GRADIENT LIQUID CHROMATOGRAPHY

Ellis Horwood Series in Analytical Chemistry
Series Editor: Dr. R. A. CHALMERS
University of Aberdeen

Founded as a library of fundamental books on important and growing subject areas in analytical chemistry, this series will serve chemists in industrial work and research, and in teaching or advanced study

Published or in active preparation

Automatic Methods in Chemical Analysis
 J. K. Foreman
 P. S. Stockwell } *Laboratory of the Government Chemist, London*

Theoretical Foundations of Chemical Electroanalysis
 Z. Galus, *Warsaw University*

Electroanalytical Chemistry
 G. F. Reynolds, *University of Reading*

Analysis of Organic Solvents
 V. Šedivec
 J. Flek } *Institute of Hygiene and Epidemiology, Prague*

Handbook of Process Stream Analysis
 K. J. Clevett, *Crest Engineering (U.K.) Inc.*

Methods of Catalytic Analysis
 G. Svehla, *Queen's University of Belfast*
 H. Thompson, *University of New York*

Organic Reagents in Inorganic Analysis
 Z. Holzbecher et alia, *Institute of Chemical Technology, Prague*

Analysis of Synthetic Polymers
 J. Urbanski et alia, *Warsaw Technical University*

Colorimetric Determination of the Elements
 Z. Marczenko, *Warsaw Technical University*

GRADIENT LIQUID CHROMATOGRAPHY

C. LITEANU
S. GOCAN
University of Cluj, Rumania
Edited by: R. A. CHALMERS,
University of Aberdeen

ELLIS HORWOOD LIMITED
Chichester

Halsted Press: a division of
JOHN WILEY & SONS Inc.
New York · London · Sydney · Toronto

First published in 1974 by

ELLIS HORWOOD LIMITED
Coll House, Westergate, Chichester, Sussex, England

Distributed by

JOHN WILEY & SONS LTD.,
Baffins Lane, Chichester, Sussex, England

Published in U.S.A. by
HALSTED PRESS, a division of

JOHN WILEY & SONS INC.,
605 Third Avenue, New York, N.Y. 10016, U.S.A.

© 1974 C. Liteanu, S. Gocan

ISBN 85312 007 2 (Ellis Horwood)

Library of Congress Cataloging in Publication Data
Liteanu, Candin.
  Gradient liquid chromatography.

  (*Ellis Horwood series in analytical chemistry*)
1. Liquid chromatography. I. Gocan, S., joint author. II. Title.

QD79.C454L5713   544'.924   74-1008
ISBN 0–470–54124–5

All rights reserved. No part of this
publication may be reproduced, stored
in a retrieval system or transmitted,
in any form or by any means, electronic,
mechanical, photocopying, recording or
otherwise, without prior permission.

Printed in Great Britain by J. W. Arrowsmith Ltd., Bristol

This work is dedicated to those who conceive and build scientific equipment, thus contributing to the progress of science no less than those who conceive and build scientific theory.

Candin Liteanu
Simion Gocan

PREFACE

Considering the wide scientific and technological applications of the different separation methods, it is not surprising that a science of separation is becoming more and more defined.

Chromatography, as a separation method, is part of a group of procedures based on the differential migration of the components in the mixture to be examined. Based on multiple partition, i.e. a repetition of the equilibration between two phases, chromatography occupies a privileged position, among not only the differential migration methods but also all separation methods. This is due to the rapidity of the partition process, which can be repeated many times in a short time interval.

The use of gradients, making it possible to vary the position of equilibrium from one equilibrium to another, by use of easily accessible experimental conditions, gives a new dimension to chromatography in general. It is therefore not surprising that there is rapid expansion in the use of mobile-phase or stationary-phase gradients, and of other types of gradient in chromatographic techniques.

The present work aims to give a full account of the present stage of use of gradients in liquid chromatography. The first part is of general character, dealing briefly with the main problems of liquid chromatography, i.e. the movement of the eluent, the movement of the solute, the reproducibility of R_f values, methods of estimating the separation efficiency, and the optimization of the chromatographic process. The four chapters in the second part deal with the use of different types of gradient: mobile-phase, stationary-phase, environmental, and combined. Various applications are also mentioned.

The work should appeal equally to chemists and research workers in chemical, biological, pharmaceutical, sanitation, alimentary, metallurgical, mining, hydrological and agricultural laboratories. It may also be useful to university students of these subjects.

C. Liteanu
S. Gocan
University of Cluj, Rumania

CONTENTS

PART I. FUNDAMENTAL PROBLEMS OF CHROMATOGRAPHY

Chapter 1 Kinetics of Eluent Migration
1.1 Introduction 3
1.2 Kinetics of eluent migration in open column chromatography 4
    1.2.1 Ascending chromatography 4
    1.2.2 Descending chromatography 7
    1.2.3 Horizontal chromatography 9
1.3 Non-uniform distribution of single-component eluents . 13
1.4 Dehomogenization of multicomponent eluents . . 19
1.5 Migration of the eluent in closed column chromatography 20

Chapter 2 Kinetics of Zone Migration
2.1 Migration rate of the zone in open column chromatography 28
    2.1.1 Isothermal chromatography 28
    2.1.2 Temperature-gradient chromatography . . 32
2.2 Relative migration rate of zone 34

Chapter 3 Factors Influencing the Reproducibility of R_f Values
3.1 Factors influencing the R_f values 39
    3.1.1 Factors determined by the composition of the stationary phase 40
    3.1.2 Factors determined by the composition of the mobile phase 41
    3.1.3 Factors determined by the medium . . . 42
3.2 Reproducibility of R_f values in paper chromatography . 46
3.3 Reproducibility of R_f values in thin-layer chromatography 47

Chapter 4 Estimation of Separation Efficiency
4.1 Separation efficiency of a closed chromatographic column 55
4.2 Separation efficiency of an open chromatographic column 67
4.3 Chromatographic selectivity 70

4.4	Resolution	76
	4.4.1 Resolution on a closed chromatographic column	76
	4.4.2 Resolution on open chromatographic columns	83

Chapter 5 Optimization of the Chromatographic Process

5.1	Adsorption and partition chromatography on closed columns	91
	5.1.1 Dimensions of the column	91
	5.1.2 Sample size, particle size of the packing, type and activity of the adsorbent	93
	5.1.3 Flow-rate of the eluent	94
	5.1.4 Viscosity of the eluent	95
	5.1.5 Structure of the bed	95
	5.1.6 Temperature	97
	5.1.7 Pressure	100
	5.1.8 Time of analysis	101
5.2	Ion-exchange chromatography on closed columns	102
	5.2.1 Size and degree of cross-linking of the particles, and the quantity of resin	102
	5.2.2 Dimensions of the column	105
	5.2.3 Pressure and flow-rate of the eluent	105
	5.2.4 Temperature	107
	5.2.5 Other factors	110
5.3	Partition and adsorption chromatography on open columns	111
	5.3.1 Choice of the mobile and stationary phases	111
	5.3.2 Choice of optimum pH	113
	5.3.3 Influence of humidity in thin-layer chromatography	115
	5.3.4 Influence of temperature	116

PART II. THE USE OF GRADIENTS

Chapter 6 Introduction to Use of Gradients

6.1	History of gradient chromatography	125
6.2	Classification of gradients	126
6.3	Types of gradient	129
6.4	Nomenclature in gradient chromatography	129

Chapter 7 Mobile-phase Gradients

7.1	Devices for achieving a mobile-phase gradient	133
	7.1.1 Closed columns	133
	7.1.1.1 Discontinuous elution gradients	133
	7.1.1.2 Concave and convex exponential gradients	137

	7.1.1.3	Linear gradient elution	156
	7.1.1.4	Concave and convex parabolic gradients	157
	7.1.1.5	Composite gradients	169
	7.1.2	Open column elution	185
	7.1.2.1	Paper chromatography	185
	7.1.2.2	Thin-layer chromatography	189
7.2	Theory of mobile-phase gradient chromatography		195
	7.2.1	Closed column chromatography	198
	7.2.1.1	Calculation of the position of the peak	198
	7.2.1.2	Resolution	221
	7.2.2	Open column chromatography	228
	7.2.3	Choice of the form of gradient	234
7.3	Uses of mobile-phase gradients		234
	7.3.1	Organic acids	234
	7.3.2	Carbohydrates	236
	7.3.3	Chlorophenols	238
	7.3.4	Aromatic hydrocarbons	238
	7.3.5	Polyprolylene glycol and polyethylene glycol	239
	7.3.6	Amino-acids	240
	7.3.7	Proteins	240
	7.3.8	Lipids	244
	7.3.9	Nucleic acids	247
	7.3.10	Steroids	248
	7.3.11	Antibiotics	250
	7.3.12	Inorganic ions	250

Chapter 8 Stationary-phase Gradient
8.1	Apparatus		258
	8.1.1	Discontinuous gradients	258
	8.1.2	Continuous gradients	258
8.2.	Resolution		260
8.3	Applications		262
	8.3.1	Composition gradient	262
	8.3.2	Impregnation gradient	264
	8.3.3	Activity gradient	265
	8.3.4	pH gradients	265

Chapter 9 Environmental Gradients
9.1	Temperature gradients		270
	9.1.1	Closed columns	270
	9.1.1.1	Apparatus	270
	9.1.1.2	Practical uses	274

                9.1.2   Open columns 278
                    9.1.2.1   Apparatus 278
                    9.1.2.2   Variation of R_f values as a function of the
                              temperature gradient 283
                    9.1.2.3   Resolution 290
                    9.1.2.4   Comparison between the parallel temperature
                              gradient and other chromatographic techniques 295
        9.2   Vapour gradient 297
                9.2.1   Apparatus 297
                9.2.2   Applications 299
        9.3   Layer thickness (section) gradient 301
        9.4   Eluent flow-velocity gradient 302

Chapter 10   Combined Gradients
        10.1   Combined mobile-phase and temperature gradient . . 306
        10.2   Combined mobile-phase gradients 313

        Appendix I 315

        Appendix II 317

        Appendix III 333

        Index 334

ACKNOWLEDGMENTS

For permission to reproduce tabular material and drawings, we are most grateful to the authors of the papers concerned, and to the other copyright holders listed below.

American Chemical Society: Figs. 1.15, 4.8, 4.9, 5.7, 5.8, 5.11, 7.1, 7.2, 7.3, 7.10, 7.21, 7.22, 7.23, 7.24, 7.25, 7.26, 7.28, 7.34, 7.36, 7.39, 7.44, 7.46, 7.47, 7.48, 7.49, 7.50, 7.57, 7.59, 7.60, 7.79, 7.89, 7.91, 7.100, 7.101, 7.102, 9.3, 9.4, 9.13, 9.29, 9.30, Tables 1.3, 7.4
Journal of Chromatography: Figs. 1.3, 1.4, 1.7, 1.8, 1.12, 4.5, 5.2, 5.6, 5.9, 5.10, 5.15, 7.4, 7.5, 7.8, 7.27, 7.31, 7.32, 7.33, 7.37, 7.38, 7.40, 7.41, 7.52, 7.53, 7.54, 7.55, 7.56, 7.64, 7.65, 7.78, 7.80, 7.86, 7.87, 7.88, 7.90, 7.92, 7.93, 7.94, 7.103, 7.104, 7.105, 8.3, 8.4, 8.10, 9.5, 9.25, 9.28, Appendix II, Tables, 7.2, 7.3, 7.5, 7.6, 7.8, 7.11
Kolloid Zeitschrift und Zeitschrift für Polymere: Figs. 1.5, 1.9, 1.10, 1.11, 1.13
Collection of Czechoslovak Chemical Communications: Figs. 4.2, 4.3
Society of Chemical Industry: Figs. 4.4, 4.12, 7.45
Separation Science: Figs. 4.6, 5.4, 9.9
Journal of Chromatographic Science: Figs. 5.1, 5.3, 7.77, 7.81, 7.82, 7.83, 7.84, 7.85, 8.9, 9.1, 9.6, 9.7, 9.20, 9.32; Table 7.9
Angewandte Chemie: Fig. 5.12
Nature: 5.13, 5.14, 7.16, 7.63, 7.96, 7.97
Chromatographia: Figs. 6.2, 8.11, 9.14
Marcel Dekker Inc: Fig. 6.3
Analytical Biochemistry: Figs. 7.11, 7.18, 7.67, 7.68
Biochimica et Biophysica Acta: Figs. 7.1, 7.12, 7.13, 7.15
Society for Analytical Chemistry: Figs. 7.14, 7.61, 7.62, 9.31
American Oil Chemists' Society: Figs. 7.17, 7.19, 7.20
Receuil des travaux chimiques des Pays-Bas: Figs. 7.42, 7.43
Experientia: Fig. 7.66
Helvetica Chimica Acta: Figs. 7.69, 7.70, 7.71, 7.72, 8.7
Archives of Biochemistry and Biophysics: Fig. 7.95
Biochemical Journal: Figs. 7.98, 7.99
U.S. Atomic Energy Commission: Fig. 7.106
Springer-Verlag: Fig. 8.5
Talanta: Figs. 4.7, 8.6, 8.8
Chimia: Fig. 9.27
Chemical Society: Fig. 10.1
Journal of Polymer Science: Figs. 10.2, 10.4
Berichte der Bunsengesellschaft für Physikalische Chemie: Fig. 10.3
Mikrochimica Acta: Table 7.10

The Authors

PART I
FUNDAMENTAL PROBLEMS OF CHROMATOGRAPHY

CHAPTER 1

KINETICS OF ELUENT MIGRATION

1.1 Introduction

The chromatographic media, whether in the form of columns, thin layers or paper, and made of powdered adsorbents, solid foams or fibres, are dispersed porous media, characterised by a statistical distribution of the size and shape of the pores.

Though in column chromatography there generally exists in the 'closed column' liquid phase a uniform distribution of the mass and composition of the eluent along the porous medium formed by the column, in the 'open column' thin-layer and paper chromatography, this distribution is usually not uniform. The hydrodynamic effect characteristic of this distribution is specific for porous media with non-uniform porosity. This is one of the main differences between column chromatography and paper or thin-layer chromatography, though all involve elution chromatography by the liquid phase. Moreover, in the case of multicomponent solvents, the presence of the solid–liquid (or even of the liquid–liquid) interface, will lead in the case of an 'open column', to changes in the composition of the mobile and the stationary phases. Concurrently there will be changes in the partition coefficients, and therefore in the column's selectivity. Knowledge of these changes will allow better understanding of the chromatographic phenomenon and improvement in separations.

During development of a chromatogram broadening of zones occurs for reasons that are well understood. The factors responsible can be varied so as to yield zones that are as symmetric and narrow as possible, which increases the resolution. Remarkable progress has been made in recent years in understanding the broadening mechanism.

The migration of a zone is also related to the chemical structure of the components present. The relative migration rate is thus useful in establishing the structure.

Chromatographic separation is affected by the migration rate, by the composition of the eluent, and the way in which it permeates the porous support (especially in the case of open columns). It is therefore natural that we should be interested in the way in which we can control the flow-rate of the eluent and its distribution in the column. The migration laws for the eluent are specific to each chromatographic technique (ascending, descending, horizontal, open and closed columns). The parameters involved are useful for characterizing the stationary phase and the conditions in the chromatographic chamber in paper and thin-layer chromatography. Great interest has arisen in these problems because of the way in which they affect the reproducibility of R_f values.

1.2 Kinetics of Eluent Migration in Open Column Chromatography

1.2.1 Ascending Chromatography

Lucas [1] had already established in 1918 an equation for the capillary ascent of a solution in a vertical paper. Then Peek and McLean [2], starting from capillary ascent in rectangular paper strips, found the law of ascendant migration of the eluent, given by

$$\frac{dz_f}{dt} = \frac{A}{z_f} - B \qquad (1.1)$$

where z_f is the distance traversed by the eluent front in time t, and A and B are empirical constants.

Wood and Strain [3] integrated Eq. (1.1) between the limits z_f and 0 and t and 0, obtaining:

$$A \ln\left(1 - \frac{B}{A}z_f\right) + Bz_f = -B^2 t \qquad (1.2)$$

By development as a series and stopping at the first term (an approximation evidently valid only for small values of z_f) Eq. (1.2) becomes

$$z_f^2 = Ct \qquad (1.3)$$

where C is an empirical constant.

The same equation is also valid for other solvents in the case of paper [3–11] and of thin layers [12–15] on condition that the migration distances are small (<20 cm) [2, 3, 16]. As in chromatographic separations on paper or thin layers migration distances longer than 20 cm are not needed, Eq. (1.3) is perfectly satisfactory. A theoretical justification of this problem was given by Ruoff et al. [9, 17] and will be discussed later in the chapter. Erdös and Vavruch [10], starting from capillary ascent and taking into consideration an effective average radius of the capillaries, also reached the same equation. For bigger migration distances, the simple law (1.3) loses its validity [2, 3, 7].

Recently, Vanhaelen [16], studying a thin layer and starting from premises analogous to those of Erdös and Vavruch [10], obtained an equation of the same form as Eq. (1.1), but gave precise physical significance to A and B, in the case of pure solvents. Thus

$$A = \frac{R\gamma}{4\eta} \quad \text{and} \quad B = \frac{R^2 dg}{8\eta} \qquad (1.4)$$

where R is the average radius of the capillaries, γ is the surface tension of the solvent, η its viscosity and d its density, and g is gravitational acceleration.

However, the theoretical values obtained for t do not estimate the experimental values sufficiently correctly, even when Eq. (1.2) is used. The difference can be expressed as a function which when added to the theoretical function for t (1.2) leads to

$$t = -\frac{1}{B}\left[z_f + \frac{A}{B}\log\left(1 - \frac{B}{A}z_f\right)\right] + a(\log z_f \cdot z_f - z_f) + bz_f^2 + cz_f \qquad (1.5)$$

which gives the experimental values correctly. Vanhaelen, trying to find a physical significance for the parameters b and c, noticed that the variation of a single factor, such as absorbent, solvent or atmosphere, led to simultaneous variation of the parameters a, b and c, besides A and B. A thorough examination showed that the changes in a, b and c are mainly dependent on the atmosphere in the chromatographic chamber, the density of the particles in the support, and the activity of the support towards the eluent.

Practical use of this relationship is extremely difficult, and usually Eq. (1.1) with parameters A and B empirically determined is to be preferred. The calculation of the average radius of the capillaries from (1.4) and empirically determined values of A and B nevertheless leads to very interesting results. Thus for two layers differing only in the density of their particles the ratio of the average radii of the capillaries ($R = 2\gamma B/gdA$) is approximately equal to the ratio of the thicknesses of the layers.

The model suggested by Vanhaelen [16] does not indeed correspond to reality, as shown by Ruoff et al. [9, 17], but is a successful attempt at finding a model for the actual phenomenon.

All these relationships refer to migration of solvents at room temperature. Gocan and Liteanu [18] have carried out studies at different temperatures and found that in isothermal chromatography the following empirical relation is valid:

$$z_f = Ct^n \qquad (1.6)$$

where C and n are empirical constants. Typical results are given in Fig. 1.1.

Equation (1.6) can be expressed in linear form as

$$\log z_f = \log C + n \log t \qquad (1.7)$$

and from graphs of $\log z_f$ vs. $\log t$, C and n can be calculated (Table 1.1).

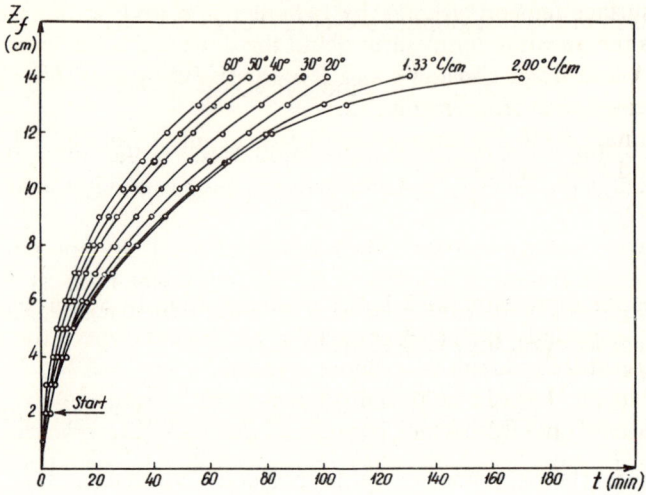

FIG. 1.1. Ascent of the eluent as a function of time. Eluent: n-butanol saturated with 4N HCl; Schleicher-Schüll paper 2040 bM [18].

This table shows that n varies only slightly over a broad temperature range, while C increases with the temperature. As shown by Lucas [1], C is proportional to the surface tension and inversely proportional to the dynamic viscosity.

TABLE 1.1. Value of constants C and n [18]

Eluent and paper	T, °C	C (cm/min^n)	n
n-Butanol + 20% 3M HCl	14	1·288	0·456
Whatman No. 2	20	1·230	0·474
	30	1·334	0·464
	40	1·531	0·445
	50	1·585	0·440
	60	1·660	0·444
	70	1·686	0·464
n-Butanol satd. with 4M HCl	20	1·513	0·475
Schleicher-Schüll 2040 bM	30	1·679	0·471
	40	1·906	0·459
	50	2·113	0·447
	60	2·239	0·446
n-Butanol satd. with 4M HCl	20	1·514	0·474
Whatman No. 4	20	1·906	0·464

The surface tension $\gamma = \mathbf{a} - \mathbf{b}T$ (where $\mathbf{a} > \mathbf{b}$, and both are constants and T is the absolute temperature) and the dynamic viscosity $\eta = Ae^{B/RT}$ (where A and B are constants and R is the gas constant) [19]. Obviously γ/η increases with temperature.

Examination of the temperature gradient effect, in Fig. 1.1, suggests that z and t are related by

$$z_f = \frac{t}{x + yt} \qquad (1.8)$$

which better represents the kinetics of eluent migration in a temperature gradient, especially in the upper half of the ascent, where the action of the temperature gradient is far more evident.

Equation (1.8) can be put into linear form by plotting $1/z_f$ vs. $1/t$, to give a series of lines from which constants x and y are determined (Table 1.2).

TABLE 1.2. Values of constants x and y for different temperature gradients [18]

Eluent and paper	Temp. (°C) Start	Temp. (°C) At spot	Temp. gradient (°C/cm)	x (min/cm)	y (cm^{-1})
n-Butanol + 20% 3M HCl	14	19	1.5	2.99	0.072
Whatman No. 2	14	33	1.9	3.11	0.077
n-Butanol satd. with 4M HCl	20	40	1.33	2.27	0.054
Schleicher–Schüll 2040 bM	20	50	2.00	2.28	0.058

Constants x and y characterise the stationary phase–mobile phase system as well as the medium conditions (temperature range and degree of saturation of the atmosphere). We shall return to the action of the temperature gradient later.

1.2.2 DESCENDING CHROMATOGRAPHY

Descending chromatography has been studied less [10, 20] owing to its less frequent use. This technique may also be used as a system of continuous chromatography, especially useful when the components have small R_f values and are only partially separated.

Erdös and Vavruch [10] studied the kinetics of migration of the solvent front in descending chromatography (see Fig. 1.2), giving a theoretical justification which is in good agreement with experimental results.

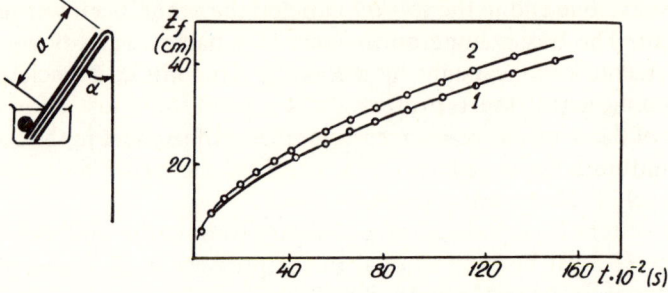

FIG. 1.2. The device used for studying descendent migration of water (1) and methanol (2) on chromatographic paper; $a = 2$ cm and $\alpha = 30°$ [10].

The rate of migration becomes constant after some time. After longer elution there is a slight decrease of the rate [20], owing to decrease of the liquid level in the feeder-tank.

To study migration of single-component eluents in descending paper chromatography, Bourdillon [20] used the device shown in Fig. 1.3. It consists of an aluminium frame containing a vessel with solvent, and with a glass rod set in holes in its sides so that the paper can be suspended

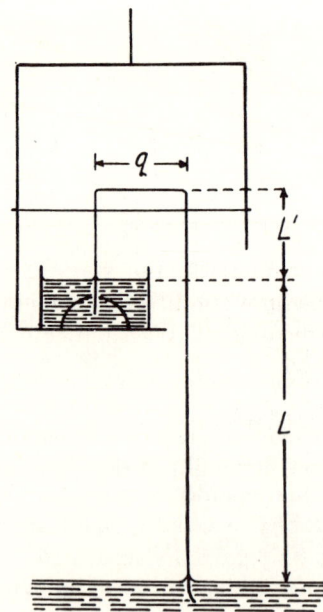

FIG. 1.3. Apparatus used for flow-rate measurements (schematic) [20].

so as to have one end in the solvent cup and the other in a lower reservoir of solvent. The box is hung on the arm of a balance, and flow-capacity measurements can be made by weighing. The whole is enclosed in a bigger container at the top of which is a small hole just allowing free passage of the connection to the balance. The atmosphere in the container is thus saturated with solvent. The length of the paper loop above the upper level of the solvent is given by $a = 2L' + q$.

If the system is considered as a capillary siphon, the fluid delivered in unit time U must be proportional to the pressure head L and inversely proportional to the total length of paper $a + L$,

$$\frac{U_0}{U} = \frac{L + a}{L} \quad \text{or} \quad \frac{1}{U} = \frac{1}{U_0} + \frac{a}{U_0 L} \tag{1.9}$$

where U_0 is the value of U when $a \equiv 0$ or $L \equiv \infty$.

Plots of $1/U$ vs. $1/L$ will give a straight line with slope proportional to a. This is in good agreement with experimental data (Fig. 1.4).

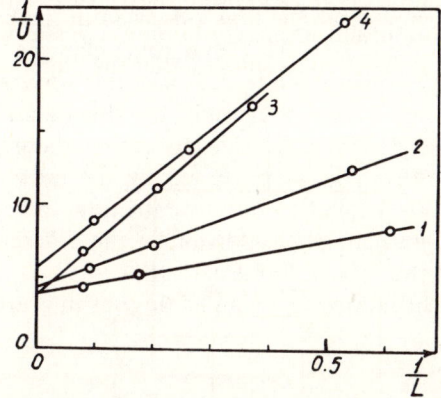

FIG. 1.4. Flow of water. Abscissae: reciprocal of pressure head in cm. Ordinates: reciprocal of flow in ml/hour/cm width. Expt. 1: length of loop $a = 2\cdot 2$ cm, $L' = 0\cdot 85$ cm. Expt. 2: $a = 4\cdot 2$ cm, $L' = 1\cdot 5$ cm. Expt. 3: $a = 8\cdot 5$ cm, $L' = 0\cdot 45$ cm. Expt. 4: $a = 8\cdot 1$ cm, $L' = 3\cdot 0$ cm. Values corrected to 20°C [20].

Thus in descending chromatography a constant rate is reached after a time which depends on the length of the siphon a and on the fall L of the level.

1.2.3 HORIZONTAL CHROMATOGRAPHY

In this case the effect of gravitation is evidently negligible. The equation describing the migration of the front is parabolic in form ($z_f^2 = Ct$) [17] as shown in Fig. 1.5.

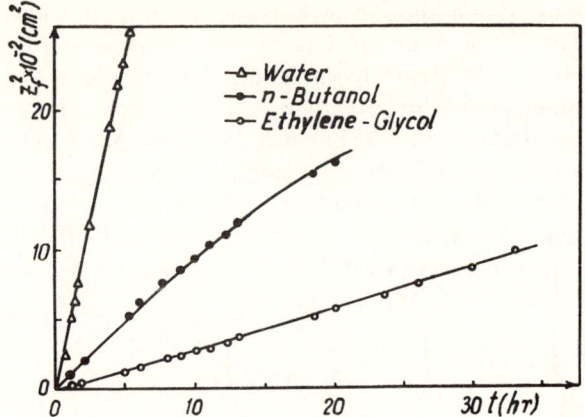

Fig. 1.5. Frontal movement of various solvents on Whatman 3 MM paper [17].

The deviations from parabolic flow observed in the case of n-butanol show, according to Ruoff et al. [17], that some changes take place in the structure of the paper owing to interaction of this solvent with the paper, and probably all deviations from linearity are due to such interactions.

The same authors [15] find that this law can be theoretically predicted by considering the drainage as governed by the laws of diffusion; the diffusion coefficient is a function of concentration and hence affected by the concentration gradient of the solvent in the different regions of the paper immersed in the solvent [see Eq. (1.3)].

In radial paper chromatography an analogous migration law has been found empirically [21–25]:

$$z_f^2 = C_r t \qquad (1.10)$$

In contrast to the other cases, here the rate changes very quickly with the distance. This sudden decrease of the rate seems to be the qualitative explanation [3] of the narrow zones. In circular chromatography good resolution is obtained as the rear edge of the zone always has a higher rate of advance than the leading edge.

Typical shapes of paper or thin layers used in radial chromatography are given in Fig. 1.6. These shapes are used for different purposes (mixtures difficult to separate by other techniques, concentration of a trace component etc.).

Ruoff and Giddings [30] have inferred a general law for the migration of the front in these cases. They start from Darcy's relationship [31] according to which the flow in a porous medium is proportional to the

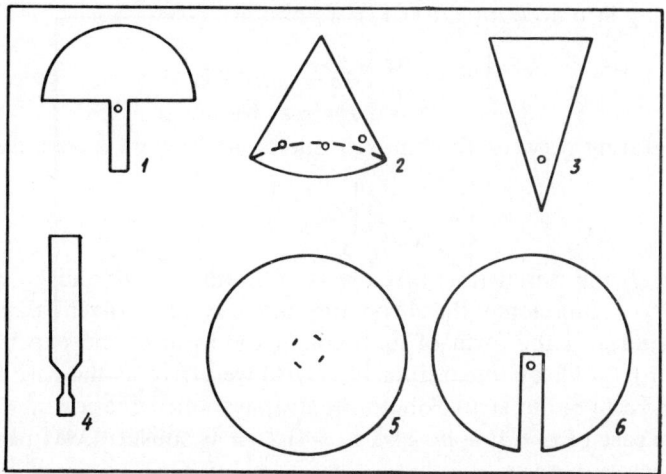

Fig. 1.6. Shapes of paper and thin layers used in chromatography: 1, 6, [25]; 2, [26]; 3, [27]; 4, [28]; 5, [29].

pressure gradient:

$$q = -K\frac{dP}{dz} \qquad (1.11)$$

where q is the mass flux of solvent [per unit breadth (L) of strip], P the pressure, z the distance and K a constant.

The total flux q_0 will be equal to qL and therefore

$$q_0 = -KL\frac{dP}{dz} \qquad (1.12)$$

By integrating along the strip from the solvent source z_0 to the solvent front z_f, we have

$$-P = \frac{q_0}{K}\int_{z_0}^{z_f}\frac{dz}{L} = \frac{qL}{K}\int_{z_0}^{z_f}\frac{dz}{L} \qquad (1.13)$$

where P is the variation of the capillary pressure between the saturated and the dry paper, and $L = f(z)$.

Supposing that the rate of advance of the front is proportional to the flow, i.e.

$$\frac{dz}{dt} = bq \qquad (1.14)$$

and taking into account Eq. (1.13), we obtain

$$\left[L \int_{z_0}^{z_f} \frac{dz}{L} \right] dz = -bPK\, dt. \tag{1.15}$$

By integrating between the limits z_0 and z_f, and t_0 and t, we have

$$\int_{z_0}^{z_f} \left[L \int_{z_0}^{z_f} \frac{dz}{L} \right] dz = \frac{1}{2} C(t - t_0) \tag{1.16}$$

where t_0 is the initial time (usually $t_0 = 0$ and $z_0 = 0$) and C (equal to $-2bKP$) is a function of the chromatographic system (solvent, paper, etc.).

L includes in the form of its function the form of the paper. If L is constant ($z = 0$), by integrating Eq. (1.16) we arrive at the case for flow through rectangular strips, obtaining the parabolic equation $z_f^2 = Ct$.

In the case of $L = a + mz$ and $z_0 = 0$, $t_0 = 0$, substitution in Eq. (1.16) and integration gives

$$\left(\frac{a}{m} + z_f \right)^2 \left[\ln\left(1 + \frac{mz_f}{a}\right) - \frac{1}{2} \right] + \frac{1}{2}\left(\frac{a}{m} \right)^2 \equiv Ct \tag{1.17}$$

Figure 1.7 shows the way in which this equation is experimentally verified when water is used on Whatman No. 1 paper at $30 \pm 0.5°C$. By use of the horizontal flow technique, the effects of gravitation are eliminated. The value of C was determined from the flow on rectangular strips. In the case of divergent flow calculated values are smaller than the experimental ones.

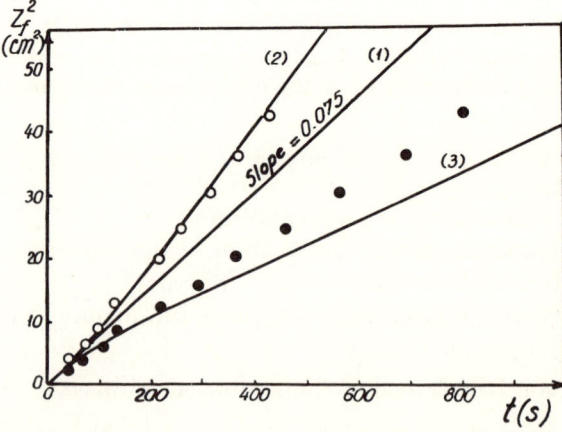

FIG. 1.7. Flow in tapered strips: (1) rectangular strips, (2) experimental and calculated converging flow ($a = 2.0$ cm, $m = -0.231$), (3) experimental and calculated diverging flow ($a = 0.32$ cm, $m = 0.258$) [30].

When discs with so-called filter drains of paper or of cotton are used, the calculation leads to:

$$z_f^2 = \frac{Ca}{2\pi L}t = C_r t \qquad (1.18)$$

where L is the drain length, a the drain width, C the flow-rate coefficient for rectangular flow, and $C_r = Ca/2\pi L$ is the radial flow coefficient.

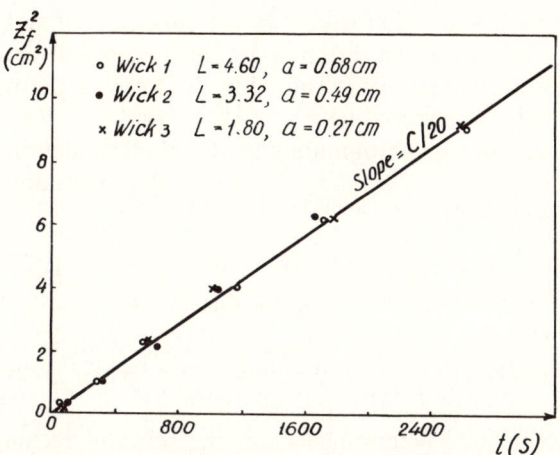

FIG. 1.8. Radial flow with wick dimensions approximately $2\pi L/a = 42$ [30].

The law is verified very well experimentally (Fig. 1.8) and if the ratio $2\pi L/a$ is kept constant, the rate practically does not vary. The measured value of C_r is approximately $C/20$ and not $C/42$ as was theoretically predicted [30]. It is therefore recommended in practice to use the relation

$$z_f^2 = \frac{Ca}{\pi L}t \qquad (1.19)$$

which agrees well with the experimental results.

These relationships permit specification of the migration rate of the solvent in horizontal chromatography with different geometries, thus giving the possibility of choosing a convenient rate for a certain separation.

1.3 Non-uniform Distribution of Single-component Eluents

In chromatographic techniques where solvent migration is due to capillarity it has been found [32–46] that the concentration of the eluent is not uniform along the whole length of the strip or column. Experiment

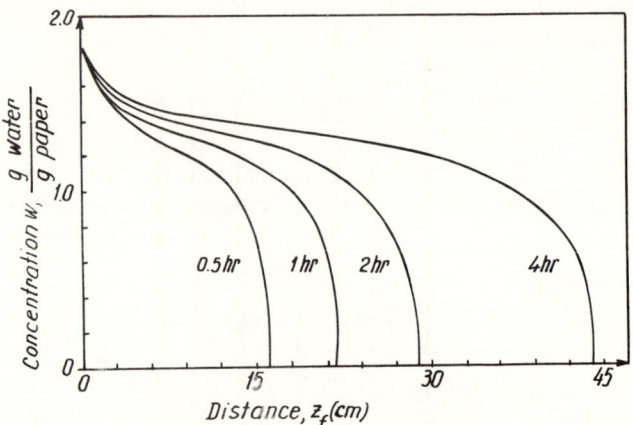

FIG. 1.9. Concentration profiles for Whatman 3 MM paper systems at various time intervals [17].

[3, 36] has led to the conclusion that the distribution is generally S-shaped, e.g. Fig. 1.9 [17].

Attempts to explain the flow mechanism of solvents through thin layers (whether in connection with chromatography or otherwise) may be divided into two categories, those based on capillary models [9, 10, 32–37], and those on analogy with diffusion [8, 9, 15]. The touchstone of these theories is the extent to which they predict the migration law for the front and the distribution of eluent concentration along the paper strip or the thin layer. Only Ruoff et al. [9–17] have succeeded in this.

Some authors [10, 37–39] have considered the thin layer or the paper as a bunch of capillaries of equal radius. In this case, though the migration law is found to agree with experiment, the distribution of the eluent is theoretically uniform, a fact which does not correspond to reality.

The model of capillaries of different radii bound in series is a better approximation to reality [2], but the arbitrary parameters used by Peak and McLean have no real physical justification. Moreover, they too did not predict a concentration gradient of the eluent. Henderson [40] takes this behaviour into account and suggests a model of capillaries of different sizes placed in parallel. This model leads to a concentration gradient, but does not correspond to reality from the physical point of view, as in porous media the channels are generally connected.

Some models of interconnected capillaries have been suggested [9, 41, 43]. According to these (Fig. 1.10) the liquid penetrates through the end of the bunch of capillaries by the action of capillary forces, which produces a decrease in the free energy. Analogously, the liquid in the

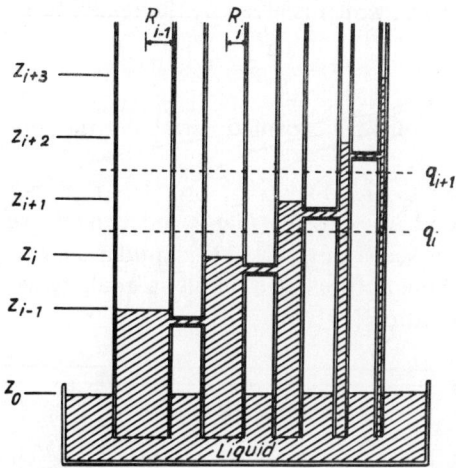

FIG. 1.10. One-dimensional interconnected capillary pore model [9].

capillaries will tend to reduce *its* free energy, migrating into an interconnected capillary of smaller radius. The displacement of the solvent front leads to a free-energy gradient along the capillary grouping [39, 41], which, in fact, is the power that ensures the penetration of the liquid.

Ruoff *et al.* [9] find that in such a model the solvent migration is analogous to diffusion. We can thus consider that the physical process on which solvent migration in chromatography is based, is a capillary migration, but that the process is more easily dealt with by means of the mathematics used for diffusion. Ruoff *et al.* [9] also suggest a three-dimensional model of chromatographic paper which, experimentally reproduced, leads to satisfactory agreement with theory.

Among the theories based on analogy with diffusion, those of Ruoff *et al.* [9, 17] are important. In the case of horizontal migration on paper [9], the use of Fick's second law [the equation of one-dimensional longitudinal diffusion] is suggested, with the difference that the diffusion coefficient is a function of the concentration:

$$\frac{\partial w}{\partial t} = \frac{\partial}{\partial z}\left[D(w)\frac{\partial w}{\partial z}\right] \qquad (1.20)$$

where w is the concentration (w/w) of the solvent in the paper, $D(w)$ is the diffusion coefficient of the solvent in aqueous medium and z the distance from the origin.

As shown by Boltzmann [44], this equation has a solution containing the concentration w as a function of a single variable called the 'Boltzmann transformed variable', which is given by the relation:

$$\lambda = \frac{z}{t^{1/2}} = C^{1/2}y \tag{1.21}$$

where λ is Boltzmann's transformed variable, t the time and $y\,(=z/z_f)$ is the reduced distance.

If y is replaced in (1.21) by z/z_f, Eq. (1.3) is immediately obtained. On the other hand, $w = f(\lambda) = f(y)$ is independent of the value of z_f, and also of time, as seen in Fig. 1.11 [17]. Similar curves were obtained for other solvents. These authors also found an analogy with diffusion in the case of radial migration [17].

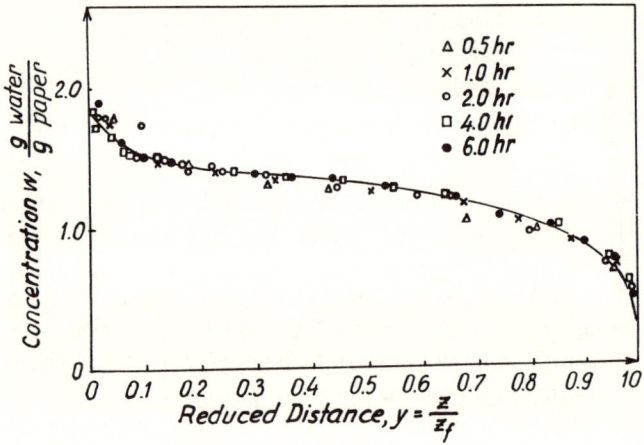

FIG. 1.11. Reduced concentration profile for a water/Whatman 3 MM paper system [17].

The superiority of the Ruoff theory [9, 17] to other diffusion theories [8, 15], lies in the fact that the diffusion coefficient is considered as a function of the concentration, the nature of which may be determined [17] from the concentration profile and the flow-rate coefficient C. Thus, on differentiating Eq. (1.3) we obtain

$$\frac{\partial z_f}{\partial t} = \frac{C}{2z_f} \tag{1.22}$$

and by introducing the reduced distance (1.21),

$$\frac{\partial z}{\partial t} = \frac{yC}{2z_f} \tag{1.23}$$

Expressing the flow as the growing weight at a given point we obtain:

$$q = \int_0^w \frac{\partial z}{\partial t} dw = \frac{C}{2z_f} \int_0^w y\, dw = \frac{C}{2z_f} A \qquad (1.24)$$

The diffusion coefficient may be expressed in terms of the flow and the concentration gradient:

$$D = \frac{-q}{\partial w/\partial z} = -\frac{z_f q}{\partial w/\partial y} = -\frac{CA/2}{\partial w/\partial y} \qquad (1.25)$$

The integral in (1.24) represents the surface A limited by the curve $w = f(y)$ and the line $w = w_y$, and the denominator gives the slope of the concentration curve at its intersection with this line, marked by an arrow in Fig. 1.12 [45].

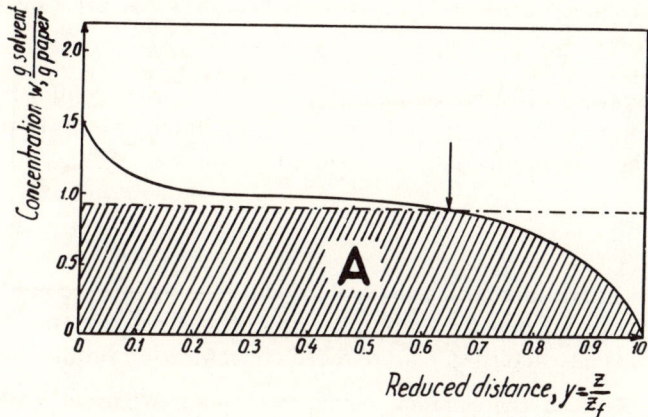

FIG. 1.12. The area term, $A = \int_0^w y\, dw$ [45].

The form of the function $D = f(w)$ is given in Fig. 1.13. The diffusion coefficient D is characteristic for each paper (or thin layer)–solvent system and may be approximated [17] by an empirical relation having the form:

$$D = \frac{D_0}{1 + \alpha w + \beta w^2} \qquad (1.26)$$

The coefficients of this formula can be used to classify chromatographic systems.

The non-uniform distribution of the solvent along the paper or thin layer causes the rate of transport of solvent through the paper to be different behind the front from that at the front.

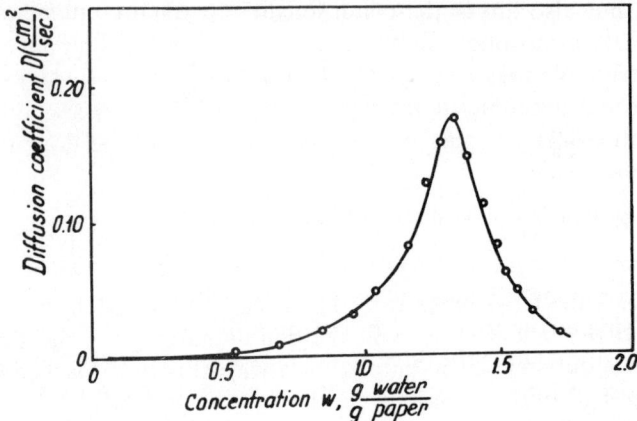

FIG. 1.13. Diffusion coefficient for a water/Whatman 3 MM paper system as a function of concentration [17].

The average flow-rate v of solvent at a certain point behind the front may be obtained, according to Giddings *et al.* [45], by means of the ratio between the solvent flux q, and its concentration w.

$$v = \frac{q}{w} \qquad (1.27)$$

In this case the flux represents g of solvent passing in 1 second through a cross-section of paper containing 1 g of dry paper per cm of length. If the flow is characterised by a reduced concentration profile, the flux is given by

$$q = v_f A = v_f \int_0^w y \, dw \qquad (1.28)$$

where v_f is the migration rate of the front and A the shaded area in the reduced concentration profile shown in Fig. 1.12.

Waksmundzki and Różyło [12–14] find that for thin layers the coefficient C of the parabolic migration law (1.3) depends on the microporous structure of the adsorbent and on the viscosity of the mobile phase. The migration rate of the mobile phase decreases with increase in the specific surface of the support, but is independent of the way in which the support is prepared, i.e. of the framework itself. They conclude that the results obtained in study of the eluent flow kinetics allow standardisation of the adsorbent-layer characteristics in thin-layer chromatography.

We consider that kinetic experiments on eluent migration tell us not only about the porous structure of the thin layer and the composition of the solvent, but also about the conditions of the medium (temperature, saturation of the atmosphere, moisture, etc.) i.e. about the whole chromatographic system. We may thus check that the working conditions of two different chromatographic chambers, or for two different chromatograms, etc. are identical.

1.4 Dehomogenisation of Multicomponent Eluents

The S-form concentration profile and the flow-rate gradient also appear when multicomponent eluents are used [45], for the same reasons as in the case of single-component ones, but in addition there is a change in eluent composition along the direction of migration, owing to the different degrees of adsorption of the eluent component on the support. This effect is usually called dehomogenisation of the eluent. In thin-layer or paper chromatography this effect is still more accentuated owing to the different volatilities of the components if the atmosphere in the chamber is not perfectly saturated.

The problem began to concern research workers using 'open column' chromatography or frontal column chromatography, after Horner et al. [46] discovered that a mixture of phenol and water, flowing through a filter paper, loses the water until the fibres are saturated with it. Only after this does the developer have a constant composition.

The problem is certainly of no interest in elution chromatography on 'closed columns' but it was studied because of the double fronts observed by numerous authors [47–55]. The problem also arose in Tiselius's [56–59] frontal analysis. There are cases when dehomogenisation is even desirable, as in the case of multizonal chromatography (the simplest elution gradient technique) [60,61]. This technique is based on the fact that in many solvent mixtures, several zones of the solvent (fronts) are formed on the surface of the layer. This is explained by the different (selective) affinity of the adsorbent for the solvent of the developer. Consequently, substances applied in different places on the surface will be differently developed, owing to the variation in the eluent composition during migration through the absorbent layer. For this reason, in the multizonal technique, the samples are applied diagonally on the chromatographic plate.

The first systematic studies of the phenomenon permitted some conclusions regarding the stationary phase in the techniques of open or closed columns [62–64], or the separation mechanism of some inorganic ions in paper chromatography [65]. This research work has not led to definitive conclusions, however, but only to some qualitative findings regarding the dehomogenisation of the eluent.

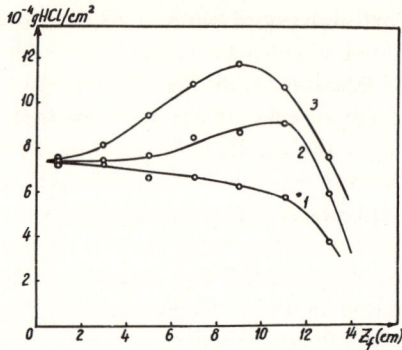

FIG. 1.14. The concentration profile of HCl in the system: propanol + HCl + H$_2$O (80:10:10 v/v). 1, Isothermal (20°C); 2, temperature gradient (0·5°C/cm), 3, temperature gradient (1·0°C/cm) [66].

We can state that during migration of the eluent a concentration gradient of the different components of the solvent mixture appears, and that the more polar a component is, the lower its concentration in the solvent front. Also, the improvement of the resolution brought about by a temperature gradient is also due to a dehomogenisation of the eluent, caused by the differential evaporation of its components (Fig. 1.14) [66].

It has recently been established [67] that even in reversed-phase chromatography with liquid ion-exchangers such a process takes place, proving the generality of this phenomenon in the case of open columns.

1.5 Migration of the Eluent in Closed Column Chromatography

As already mentioned, a chromatographic column is a porous medium, and we must therefore appeal to the knowledge of the flow of liquids through porous media [31, 68, 69] to understand the migration of the eluent through the column. The complexity of the problem makes it necessary to use empirical laws. Nevertheless attempts have been made to find a simple model of the porous medium that would allow integration of the flow equations of viscous liquids, with adequate limit conditions. Hence the models of porous media, the simplest of which is a bunch of capillaries of circular section, parallel and having the same diameter. This model, known as an 'ideal porous medium', verifies Darcy's law but suffers from serious drawbacks. Thus, as all the capillaries are parallel, the flow is one-dimensional and only the permeability in a single direction may be obtained. If the model is changed, placing a third of the capillaries along each of the Cartesian coordinates, a relation between porosity and permeability is obtained, which also includes the specific surface. An oversimplification of this model is for all the tubes to have the same diameter, representing the average diameter of the pores. As there is really a distribution of pore dimensions, the capillary model must be modified. Therefore all the capillaries allowing the flow in a certain direction are parallel to it, but with different diameters, so that the distribution of the pore dimensions should be identical to the one in a real porous medium. In these models, known as parallel-type models, the relation between porosity and permeability is the same as in the simple capillary models and cannot be applied directly to real porous media.

The parallel-type models generally include a simplification far enough from reality, as they suppose that all the pores go straight from one face of the porous medium to the other. The fluid particles actually do not cover the shortest distance, represented by a straight line between those two faces, but a markedly longer distance. This is taken into consideration by including the notion of tortuosity of the porous medium, which exactly expresses the ratio between the average path length traversed by the fluid particles and the length of the porous medium. Such models are known as 'serial-type models', as they are made of capillaries of different diameters placed in series, one behind the other, and in this case a relationship between porosity and permeability can be deduced, and if the real value of the tortuosity could be included, the serial model could be considered as satisfactory.

These types of models assume that the porous medium is a solid material with inner gaps (pores). From the opposite standpoint, the porous medium is a space filled with fluid and having a certain number of solid obstacles inside. The resistance opposed by the porous medium to the flow of the fluid is obtained by adding the resistance of all the inner obstacles. Such a model may be considered as satisfactory for describing flow through a very porous medium.

Nevertheless none of these models is really adequate since none of the different relationships between porosity and permeability found for the models may be applied to actual porous media.

Another type of model, proposed by Scheidegger [68], is the statistical one. The basic idea is that any type of capillary model has too great a regularity to be able to represent a natural porous medium, which is essentially disordered. It is therefore necessary to use statistical mechanics to describe the phenomenon of flow through these media. Notwithstanding the fact that interesting results were reached by this method, this model is still unable to give a completely satisfactory explanation of the different phenomena appearing in the flow through porous media. It nevertheless appears that the statistical approach is the way to reach an adequate description of the phenomena.

The problem of the flow of a fluid through a porous medium should be studied by means of the equations of motion of real fluids. In the case of laminar flow, these are Navier and Stokes's well-known equations. For a compressible fluid with a variable viscosity and laminar flow, we have [31] for projection on the $0z_i$ axis.

$$\rho\left(\frac{\partial v_i}{\partial t} + \frac{\partial v_i}{\partial z_j}v_i\right) = \rho F_i - \frac{\partial P}{\partial z_i} + \frac{\partial}{\partial z_j}\left[\eta\left(\frac{\partial v_i}{\partial z_j} + \frac{\partial v_j}{\partial z_i}\right)\right] - \frac{2\partial}{3\partial z_i}\left(\eta\frac{\partial v_j}{\partial z_j}\right) \quad (1.29)$$

where v_i represents the components of the velocity v, and F_i those of the mass force per unit mass, and ρ is the density of the fluid.

If the viscosity η is constant, we obtain the simpler form:

$$\frac{\partial v_i}{\partial t} + \frac{\partial v_i}{\partial z_j}v_j = F_i - \frac{1}{\rho}\frac{\partial P}{\partial z_i} + \frac{\eta}{\rho}\frac{\partial^2 v_i}{\partial z_j \partial z_j} + \frac{1}{3}\frac{\eta}{\rho}\frac{\partial}{\partial z_i}\left(\frac{\partial v_j}{\partial z_j}\right) \quad (1.30)$$

and finally, if the fluid is incompressible ($\partial v_j/\partial z_j = 0$) there results

$$\frac{\partial v_i}{\partial t} + \frac{\partial v_i}{\partial z_j}v_j = F_i - \frac{1}{\rho}\frac{\partial P}{\partial z_i} - \frac{\eta}{\rho}\frac{\partial^2 v_i}{\partial z_j \partial z_j} \quad (1.31)$$

In any of these situations, however, solution of flow problems in porous media by means of these equations is impossible, because of the extremely complicated geometry of the porous media, which cannot be described quantitatively (on the microscale). The limits of the flowing system cannot be specified, and the irregular forms of the pores and the multiple links between them render impossible any attempt to integrate the Navier–Stokes equations. Several attempts have been made to infer Darcy's law from these equations, but the demonstrations are not convincing, so Darcy's law must be accepted as an empirical law, as will be seen below.

Giddings [70] shows that the inner space of a chromatographic column is made of two regions: a solid region formed by the support particles, and the free portion between the particles and their pores. Both regions are determined by a continuous surface common to both, namely the surface created by the contact of contiguous particles. By thus describing one of the two regions, we specify the other. The geometry of this common surface determines in fact the structure of the column filling. The complexity of this geometry prevents us from using it to characterise a column and therefore other characteristics are used, such as the free volume of the column, the porosity of the mixture or the distribution of the pores.

The fraction (Φ) of free volume of the column is an important parameter:

$$\Phi = f_0/f \tag{1.32}$$

where f_0 is the porosity between the particles (the free space between the particles), and f the total porosity. A well-filled chromatographic column has $f_0 \sim 0.4$.

As the migration of the eluent during column chromatography practically always corresponds to laminar flow, we shall begin by examining such a flow through porous media, and use Darcy's empirical law [31], which is essentially Ohm's law [70] applied to fluid flow

$$q = -K\frac{dP}{dz} \tag{1.33}$$

and shows that the fluid flow q per unit cross-section is proportional to the pressure drop per unit length (the pressure gradient dP/dz is negative), and the proportionality constant K is an empirical parameter. The law is valid only for low flow-rates where the resistance to flow is due to viscosity. By means of Darcy's law we may forecast the total flow capacity through a known medium under given hydrostatic conditions.

It is evident that at constant pressure, increase of the fluid viscosity η will lead to decrease of the volumetric flow of the fluid per unit surface area, so that Eq. (1.30) may be written:

$$q = -\frac{K_0}{\eta}\frac{dP}{dz} \tag{1.34}$$

where K_0, the specific permeability, has the dimensions of a surface, a fact resulting from the dimensions of Eq. (1.34) and is expressed in terms of a unit called the Darcy (1 Darcy = 9.87×10^{-9} cm^2). The permeability depends only on the properties of the porous medium, being a characteristic dimension of it.

The flow-rate v of the eluent is more interesting in chromatography than the flux q, and between the two dimensions there exists the relation

$$v = \frac{q}{f} \tag{1.35}$$

and therefore:

$$v = -\frac{K_0}{\eta f}\frac{dP}{dz} \tag{1.36}$$

However, it may be that the most important factor controlling the flow of the mobile phase in chromatography is the size of the particles, a dimension closely connected to the efficiency of the column. The size of the particles is included in the value of K_0 in Eq. (1.36). Thus, according to Giddings [70]:

$$K_0 = \frac{f \Phi d_p^2}{2\phi'} = \frac{f_0 d_p^2}{2\phi'} \qquad (1.37)$$

where ϕ' is an empirical constant (~ 300), $\Phi = f_0/f$ is the fraction of free volume between particles, f_0 is the free space between particles, f is the total porosity of the column, and d_p is the diameter of the particles.

The parameter ϕ' is directly connected to the free space f_0 between the particles by means of the Kozeny–Carman equation [71, 72]:

$$K_0 = \frac{d_p^2}{180} \frac{f_0^3}{(1-f_0)^2} \qquad (1.38)$$

which combined with (1.37) leads to

$$\phi' = \frac{90(1-f_0)^2}{f_0^2} \qquad (1.39)$$

As shown by various authors [73, 74] the porosity between particles (f_0) is generally 0.4 ± 0.03 for a well-packed chromatographic support.

The porosity of the particles may also influence the migration of the eluent through the column. If the granules are small, we may consider the eluent in these internal channels as stationary, but if the diameter d_c of the inner channels is somewhere near the diameter d_p of the channels between particles ($10 d_c > d_p$) [70], the flow will take place in equal measure between the particles and their inner channels. This is the case with some porous adsorbents, as for instance Chemosorb W (diatomaceous earth). There exist many sources of variation of the local flow-rate in a column, especially in the direction transverse to the direction of flow. The importance of these effects is very great in preparative chromatography on wide columns. The importance of these non-uniformities decreases in the case of narrow columns (radius ~ 4 mm).

In the case of liquid chromatography at moderate pressures, the rate v is constant, thus $-dP/dz$ is constant, and therefore P varies linearly with the distance. Nevertheless it has been found experimentally that if the rate increases (under increase in the pressure) its relationship to the pressure ceases to be linear. At the same time the fluid flow changes from laminar to turbulent. The turbulence is characterized by rapid fluctuations of rate, direction of flow and pressure at a given point. Increase of turbulence leads to increase of lateral transport and of effective diffusion rate. Reynolds number Re, quantitatively characterising the degree of turbulence in granular materials, is given by Re = $\rho v d_p/\eta$ and is a non-dimensional quantity proportional to the flow-rate. As is well known, in the case of straight tubes, turbulence appears suddenly when Re becomes > 2300.

For porous media [68] turbulence becomes increasingly important as Re rises from 1 to 100. The gradual increase in turbulence in the granular material is undoubtedly due to the fact that channels of different sizes and configurations are involved. At Re ~ 1 we may consider that turbulence is produced only in some of the big channels. As Re grows, the turbulence gradually spreads to the other channels.

Considering the analogy between flow in circular tubes and in porous media, it might be expected that the critical Reynolds numbers should be close. The fact that for flow in porous media the linear relationship between flow-rate and pressure ceases to be valid for very small Reynolds numbers, seems to be due not to the appearance of turbulence, but to that of inertial forces [31]. For flow of a fluid through a straight tube, the terms representing the inertial forces disappear from the equation of motion, but in the case of a curved tube, the effects of inertia also show in the laminar flow, the equation of motion losing its linear character. Since the pores of a porous medium frequently change direction, it therefore ensues that this is the reason limiting the validity range of Darcy's law at such small Reynolds numbers. Other possible reasons are the numerous changes of section of the pores as well as the marked irregularity of the surfaces.

Though most chromatographic separations are done with laminar flow, the theoretical developments in recent years show that much is

TABLE 1.3. Typical flow-rates and analysis times for liquid chromatography in laminar and turbulent systems at different values of the column radius (r_0) for a concentration distribution coefficient $K = 100$ [77]

	Laminar		Turbulent	
r_0 (cm)	U_0 (cm/s)	t (s)	U_0 (cm/s)	t (s)
0.01	0.06	1.51×10^1	3000	2.80×10^{-3}
0.05	0.01	3.78×10^2	600	7.00×10^{-2}
0.08	0.008	9.65×10^2	375	1.80×10^{-1}
0.1	0.006	1.51×10^3	300	2.80×10^{-1}
0.2	0.004	6.02×10^3	150	1.12
0.3	0.002	1.36×10^4	100	2.52
0.5	0.001	3.75×10^4	60	7.00

gained by using chromatographic columns with very high flow capacity. Turbulence can then play a positive role in chromatographic analysis, as many authors have shown.

Pretorius and Smuts [77] show that in the case of liquid–liquid chromatography the flow-rates in a turbulent system must be considered as approximately 10^4 times those in a laminar system, and the time for analysis decreases by approximately the same factor (Table 1.3).

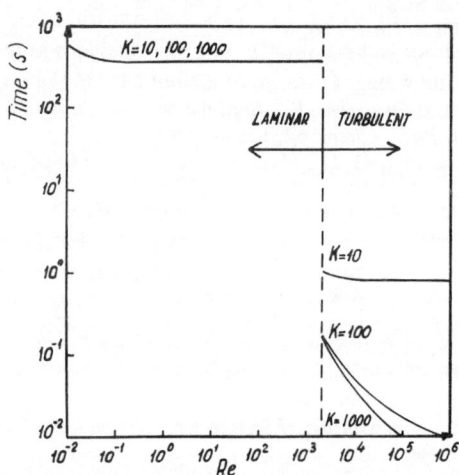

FIG. 1.15. Effect of concentration partition coefficient on analysis time at various Reynolds numbers for column radius $r_c = 0.05$ cm [77].

Figure 1.15 [77] gives the variation of the analysis time with Reynolds number for different values of the concentration distribution coefficient K for an open tubular column, $r_0 = 0.05$ cm, and with the coefficients for mass-transport by molecular diffusion the same in the mobile and stationary phases, $D_M^m = D_S^m = 10^{-5}$ cm^2/s.

Undoubtedly the biggest effect of turbulence in chromatography is the increase of the mass transfer rate, a fact leading to reduction of the time of analysis.

REFERENCES

1. Lucas, R., *Kolloid Z.* **23**, 15 (1918).
2. Peek, R. L., Jr. and McLean, D. A., *Ind. Eng. Chem., Anal. Ed.* **6**, 85 (1934).
3. Wood, S. E. and Strain, H. H., *Anal. Chem.* **26**, 260 (1954).
4. Fujita, H., *J. Phys. Chem.* **56**, 625 (1952).
5. Kowkobany, G. N. and Cassidy, A. G., *Anal. Chem.* **22**, 817 (1950).
6. Müller, R. H. and Clegg, D. L., *Anal. Chem.* **23**, 396 (1951).
7. Nakagaki, M. and Osagawa, K., *Bull. Chem. Soc. Japan* **32**, 344 (1959).
8. Rudd, D. F., *J. Phys. Chem.* **64**, 1254 (1960).
9. Ruoff, A. L., Stewart, G. H., Hyung Kyu Shin and Giddings, J. C., *Kolloid Z.* **173**, 14 (1960).
10. Erdös, E. and Vavruch, J., *Chem. Listy* **50**, 29 (1956).
11. Eliseeva, G. D., *Tr. Komis po Analit. Khim.* **6**, 439 (1955).
12. Waksmundzki, A. and Różyło, J., *Chem. Analit. (Warsaw)* **13**, 115 (1968).
13. Waksmundzki, A. and Różyło, J., *Chem. Analit. (Warsaw)* **13**, 104 (1968).
14. Waksmundzki, A. and Różyło, J., *Chem. Analit. (Warsaw)* **15**, 747 (1970).
15. Rachinskii, V. V. in *Stationary Phase in Paper and Thin Layer Chromatography*, Proc. 2nd Intern. Symp. Lieblice 1964, p. 284, K. Macek and I. M. Hais (eds.), Elsevier (1965).
16. Vanhaelen, M., *Chromatog. Electrophor. Symp. Int. 4th (Bruxelles)* 1966 (Publ. 1968), p. 160.
17. Ruoff, A. L., Prince, D. L., Giddings, J. C. and Stewart G. H., *Kolloid. Z.* **166**, 144 (1959).
18. Gocan, S. and Liteanu, C., *Bull. Soc. Chim. France* 1409 (1967).
19. Moelwyn-Hughes, E. A., *Physical Chemistry*, Pergamon Press, London (1961).
20. Bourdillon, J., *J. Chomatog.* **6**, 461 (1961).
21. Müller, R. H. and Clegg, D. L., *Anal. Chem.* **21**, 1429 (1949).
22. LeStrange, R. J. and Müller, R. H., Anal. Chem. **26**, 953 (1954).
23. Irreverre, F. and Martin, W., *Anal. Chem.* **26**, 257 (1954).
24. Hendrickson, M. J., Berueffy, R. R. and McIntyre, A. R., *Anal. Chem.* **29**, 1810 (1957).
25. Rutter, L., *Nature* **161**, 435 (1948); *Analyst* **75**, 37 (1950).
26. Osawa, Y., *Nature* **180**, 705 (1957).
27. Prey, V., Berbalk, H. and Kausz, M., *Mikrochim. Acta* 968 (1961).
28. Mathias, W., *Naturwiss.* **41**, 17 (1954).
29. Stahl, E., *Parfumen-Kosmetik* **39**, 564 (1958).
30. Ruoff, A. L. and Giddings, J. C., *J. Chromatog.* **3**, 438 (1960).
31. Oroveanu, T., *Scurgerea fluidelor prin medii poroase neomogene*, Ed. Academiei RPR, Bucharest (1963).
32. Dahn, H. and Fuchs, H., *Helv. Chim. Acta* **45**, 261 (1962).
33. Krulla, R., *Z. Phys. Chem.* **66**, 307 (1909).
34. Schmidt, H., *Kolloid Z.* **24**, 49 (1919).
35. Liteanu, C. and Gocan, S., *Rev. Chim. Acad. RPR* **7**, 1041 (1962).
36. Ackerman, B.-J. and Cassidy, H. G., *Anal. Chem.* **26**, 1874 (1954).
37. Lucas, R., *Kolloid Z.* **23**, 15 (1918).

38. Washburn, E. W., *Phys. Rev.* **17**, 276 (1921).
39. Bosanquet, C. H., *Phil. Mag. Ser.* 6, **45**, 525 (1923).
40. Henderson, H. J., *Producers Monthly* **14** (1), 32 (1949).
41. Childes, E. C. and Collis-George, N., *Proc. Roy. Soc., A* **201**, 392 (1950).
42. Fatt, I., *Petroleum Technology* **8**, 144 (1956).
43. Stewart, G. H., Chap. 3 in *Advances in Chromatography*, Vol. 2, J. C. Giddings and R. A. Keller (eds.), Arnold, London (1960).
44. Boltzmann, L., *Ann. Phys.* **53**, 959 (1894).
45. Giddings, J. C., Stewart, G. H. and Ruoff, A. L., *J. Chromatog.* **3**, 239 (1960).
46. Horner, L., Emrich, W., and Kirschner, A., *Z. Elektrochem.* **56**, 987 (1952).
47. Lederer, M., *Nature* **162**, 776 (1948).
48. Boman, H. G., *Nature* **170**, 703 (1952).
49. Munier, R., Macheboeuf, M. and Cherbier, N., *Bull. Soc. Chim. Biol.* **34**, 204 (1952).
50. Kowkabany, G. and Cassidy, H. G., *Anal. Chem.* **24**, 643 (1952).
51. Macek, K., *Chem. Listy* **48**, 1181 (1954).
52. Pollard, F. H., McOmie, J. F. W. and Jones, D. J., *J. Chem. Soc.* 4337 (1955).
53. Diximer, J., Dupuis, P. and Nortz, M., *Chim. Anal. Paris* **38**, 129 (1956).
54. Bouzková, J., Hejtmánek, M. and Vavruch, I., *Collection Czech. Chem. Commun.* **22**, 1219 (1957).
55. Bungenberg de Jong, H. G. and Hoogeveen, J. Th., *Koninkl. Ned. Akad. Wetenschap. Proc.* **64 B**, 1 (1961).
56. Tiselius, A., *Arkiv. Kemi Mineral Geol.* **14 B**, No. 22 (1940).
57. Claesson, S., *Arkiv. Kemi Mineral. Geol.* **20 A**, No. 3 (1945).
58. Griffiths, J., James, D. and Phillips, C., *Analyst* **77**, 897 (1952).
59. Fatt, I. and Selim, M. A., *J. Phys. Chem.* **63**, 1641 (1959).
60. Niederwieser, A. and Honegger, C. C., in *Advances in Chromatography*, Vol. 2, p. 123, J. C. Giddings and R. A. Keller (eds.), Arnold, London (1966).
61. Brenner, M., in *Stationary Phase in Paper and Thin-Layer Chromatography Proc. 2nd Intern. Symp. Lieblice* 1964, K. Macek and I. M. Hais (eds.), Elsevier (1965).
62. Martin, E. C., *J. Chromatog.* **10**, 338, 347 (1965).
63. Michal, J. and Ackermann, G., *Talanta* **11**, 441 (1964); **12**, 171 (1965).
64. Michal, J. and Ackermann, G., *J. Chromatog.* **33**, 38 (1968).
65. Janardhan, S. and Paul, A., *Indian J. Chem.* **5**, 297 (1967).
66. Liteanu, C. and Gocan, S., *Bull. Soc. Chim. France* 3836 (1967).
67. Bark, L. S. and Duncan, G., *J. Chromatog.* **49**, 278 (1970).
68. Scheidegger, A. E., *The Physics of Flow through Porous Media*, Revised Edition, University of Toronto Press (1960).
69. Collins, R. E., *Flow of Fluids through Porous Materials*, Reinhold, New York (1961).
70. Giddings, J. C., *Dynamics of Chromatography*, Part 1, p. 195, Dekker, New York (1965).
71. Kozeny, J. S. B., *Akad. Wiss. Wien. Abt. IIa* **136**, 271 (1927).
72. Carman, P. C., *Trans. Inst. Chem. Eng. London* **15**, 150 (1937).
73. Dal Negare, S. and Juvet, R. S., Jr., *Gas–Liquid Chromatography*, p. 135, Interscience, New York (1962).
74. Boremen, J. and Purnell, J. H., *J. Chem. Soc.* 360 (1961).
75. Bernard, R. A. and Wilhelm, R. H., *Chem. Eng. Progr.* **46**, 223 (1950).
76. Sternberg, J. C. and Poulson, R. E., *Anal. Chem.* **36**, 1492 (1964).
77. Pretorius, V. and Smuts, T. V., *Anal. Chem.* **38**, 274 (1966).

CHAPTER 2

KINETICS OF ZONE MIGRATION

2.1 Migration Rate of the Zone in Open Column Chromatography

2.1.1 Isothermal Chromatography

In elution chromatography, the components of a mixture migrate as zones or spots, at rates changing from zero to that of the eluent, the movement occurring only while the components are in the mobile phase. This means that the law for migration of the zone has the form of the equation for eluent migration. This was experimentally confirmed by Eliseeva [1] for isothermal chromatography at 20°. Similar results were obtained by Gocan and Liteanu [2] by studying the migration of Cu^{2+}, Ni^{2+} and Bi^{3+} at different temperatures (e.g. Fig. 2.1).

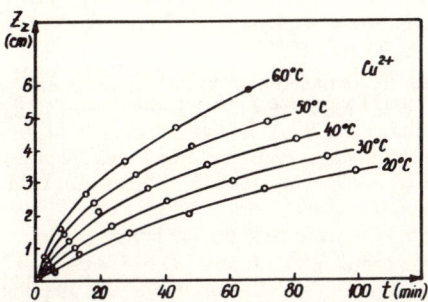

Fig. 2.1. Zone migration as function of time in isothermal conditions; eluent: n-butanol satd. with 4M HCl, Schleicher-Schüll paper 2040 bM [2].

The curves in Fig. 2.1 suggest that the zones in isothermal chromatography migrate according to a relation of the form

$$z_z = At^m \qquad (2.1)$$

or

$$\log z_z = \log A + m \log t \qquad (2.2)$$

Plots of $\log z_z$ vs. $\log t$, give a series of lines for different temperatures as in Fig. 2.2. Table 2.1 records the values obtained for A and m, and also

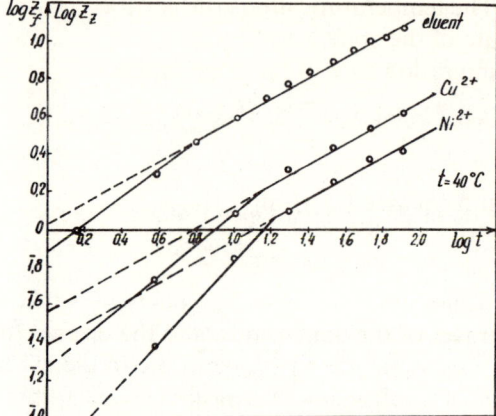

FIG. 2.2. Curves of zone migration [Equation (2.2)] at 40°C, eluent: n-butanol satd. with 4M HCl, Schleicher-Schüll paper 2040 bM, and of the same eluent [Eq. (1.7)] also at 40°C [2].

TABLE 2.1. The values of coefficients A, C and of exponents m and n for different temperatures and different ions (from [2] by permission of the copyright holders)

Eluent and paper	Temp. °C	m			n	A (cm/min^m)			C (cm/min^n)
		Cu^{2+}	Ni^{2+}	Bi^{3+}		Cu^{2+}	Ni^{2+}	Bi^{3+}	
n-Butanol satd. with 4N HCl Schleicher– Schüll 2040 bM	20	0·61	0·60		0·59	0·21	0·15		0·79
	30	0·56	0·56		0·54	0·31	0·21		1·06
	40	0·57	0·56		0·56	0·37	0·34		1·07
									1·07
	50	0·54	0·55		0·54	0·31	0·31		1·26
	60	0·51	0·50		0·51	0·67	0·41		1·46
n-Butanol satd. with 4N HCl Whatman No. 4	20			0·50	0·50			0·76	1·32
	40			0·51	0·50			0·85	1·50

the constants C and n for the eluents [see Eq. (1.6)] the origin of the graphs being taken as the starting line.

The constants A and m depend on the properties of the migrating substance and also on those of the stationary and mobile phases and on the other conditions used, i.e. ultimately on the partition coefficient between the two phases. The temperature affects the partition coefficient and hence the migration rate of the zone.

The average migration rate of the zone centre is

$$\bar{v}_z = \frac{z_z}{t} = At^{m-1} \tag{2.2}$$

and of the eluent front [see Eq. (1.6)]:

$$\bar{v}_f = \frac{z_f}{t} = Ct^{n-1} \tag{2.3}$$

The rates of travel of the zone centre and the eluent front are

$$v_z = \frac{dz_f}{dt} = mAt^{m-1} \tag{2.4}$$

and

$$v_f = \frac{dz_f}{dt} = nCt^{n-1} \tag{2.5}$$

The R_f value is defined as the ratio of the average migration rates $\bar{v}_z$ of the zone centre and $\bar{v}_f$ of the eluent front:

$$R_f = \frac{\bar{v}_z}{\bar{v}_f} = \frac{A}{C}t^{m-n} \tag{2.6}$$

Figure 2.1 shows that at a certain height, approximately 4–6 cm, the slopes n and m of the two lines become approximately equal, which means that the R_f value becomes

$$R_f = \frac{A}{C} \tag{2.7}$$

i.e. practically independent of the development time.

Table 2.2 gives the R_f values calculated by means of Eq. (2.7) and Table 2.1. The results are in good agreement with each other and with the research of Giddings *et al.* [4] which has shown that the R_f values have an initial period of change during which they approach the limiting values. This is illustrated in Fig. 2.3.

The theoretical approach to the problem of zone migration in ascending chromatography requires the introduction of an additional term (owing to the gravitational field) in the diffusion equation. This prevents a simple solution of the differential equation by means of Boltzmann's transformation [see Eq. (1.21)]. A solution is nevertheless possible by numerical calculus [3].

TABLE 2.2. R_f values calculated by means of Eq. (2.7) and of the experiment for Cu^{2+}, Ni^{2+} and Bi^{3+} and for different temperatures (from [2] by permission of the copyright holders)

Eluent and paper	Temp. °C	$R_f = A/C$			R_f (experimental)		
		Cu^{2+}	Ni^{2+}	Bi^{3+}	Cu^{2+}	Ni^{2+}	Bi^{3+}
n-Butanol satd. with 4N HCl Schleicher–Schüll 2040 bM	20	0.26	0.19		0.26	0.19	
	30	0.29	0.20		0.30	0.20	
	40	0.34	0.29		0.37	0.23	
	50	0.40	0.25		0.41	0.25	
	60	0.46	0.28		0.47	0.27	
n-Butanol satd. with 4N HCl Whatman No. 4	20			0.58			0.60
	40			0.57			0.60

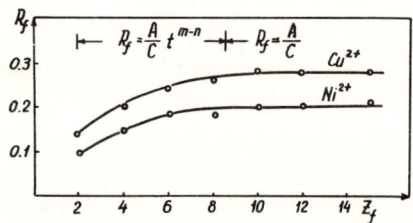

Fig. 2.3. Plot of R_f values as function of the height of the eluent migration in isothermal conditions at 20°C, eluent: n-butanol satd. with 4M HCl, Schleicher–Schüll paper 2040 bM [4]. (By permission of the copyright holders.)

2.1.2 Temperature-Gradient Chromatography

In the case of chromatography with a temperature gradient, it was established [2] that the centre of the zone shifts according to a law of the form:

$$z_z = bt^m \quad (2.8)$$

and the eluent front according to

$$z_f = \frac{t}{A + Bt} \quad (2.9)$$

The average rates of ascent of the zone centre and of the eluent front are

$$\bar{v}_z = \frac{z_z}{t} = bt^{m-1} \quad (2.10)$$

and

$$\bar{v}_f = \frac{z_f}{t} = \frac{1}{A + Bt} \quad (2.11)$$

The rates at any time t are given by

$$v_z = \frac{dz_z}{dt} = mbt^{m-1} \quad (2.12)$$

$$v_f = \frac{dz_f}{dt} = \frac{A}{(A + Bt)^2} \quad (2.13)$$

For R_f we obtain

$$R_f = \frac{\bar{v}_z}{\bar{v}_f} = b(At^{n-1} + Bt^m), \quad (2.14)$$

an equation showing that in the case of temperature-gradient chromatography, the values of R_f increase with the development time. The R_f values calculated by means of Eq. (2.14) and the experimental values are given in Table 2.3. There is good agreement between these values.

TABLE 2.3. R_f values calculated by means of Eq. (2.14) and from experiment. Eluent: n-butanol saturated with $4N$ HCl. Paper: Schleicher–Schüll 2040 bM [2]

Temp. gradient (°C/cm)	t (min)	R_f calc.		R_f exper.	
		Cu^{2+}	Ni^{2+}	Cu^{2+}	Ni^{2+}
0·66	50	0·26	0·20	0·28	0·20
	76	0·27	0·22	0·28	0·21
	108	0·30	0·24	0·28	0·23
1·33	63	0·27	0·20	0·27	0·20
	78	0·28	0·22	0·29	0·21
	128	0·30	0·25	0·32	0·24
2·00	51	0·29	0·20	0·25	0·18
	80	0·29	0·22	0·29	0·20
	105	0·30	0·23	0·29	0·21
	168	0·34	0·27	0·35	0·27

2.2 Relative Migration Rate of the Zone

The relative migration rate of the zone is characterised by the parameter R_f. Though defined in different ways, it may be considered as the fraction of the component that is dissolved in the eluent at a certain moment [5]. For instance if $R_f = 1/3$, we may consider that 1/3 of the molecules are to be found in the mobile phase and $1 - R_f = 2/3$ in the stationary phase.

If the mobile phase migrates at an average rate $\bar{v}$, the fraction R_f of the component, while dissolved in the mobile phase, migrates at the same rate $\bar{v}$. The rest is undoubtedly stationary. The average rate of the zone centre is $\bar{v}_z = R_f \bar{v}$, hence $R_f = \bar{v}_z/\bar{v}$ represents the relative migration rate of the zone centre.

The parameter R_f also gives information about the distribution of the component between the two phases, which takes place according to a reversible thermodynamic equilibrium. Thus, in every volume element of the column, the quantity of component in the mobile phase relative to the quantity in the stationary phase will be $A_M C_M/A_S C_S = A_M/\alpha A_S$ (where $\alpha = C_S/C_M$ is the partition coefficient, and A_M/A_S is the volume ratio of the mobile and stationary phases and C_M and C_S are the concentrations of the component in those phases. We can therefore write

$$\frac{R_f}{1 - R_f} = \frac{A_M}{A_S} \tag{2.15}$$

and

$$R_f = \frac{A_M}{A_M + \alpha A_S} \tag{2.16}$$

a relation equivalent to the one obtained by Martin and Synge [6].

Sometimes the relation may be produced by several mechanisms [7, 8] and in this case:

$$R_f = \frac{A_M}{A_M + \Sigma \alpha_i A_{S_i}} \tag{2.17}$$

Equation (2.16) can also be written in the form:

$$R_f = \frac{1}{1 + \alpha A_S/A_M} = \frac{1}{1 + k} \tag{2.18}$$

where $k = \alpha A_S/A_M$ is the capacity coefficient.

Another way of dealing with the problem is to consider the migration of the zone on the molecular scale. Let us suppose that t_a (adsorption time) is the average time elapsed between the adsorption of a molecule and a new adsorption, a time in which the molecule will move with the rate v of the eluent, and that t_d (desorption time) is the average time spent by an adsorbed molecule in the stationary phase till desorption. Giddings [9], starting from the equation given by LeRosen [10], writes that the ratio between the fraction R_f of molecules in the mobile phase, and the fraction $1 - R_f$ of molecules in the stationary phase is proportional to the ratio between the mean time needed for adsorption t_a and the mean desorption time t_d, i.e.

$$\frac{R_f}{1 - R_f} = \frac{t_a}{t_d} \tag{2.19}$$

or

$$R_f = \frac{t_a}{t_a + t_d} \tag{2.20}$$

In paper and thin-layer chromatography the relative migration rate of the zone centre is practically determined by the ratio

$$R_f = \frac{\bar{v}_z}{\bar{v}_f} = \frac{\text{Distance covered by the centre of the zone}}{\text{Distance covered by the eluent front}} \tag{2.21}$$

If the rate of the mobile phase at the position of the zone is equal to the rate of the front, and the ratio A_M/A_S constant over the whole length of the chromatographic strip, we can consider Eq. (2.16) as being rigorously valid also in the case of paper and thin-layer chromatography.

The Eq. (2.16) given by Martin and Synge [6] is certainly an approximation. This was discussed at length by Giddings and co-workers [3]. Because of the eluent's concentration profile, as well as the change in its composition (in the case of multicomponent eluents) the value of R_f depends on the migration distance of the zone. Generally the initial portion differs the most from the rest, owing to the S-form distribution of eluent concentration along the support (see Chap. 1). Moreover, the migration rate of the zone is not constant [1, 2, 11–13].

Bark and Duncan [14] have recently suggested an empirical relation for the dependence of R_f on α:

$$\alpha = A(1/R_f - 1) - B \tag{2.22}$$

where α is the experimental partition coefficient, and A and B are constants.

In the case of adsorption chromatography another approximate relation [15] for R_f can be given in the form

$$R_f = \frac{1}{1 + \alpha^* W_s/V_M} \tag{2.23}$$

where α^* is the partition coefficient [taken as the ratio of concentration of the component in the adsorbed phase (g/g) to that in the eluent (g/ml)], W_s is the quantity of adsorbent (g), and V_M is the volume of the mobile phase (ml). The ratio W_s/V_M is considered constant at every point behind the eluent front (at any moment during the separation time).

Generally, the experimental R_f value has proved to be smaller than the calculated value by a factor of 1·1–1·2. Therefore, in the systems of thin-layer adsorption chromatography where solvent gradients are formed (and of course when multicomponent solvents are used) the following relation is valid [15, 16]:

$$1 \cdot 1 R_f = \frac{1}{1 + \alpha^* W_s/V_M} \qquad (2.24)$$

The correlation between the relative migration rate of the zone in strip chromatography and in circular chromatography was studied by LeStrange and Müller [17] as well as by Giddings and co-workers [3]. These studies reached the conclusion that the R_f values obtained for strip chromatography are equal to the square of the R_R values obtained by circular chromatography:

$$R_f = R_R^2 \qquad (2.25)$$

a relationship experimentally checked by Ganguli [18].

Along with the usual chromatographic techniques, one-dimensional multiple chromatography (UMC) was evolved, and is one of the group of repeated chromatography methods. This procedure consists in the repeated irrigation of the same support (paper, thin layer) with the same eluent or with different eluents, in the same direction. An increase of R_f values is obtained by this procedure. UMC was studied from a theoretical standpoint in a series of works [19–26].

The distance covered by the centre of the zone after the first development is

$$z_z^{(1)} = LR_f$$

and the distance between the centre of the zone (the new starting line) and the front of the eluent is $L - LR_f$.

In the second development, the centre will shift by

$$z_z^{(2)} = LR_f(1 - R_f)$$

and in the third development by

$$z_z^{(3)} = LR_f(1 - R_f)^2,$$

and so on till the nth:

$$z_z^{(n)} = LR_f(1 - R_f)^{n-1} \tag{2.26}$$

The total distance z_z covered by the centre of the zone after n developments will be:

$$z_z = \sum_{i=1}^{n} z_z^{(i)} = L[1 - (1 - R_f)^n] \tag{2.27}$$

Thus, the value $R_f^{(n)}$ after n developments is

$$R_f^{(n)} = \frac{z_z}{L} = 1 - (1 - R_f)^n \tag{2.28}$$

By this technique a differentiated increase of the R_f values is obtained, and therefore an increased resolution on the chromatographic strip. The UMC technique has been used to separate mixtures of substances otherwise separable with difficulty in normal conditions [19, 22, 26–35].

REFERENCES

1. Eliseeva, G. D., *Tr. Komis. po Analit. Khim.* **6**, 439 (1955).
2. Gocan, S. and Liteanu, C., *Bull. Soc. Chim. France* 1409 (1969).
3. Giddings, J. C., Stewart, G. H. and Ruoff, R. L., *J. Chromatog.* **3**, 239 (1960).
4. Liteanu, C. and Gocan, S., *Bull. Soc. Chim. France* 4527 (1970).
5. Keller, R. A. and Giddings, J. K., in *Chromatography*, 2nd Ed., E. Heftmann (ed.), Reinhold, New York (1967).
6. Martin, A. J. P. and Synge, L. M., *Biochem. J.* **35**, 91 (1941).
7. Keller, R. A. and Stewart, G. H., *Anal. Chem.* **34**, 1834 (1962).
8. Knudson, G., Romaley, L. and Keller, R. A., *Anal. Lett.* **2**, 433 (1969).
9. Giddings, J. C., *Dynamics of Chromatography*, Part 1, *Principles and Theory*, Vol. I, Dekker, New York (1965).
10. LeRosen, A. L., *J. Am. Chem. Soc.* **67**, 1683 (1945).
11. Wood, D. E. and Strain, H. H., *Anal. Chem.* **26**, 260 (1954).
12. Rudd, D. F., *J. Phys. Chem.* **64**, 1254 (1960).
13. Stewart, G. H. and Gierke, T. D., *J. Chromatog. Sci.* **8**, 129 (1970).
14. Bark, L. S. and Duncan, G., *J. Chromatog.* **49**, 278 (1970).
15. Brenner, M., Niederwieser, A., Pataki, G. and Weber, R., in *Thin-Layer Chromatography*, E. Stahl (ed.), Academic Press, New York (1965).
16. Dallas, M. S. J., *J. Chromatog.* **17**, 267 (1965).
17. LeStrange, R. J. and Müller, R. H., *Anal. Chem.* **26**, 953 (1954).
18. Ganguli, N. C., *Anal. Chim. Acta* **12**, 335 (1955).
19. Jeanes, A. J., Wise, C. S. and Dimler, R. J., *Anal. Chem.* **23**, 415 (1951).
20. Thoma, J. A., *Anal. Chem.* **35**, 214 (1963).
21. Thoma, J. A., *J. Chromatog.* **12**, 441 (1963).
22. Chakrabortty, H. and Burma, D. P., *Anal. Chim. Acta* **15**, 451 (1956).
23. Stárka, L. and Hampl, R., *J. Chromatog.* **12**, 347 (1963).
24. Rüdiger, R. and Rüdiger, H., *J. Chromatog.* **17**, 186 (1965).
25. Lenk, H. P., *Z. Anal. Chem.* **184**, 107 (1961).
26. Goldstein, G., *Anal. Chem.* **42**, 140 (1970).
27. Franks, F., *Analyst* **81**, 384 (1956).
28. Giri, K. V., *Nature* **173**, 1194 (1954).
29. Severin, S. E. and Fedorova, V. N., *Dokl. Akad. Nauk SSSR* **82**, 443 (1953).
30. Chu, T. C., Green, A. A. and Chu, E. I., *J. Biol. Chem.* **190**, 643 (1951).
31. Pollard, F. W., McOmie, J. F. W. and Elbeih, I. I. M., *J. Chem. Soc.* 470 (1951).
32. Macek, K. and Vaněček, S., *Pharmazie* **10**, 422 (1955).
33. Jepson, J. B. and Smith, I., *Nature* **172**, 1100 (1953).
34. Mefferd, R. B., Summers, R. and Fernandez, J. G., *Anal. Lett.* **1**, 279 (1968).
35. Petrowitz, H. J., *Chem. Ztg., Chem. App.* **93** (9), 329 (1969).

CHAPTER 3

FACTORS INFLUENCING THE REPRODUCIBILITY OF R_f VALUES

In paper and thin-layer chromatography the R_f value permits identification of the analysed components as well as the retention parameters (time, volume, retention index) of gas and liquid-phase column chromatography. The difference between the R_f values of different components of a mixture in the same chromatographic system are due to their particular structural features, leading to differences between the values of the partition coefficients. The chromatographic conditions (stationary and mobile phase) will be chosen in such a way as to obtain the largest possible differences between the partition coefficients, so that the separation is as clear as possible.

The method of qualitative chromatographic analysis suggested by Pollard et al. [1, 2] and extended by Elbeih and Abou-Elnaga [3] uses comparison of R_f values, the constancy of which becomes an important problem. Galanos and Kapoulas [4] suggest a method of calculation of R_f values, to be applied to paper chromatography separations of any organic substances, making use of two reference components for the identification of an unknown substance.

In thin-layer chromatography Stahl [5] uses some dyes as reference substances. Other research-workers [6–8] recommend use of the ratio of the migration distances of the sample and a standard substance (i.e. ratio of the R_f values of the component and a standard). Turina and Horváth [9] recommend use of the ratio of the HETP for the unknown component and the reference.

It is therefore important to obtain reproducible values of R_f, and accordingly it is necessary to control all the factors influencing this parameter. The reproducibility of R_f values in thin-layer chromatography is poorer than in paper chromatography because it is more difficult to make the preparation of the chromatographic plates reproducible, notwithstanding standardisation of the working conditions.

3.1 Factors Influencing the R_f Values

A series of factors act differently during the chromatographic process and any modification of these factors will also modify the R_f values, so knowledge of these factors is essential. The factors may be grouped in four categories according to whether they are determined by (a) the chemical structure of the substance to be chromatographed, (b) the composition of the stationary phase, (c) the composition of the mobile phase, or (d) the medium. We shall deal below only with the last three categories, as they include the factors which can be modified during the chromatographic process and from one experiment to another.

3.1.1 Factors Determined by the Composition of the Stationary Phase

From research done by a series of authors [10–12] on comparison of different sorts of chromatographic paper it is evident that the paper must be chosen according to the mixture to be separated. Both the mixture of standard substances and that of unknown substances must be separated on the same type of chromatographic paper. The Whatman No. 1 and Schleicher-Schüll 2043 b MG1 papers, which have very similar qualities, may be recommended as standard papers, as a great many substances have been separated on them with good results. This is a decided advantage in comparison of the results of different authors.

Sommer [13] has studied the behaviour of a large number of inorganic ions on three sorts of Schleicher-Schüll chromatographic paper (2040 b, 2043 b and 2045 b), using as eluent a mixture of methanol, hydrochloric acid and water (80:10:10 v/v). This study shows different R_f values for the same ion on the different papers, the differences in R_f sometimes being as much as 0·14.

Hais [14] also draws attention to the role of the paper as a factor influencing R_f values. Churáček [15] has studied the source of deviations of R_f values in impregnated-paper chromatography.

Ackermann and Frey [16] have studied the effect of ion-exchange on chromatographic paper, deducing that in different conditions this effect may explain the formation of the chromatographic profile. They have also shown that the protons released by exchange can be detected on paper by different methods. The exchange capacity of cellulose is influenced by the lactone groups as well as by the dissociation of the acid groups, depending on the pH.

In the case of thin-layer chromatography, the quality of the adsorbent [17, 18] may even vary from one lot to another, and it is therefore advisable that the same adsorbent should be used for the same type of experiment.

The problem of the stationary phase in paper and thin-layer chromatography was the subject of the Liblice Symposium in 1961.

Waksmundzki and Różyło have studied the effect of the adsorbent structure on R_f values in thin-layer chromatography [20-23]. From these studies the conclusion is drawn that the R_f values usually increase with the decrease of specific area of the adsorbent for solvents such as water and carbon tetrachloride. For other solvents the effect is negligible. These variations may be explained in terms of the interaction between component, solvent and adsorbent. Also, the flow-rate of a solvent mixture (mobile phase) decreases with the increase of the specific surface of a silica gel stationary phase [24].

3.1.2 Factors Determined by the Composition of the Mobile Phase

The migration rate of inorganic substances varies with the nature and concentration of the acid added to the organic solvent [13, 25-34]. The R_f values generally increase with the concentration of the added acid, up to a certain concentration, after which they begin to decrease.

In the case of partition chromatography, the stationary phase is considered as a water-cellulose complex, and therefore the water in the eluent plays an important role in the chromatographic migration process of the ions [13, 30, 35, 36].

For the separation of ionisable substances in the aqueous phase and the non-ionisable ones in the organic phase in the partitition process, a double state of equilibrium will appear [37]. For this case an apparent partition coefficient α^* was defined, in the form $\log \alpha^* = \log \alpha - pK_A + pH$. If the pH of the eluent is changed, the partition coefficient of the component changes, and so does the R_f value [38].

The problem of the optimum pH for obtaining the best separations was discussed in a series of papers [39-42]. There is a simple method [43] of presenting the theoretical curves for variation of R_f value with pH in buffered-paper chromatography.

The effect of the mobile phase pH on the partition of penicillin and cephalosporins between a polar mobile phase and a non-polar stationary phase, was used in reverse-phase thin-layer chromatography. The R_M values of the compounds generally decrease with the rise in the pH [44].

Kraus and Dumont [45] have studied the influence of the pH value on the stationary phase after acclimatisation with different concentrations of acid or ammonia solutions, as well as the influence of the mobile phase pH in the thin-layer chromatography of caffeine, theophylline and theobromine. This study has made it clear that the separation of the three xanthines is influenced by the pH and by the water content of the stationary phase.

The influence of esterification on the reproducibility of results in paper chromatography has been little studied. Grüne [46] has studied the eluent made of n-butanol, acetic acid and water (80:20:20 v/v). The use of this eluent (kept at 20° for several days) has led to different values of R_f. When an eluent is made of solvents which do not react with one another, the mixture can be kept for some time. Sommer [47] has studied the behaviour of the eluent made of methanol, hydrochloric acid and water (80:10:10 v/v), using it immediately after preparation, after 5 days, and after 45 days, for a series of inorganic ions. The deviations of the R_f values were within experimental error.

Esterification of the eluent in thin-layer chromatography changes the polarity of the solvent and hence alters the R_f values.

The influence of the eluent composition on the form and position of the spots in paper chromatography was studied by Parissakis and Vrandi-Piscou [48] for a large number of inorganic ions with butanol–water–hydrochloric, nitric or acetic acid as eluent. This study showed that the R_f values increased with the increase in the acid or water concentration, or with the strength of the acid.

3.1.3 Factors Determined by the Medium

The effect of temperature on the values of $R_f = 1/(1 + \alpha V_S/V_M)$ appears in the change of the partition coefficient $\alpha = \text{const.} \, e^{-\Delta H^0/RT}$, where ΔH^0 is the sorption enthalpy change of the component, R the gas constant, and T the absolute temperature, as well in change of the ratio of the phase volumes V_S/V_M.

Some quantitative relations between R_f values and the composition of some types of solvent mixture have been deduced for paper chromatography [49–52].

Waksmundzki and Różyło [53] have studied the variation of the R_f values of substances of various chemical structure vs. the composition of the solvent, plotting diagrams which can be used in practical chromatography. The ascending technique was used, on silica gel as adsorbent.

Souto et al. [54] have shown that the information given by chromatographic data will chiefly depend on three factors: (a) the distribution of R_f values of the reference substances in every solvent, (b) the relation between the R_f values of a substance in different solvents, (c) the experimental error.

From all this, the general conclusion is that every change in the eluent composition during the chromatographic process or from one experiment to another, can change the R_f values.

The influence of temperature on R_f values, and the use of this factor to improve separations, was studied from the beginning of chromatography. Hais and Macek [55] discuss the results of the first work [56–61] on the influence of temperature in the chromatography of organic substances, when it was found that in separating amino-acids, with collidine and water as eluent, a sudden change of R_f values appears at temperatures between 10 and 20° [56]. This is explained by the fact that the critical solubility temperature of water in collidine occurs in this temperature interval. Neiman et al. [62] showed that in every case the R_f values at 40° are higher than at 20°.

Lederer [63] showed that a series of equilibrium processes change with the temperature: the dissociation and association of substances chromatographed, the partition equilibria, and the degree of hydration of the cellulose.

The temperature effect is used in the separation of amino-acids on paper [64] impregnated with ion-exchangers [65]. French et al. [66] separated the oligosaccharides of amidone by paper chromatography at high temperature. Tirzite et al. [67] studied the correlation between R_f values and temperature (10–30°) in the paper chromatography of nucleotide bases.

Brenner et al. [18] showed that the effect of temperature on amino-acid separation is bigger for paper chromatography than for thin-layer chromatography. Geiss and Schlitt [68] showed that the degree of activity of the adsorbent changes with rise of temperature (the amount of water retained by the support decreases), which changes the R_f values. Lederer and co-workers [69] have also studied the effect of temperature variation in partition and ion-exchange paper chromatography for a few inorganic ions and have found that temperature variations have an important effect in partition chromatography, but that the effect is very slight in ion-exchange paper chromatography.

Sommer [70] studied the temperature effect on R_f values for alkaline and alkaline-earth metals. For these the R_f values increase with rise in temperature. Tewari [71–73] using circular development on paper, found a decrease of R_f with rise of temperature (in the 10–40° interval) for the cations of the first, second and third analytical groups. Using a mixture of acetone–water–hydrochloric acid (80:20:10 v/v) as eluent, Majumdar and Chakrabarty [74] noted an increase of R_f with temperature for the cations Cu^{2+}, Pb^{2+}, Co^{2+} and Ni^{2+}.

The effect of temperature on the relative migration rate of some cations was also noticed by Ritchie [75], and Ossicini [76] has studied the effect of low temperatures.

Lederer [77] found that increasing temperature of separation of some rare earths in the system acetylacetone–acetone–water or acetylacetone–acetic acid–water (30:10:10 v/v) made the spots narrower. Parissakis and Vandi-Priscou [78] studied the effect of temperature on the form and position of the spots for a series of salts. The paper chromatograms were run at 30, 35 and 40° in a butanol–acetic acid (30:10 v/v) solvent. The spots generally became narrower in the direction of travel, as the temperature was increased.

Liteanu and Gocan [79–83] have studied the influence of temperature on the separation of some inorganic ions on paper with various eluents and found that temperature acts differentially on the migration rate of the components. The temperature is therefore a parameter which must be taken into account in the chromatographic process.

The effect of the temperature on R_f in thin-layer chromatography has also been studied [84–88] and the effect found to be small. Stahl [89] recommends the use of low temperatures in some cases to improve the resolution.

Another important factor is the influence of the composition of the gaseous phase in the chromatographic chamber. We shall discuss below the degree of saturation of the atmosphere in the chromatographic chamber, the composition of the vapour in the chamber and the water content of the adsorbent layer or of the paper, and the way in which these factors influence the R_f value. The role of these factors must be considered in correlation with the stationary phase, or more precisely with its formation. The problem is of high interest in thin-layer chromatography, and in paper chromatography considered as a special case of thin-layer chromatography. The importance of the influence of the vapour phase on the R_f values in thin-layer chromatography was amply discussed by Reimers [90].

In adsorption chromatography, the active centres on the adsorbent at the gas–solid interface, are occupied or blocked by the components of the surrounding gaseous mixture, more or less according to their concentrations. Geiss *et al*. [91] have thus found a threefold bigger variation of R_f values in experiments in unsaturated atmospheres than in those done in a saturated atmosphere. Stahl [92] has proved that in an unsaturated chamber, the so-called 'edge effect' appears, owing to the more intense evaporation at the edge of the plate. This effect may be avoided by complete saturation of a normal chamber with eluent vapour, or by using a chamber with a very small volume, as for instance a sandwich chamber [93] or analogous devices [94–100].

The fact that the chamber is unsaturated is not unfavourable to separations, and indeed as de Zeeuw [101–105] has shown, in some cases it has improved the separation, the reproducibility being the same as in a saturated chamber, provided that the working conditions are the same each time. The fact of not being saturated leads not only to the passage of a supplementary volume of eluent [102], but also, owing to the preferential adsorption of the polar component of the solvent mixture evaporated into the gas phase, during the eluent migration, an activity gradient on the stationary phase. This is due to the fact that the upper portion of the plate is longer in contact with the vapour mixture in the chamber, which is then more saturated than at the beginning of the migration. These findings have led to the chamber with vapour gradient [106]. Though it might seem that the sandwich chamber would lead to impossibility of diffusion of eluent vapour towards the top of the plate in the tank, it was proved [90] that adsorption of the polar component also takes place here, evidently in a slightly smaller proportion than in a normal chamber [90, 107, 108]. The conditions of total lack of saturation with eluent vapour are obtained in an unsaturated sandwich chamber designed by Geiss and Schlitt [109].

Vanhaelen [110] also shows that the R_f values and the shapes of the spots in thin-layer chromatography depend on the saturation of the developing chambers.

The effect of humidity on the stationary phase, and hence on the R_f values, is a problem of importance in adsorption chromatography, especially when non-polar eluents are used [18, 84, 86, 91, 105, 106, 111–116]. It was proved that silica gel G adsorbs in 3 min 50 per cent of the equilibrium quantity of moisture and that the final equilibrium is reached in approximately 15 min in an atmosphere of 57 per cent relative humidity. In an atmosphere of 80 per cent relative humidity, alumina reaches equilibrium in 4 min [91]. This means that after the time that the spot is applied, the plate is dried and introduced into the chamber, the absorbent will contain nearly as much moisture as it would if, after application of the dry layer, it had been left in air at room temperature (so activation in this case is useless). Moreover even the moisture usually contained by the chamber atmosphere (supposing that the solvent does not contain water) will considerably diminish the activity [111]. Devices have been made which keep the humidity constant [84, 86, 109, 111, 118] in the chromatographic chamber, or create an orthogonal humidity gradient [119], the so-called 'KS-Vario' chambers [120], meant to allow the finding of optimum working conditions. The influence of humidity on the most common adsorbents was studied [91], for different solvents. Up till the appearance of this study it was known that when non-polar solvents were used with polar adsorbents for separation of certain lipophilic components, increase in the humidity usually led to an increase of the R_f values, owing to decrease in the adsorption power of the solvent [84, 87, 91, 111, 112, 115]. It also results that the humidity has smaller effects when the polarity of the eluent and of the component increase.

In the case of paper, which can be placed in the category of polar adsorbents because of the hydrophilic nature of cellulose and the large surface, increase in humidity will lead to decrease of the R_f values [121–

127]. This is explained by the fact that the volume of the stationary phase increases with increase in humidity[1] and according to the definition $R_f = 1/(1 + \alpha V_S/V_M)$ will lead to diminution of the R_f value.

However, non-polar adsorbents, as for instance the polyamides, active carbon and graphite, are not influenced by variations in the water vapour concentration as they retain water to a lesser degree, 'unspecifically', and this does not deactivate the adsorbent, even when ethanol or methanol are used as eluents.

Of course, if only the separation itself is important, we may neglect the humidity and even the saturation, at the risk of a lower reproducibility, but in the study of the effect of different factors, lack of control of the humidity can bring about some inconsistencies. Thus for instance Pataki and Keller [128] have found the value of R_f to increase with the thickness of the layer of adsorbent, while Dallas [129] and Jänchen [130] have proved that this effect is due to the decrease of saturation of the adsorbent through the increase of its thickness.

The R_f values are also influenced by the particle size of the adsorbent [20–23, 131, 132], increasing as the particles become finer.

Honegger [133] recommends the use of the narrow saturated SK chamber for the study of theoretical problems. The technique of horizontal chromatography in narrow saturated chambers is recommended for the determination of absolute R_f values. If the atmosphere in the chromatographic chamber is not saturated, the solvents on the layer evaporate until the atmosphere *is* saturated, so that a bigger quantity of developer becomes necessary to cover a given distance. The development time thus increases [134]. Owing to the supplementary volume of preferentially adsorbed polar component and to the dehomogenisation of the eluent, the values of R_f will increase [102, 135, 136].

3.2 Reproducibility of R_f Values in Paper Chromatography

Taking into account the factors determining the variation of R_f values, a series of works [13–15, 71, 137–150] recommend the following working conditions to obtain the most reproducible values:

(*a*) use of the same chromatographic paper
(*b*) uniform distribution of the stationary phase
(*c*) same degree of impregnation of the paper
(*d*) absence of any secondary reactions between the solvents composing the eluents
(*e*) constant volume ratio of stationary phase to mobile phase
(*f*) constant temperature during development
(*g*) the balancing of humidity of the paper with the humidity of the atmosphere in the chromatographic chamber

(h) same degree of saturation of the chamber
(i) absence of foreign substances
(j) the quantity of substances applied to be approximately the same
(k) constant time for development
(l) the distance between the start-line and the final position of the eluent front to be the same
(m) heating of the substances used, at a suitable temperature, before their chromatography.

Apart from the result of neglect of the conditions above, deviations of the R_f values may also appear, as shown by Grüne [151], as the consequence of isomeric changes or dissociation or association during chromatography. Also too high a heating of amino-acids on the chromatogram leads to their reacting with the paper, and therefore to errors.

Fairbairn and Relph [152] showed that in quantitative paper and thin-layer chromatography hand-spotting may be a source of errors. This can be greatly reduced by mechanical spotting. They also showed that a bigger source of errors is to be found in the different chromatographic behaviour from one sheet of paper to another. These differences can be corrected by using standard solutions and test solutions for each sheet.

Finley [153] showed that in standard working conditions, standard deviations between 0·010 and 0·032 can be obtained for the values of R_f. Pokorný and Hladík [154], chromatographing oxidised alkyl oleates on paper, obtained an estimated standard deviation between 0·0001 and 0·020. In separation of cholesterol esters on paper, Tichý and Dencker [155] obtained for 30 determinations a standard deviation between 0·026 and 0·036.

Liteanu and Gocan [156, 157] demonstrated that in isothermal chromatography, as in temperature gradient chromatography, the distribution of the R_f values is normal. Once it is established that the R_f values for a certain chemical species obey the Gauss normal distribution law, the statistical processing of these results by use of the characteristic parameters of this distribution, $\bar{X}$, σ and z, or those of the 'Student' distribution $\bar{X}$, s and t, is fully justified. A total of 200 measurements was made for the Fe^{3+} ion, with ascending isothermal development at 20°, in n-butanol saturated with $4N$ hydrochloric acid, on Whatman No. 4 paper. An average $R_f = 0.084$ and an estimated standard deviation of 0·010 were obtained.

3.3 Reproducibility of R_f Values in Thin-layer Chromatography

At the beginning of thin-layer chromatography Kirchner and Miller [158, 159] showed that through a strict control of the working conditions (standardisation) the R_f values may be reproduced to within ± 0.05.

The factors favouring high reproducibility of the R_f values were discussed in a series of papers [18, 84, 87, 91, 103, 131, 133, 160–178], and it was shown that the following conditions must be observed:
  (a) careful preparation and uniform spreading of the layer
  (b) the quality of the adsorbent
  (c) the quality of the eluent, the purity and age of the mixture
  (d) the working temperature
  (e) the degree of saturation of the chamber
  (f) the degree of activation of the layer
  (g) the granulation of the adsorbent
  (h) the quantity of substance put on the plate
  (i) the distance of development and the distance between the level at immersion and the start line
  (j) no condensation should be produced on the plate covering the chamber
  (k) the edges are not taken into consideration
  (l) for the ascending technique the migration time should not be longer than 45 min
  (m) narrow chambers should be used.

Thus Bark et al. [178] described a technique for the multiple simultaneous spotting of thin-layer chromatographic plates. Benzene solutions of Sudan Red G, indophenol blue and p-dimethylazobenzene were used as test substances. The elution was done with a mixture of petrol ether (b.p. 40–60°) and diethyl ether (70:30 v/v) in a saturated chamber. The results show that the multiple simultaneous spotting method gives smaller variations of the R_f values, the reproducibility being ± 0.01.

Ebing [176] gave a method of chromatographic analysis of chlorinated hydrocarbon pesticides offering a high reproducibility of R_f values. He preferred to use single-component solvents. The maximum deviation from the average R_f value was not higher than 0.03, the average deviations were 0.008, and the standard deviations 0.012.

Tichý et al. [155] obtained for different cholesterol esters chromatographed on a thin layer of silica gel G activated for an hour at 120°, a standard deviation between 0.015 and 0.023 for 100 measurements. It was also shown that for 30 measurements by paper chromatography the standard deviation was between 0.026 and 0.036. Other research workers [18, 154] have obtained similar values for standard deviations in paper and thin-layer chromatography.

Macek [179] showed that the non-reproducibility of R_f values is chiefly due to the existence of gradients during the separation. The first problem to be solved is thus to diminish the formation of the gradient of the stationary phase. This can be achieved by applying the sample on an open chromatographic column which is already working. In this case, the stationary phase being already in equilibrium with the mobile phase, the situation is similar to the one in column chromatography. The second problem is that of ensuring the reproducibility of the gradients, by standardising the working conditions. One solution would be to apply the standard on the same chromatogram as the substance to be analysed. In this way the gradient of the solvent's system will influence equally the movement of all the components.

In conclusion, it can be said that nowadays, through the standardisation of working conditions, thin-layer chromatography has reached the standards of paper chromatography regarding the reproducibility of R_f values.

REFERENCES

1. Pollard, F. H. McOmie, J. F. W. and Elbeih, I. M., *J. Chem. Soc.* 471 (1951).
2. Pollard, F. H., McOmie, J. F. W. and Stevens, H. M., *J. Chem. Soc.* 771 (1951).
3. Elbeih, I. I. M. and Abou-Elnaga, M. A., *Anal. Chim. Acta* **17**, 397 (1957).
4. Galanos, D. S. and Kapoulas, V. M., *J. Chromatog.* **13**, 128 (1964).
5. Stahl, E., *Arch. Pharm.* **292/64**, 411 (1959).
6. Peereboon, J. W. C., *J. Chromatog.* **3**, 323 (1960).
7. Dhont, J. H. and Rooy, C., *Analyst* **86**, 74, 527 (1961).
8. Lisboa, B. P. and E. Diczfalusy, *Acta Endocrinol.* **40**, 60 (1962).
9. Turina, S. and Horváth, L., *J. Chromatog.* **33**, 402 (1968).
10. Kowkabany, G. N. and Cassidy, H. G., *Anal. Chem.* **22**, 817 (1950).
11. Macek, K. and Hacaperková, J., *Chem. Listy* **51**, 895 (1957).
12. Pohloudek-Fabini, R. and Wollmann, H., *Pharmazie* **15**, 590 (1960).
13. Sommer, G., *Z. Anal. Chem.* **147**, 241 (1955).
14. Hais, I. M., *J. Chromatog.* **33**, 25 (1968).
15. Churáček, J., *Sb. Ved. Pr., Vys. Str. Chemicko-technol., Pardubice* **21** (Pt. 3), 51 (1969).
16. Ackermann, G. and Frey, H.-P., *Z. Anal. Chem.* **233**, 321 (1968).
17. Tschesche, R., Lampert, F. and Snatzke, G., *J. Chromatog.* **5**, 217 (1961).
18. Brenner, M., Niederwieser, A., Pataki, G. and Fahmy, A. R., *Experientia* **18**, 101 (1962).
19. Macek, K. and Hais, I. M. (eds.), *Stationary Phase in Paper and Thin-Layer Chromatography*, Elsevier, Amsterdam (1964).
20. Waksmundzki, A. and Różyło, J., *Ann. Univ. Marie Curie-Sklodowska, Sect. A* **20**, 93 (1965) (Pub. 1967).
21. Waksmundzki, A. and Różyło, J., *Chim. Analit.* (*Warsaw*) **13**, 715 (1968).
22. Waksmundzki, A. and Różyło, J., *J. Chromatog.* **33**, 90 (1968).
23. Waksmundzki, A. and Różyło, J., *J. Chromatog.* **33**, 96 (1968).
24. Waksmundzki, A. and Różyło, J., *Chim. Analit.* (*Warsaw*) **13**, 1041 (1968).
25. Elbeih, I. I. M., McOmie, J. H. W. and Pollard, F. H., *Disc. Faraday Soc.* **7**, 183 (1949).
26. Lederer, M., *Microchim. Acta* 43 (1956).
27. Kertes, S. and Lederer, M., *Anal. Chim. Acta* **15**, 543 (1956).
28. Almássy, Gy. and Dezsö, I., *Magy. Kem. Foly.* **62**, 60 (1956).
29. Carvalho, R. G., *Anal. Chim. Acta* **16**, 555 (1957).
30. Carvalho, R. G., *J. Chromatog.* **4**, 353 (1960).
31. Născuţiu, T., *Rev. Roumaine Chim.* **9**, 273 (1964).
32. Născuţiu, T., *Rev. Roumaine Chim.* **10**, 989 (1965).
33. Născuţiu, T., *Rev. Roumaine Chim.* **12**, 839, 845 (1967).
34. Născuţiu, T. and Iliescu, V., *Rev. Roumaine Chim.* **12**, 853 (1967).
35. Chakrabarty, S. and Burma, D. P., *Sci. Culture* (*Calcutta*) **16**, 485 (1951).
36. Ćelap, M. B., Janjić, T. J. and Urosevica, S. E., *Glas. Hem. Drus.* (*Beograd*), **25/26**, 393 (1960–61).

References

37. Golumbic, C., Orchin, M. and Weller, S., *J. Am. Chem. Soc.* **71**, 2624 (1949).
38. Carless, J. E. and Woodhead, H. B., *Nature* **168**, 203 (1951).
39. Scozewiński, E., *Nature* **188**, 391 (1960).
40. Waksmundzki, A. and Soczewiński, E., *Roczniki Chem.* **32**, 863 (1958).
41. Waksmundzki, A. and Soczewiński, E., *Roczniki Chem.* **33**, 1423 (1959).
42. Andreev, L. G., *Sb. Nauk. Tr. Tsentr. Nauk Issled. Ap. Techn. Inst.* **5**, 191 (1964).
43. Waksmundzki, A. and Scozewiński, E., *J. Chromatog.* **3**, 252 (1960).
44. Biagi, G. L., Barbaro, A. M. and Guerra, M. C., *J. Chromatog.* **51**, 548 (1970).
45. Kraus, L. J. and Dumont, E., *J. Chromatog.* **48**, 96 (1970).
46. Grüne, A., *Allg. Papierdsch.* **677**, 716 (1956).
47. Sommer, G., *Z. Anal. Chem.* **151**, 336 (1956).
48. Parissakis, G. and Vrandi-Piscou, D., *Journées Hellénes des séparation immédiate et de chromatographie*, p. 43 (III-émes J.I.S.I.C.) (1965).
49. Soczewiński, E., *J. Chromatog.* **8**, 119 (1962).
50. Soczewiński, E. and Wachtmeister, C. A., *J. Chromatog.* **7**, 311 (1962).
51. Connors, K. A., *Anal. Chem.* **40**, 1386 (1968).
52. Perisho, C. R., *Anal. Chem.* **40**, 551 (1968).
53. Waksmundzki, A. and Różyło, J., *J. Chromatog.* **49**, 313 (1970).
54. Souto, J. and Gonzalez de Valesi, A., *J. Chromatog.* **46**, 274 (1970).
55. Hais, T. M. and Macek, K., *Paper Chromatography*, p. 105, Publishing House of the Czechoslovak Academy of Sciences, Prague (1963).
56. Kowkabany, G. and Cassidy, H. G., *Anal. Chem.* **24**, 643 (1953).
57. Counsell, J. N., Haugh, L. and Wadman, W. H., *Research (London)*, **4**, 143 (1951).
58. Jörgenson, B., *Univ. Texas Pubs.* 5109, 56 (1951).
59. Paladini, A. C. and Leloir, L. F., *Anal. Chem.* **24**, 1024 (1952).
60. Burma, D. P., *Nature* **168**, 565 (1951).
61. Boscott, R. J., *Biochem. J.* **51**, XIV (1952).
62. Neiman, M. B., Levkovskii, V. N. and Lukovnikov, A. F., *Dokl. Akad. Nauk SSSR* **81**, 841 (1951).
63. Lederer, M., *Electrical Phenomena and Solid/Liquid Interface, Proceedings of the Second International Congress of Surface Activity*, p. 506, Butterworths, London (1957).
64. Silbalić, S. M. and Radej, N. V., *Anal. Chem.* **33**, 1223 (1961).
65. Knight, C. S., *Nature* **200**, 1316 (1963).
66. French, D., Mancusi, J. L., Abdullah, M. and Brammer, G. L., *J. Chromatog.* **19**, 445 (1965).
67. Tirzite, G., Durburs, G. and Abolina, F., *Khim. Geterotsikl. Soedin.* 940 (1968).
68. Geiss, F. and Schlitt, H., *J. Chromatog.* **33**, 208 (1968).
69. Lederer, M., Mancini, P. and Ossicini, L., *J. Chromatog.* **12**, 89 (1963).
70. Sommer, G., *Z. Anal. Chem.* **147**, 241 (1955).
71. Tewari, S. N., *Naturwiss* **41**, 229 (1954).
72. Tewari, S. N., *Kolloid Z.* **138**, 178 (1954).
73. Tewari, S. N., *Z. Anal. Chem.* **141**, 401 (1954).
74. Majumdar, A. K. and Chakrabarty, M. M., *Anal. Chim. Acta* **17**, 315 (1957).
75. Ritchie, A. S., *J. Chromatog.* **10**, 281 (1963).
76. Ossicini, L., *J. Chromatog.* **42**, 159 (1969).
77. Lederer, M., *Compt. Rend.* **236**, 1557 (1923).

78. Perissakis, G. and Vrandi-Piscou, D., *Journées Hellénes des séparation immediate et de chromatographie* (III-émes JISIC) (1965), p. 61.
79. Liteanu, C. and Gocan, S., *Rev. Chim. Acad. RPR* **7**, 1041 (1962).
80. Liteanu, G., Gocan, S. and Onişor, M., *Stud. Univ. Babes-Bolyai, Chem.* **11**, 79 (1966).
81. Liteanu, C. and Gocan, S., *Bull. Soc. Chim. France* 1416 (1969).
82. Gocan, S. and Liteanu, C., *Bull. Soc. Chim. France* 1409 (1969).
83. Liteanu, C., Gocan, S. and Hodişan, T., *Rev. Roumaine Chim.* **15**, 1751 (1970).
84. Dallas, M. S. J., *J. Chromatog.* **17**, 267 (1965).
85. Honegger, C. G., *Helv. Chim. Acta* **46**, 1772 (1969).
86. Geiss, F., Schlitt, H., Ritter, F. J. and Weimar, W. M., *J. Chromatog.* **12**, 469 (1963).
87. Cone, N. J., Miller, R. and Neuss, N., *J. Pharm. Sci.* **52**, 688 (1963).
88. Furukawa, T., *J. Sci. Hiroshima, Univ. Ser. A* **21**, 285 (1958).
89. Stahl, E., *Angew. Chem. (Intern. Ed. Engl.)* **3**, 784 (1964).
90. Reimers, F., *Dansk. Tidsskr. Farm.* **42**, 204 (1968).
91. Geiss, F., Schlitt, H. and Klose, A. *Z. Anal. Chem.* **213**, 331 (1965).
92. Stahl, E., *Pharm. Rundsch.* **1**, (2), 1 (1959).
93. Stahl, E., *Dünnschicht Chromatographie*, 2 Aufl. p. 70, Springer-Verlag, Berlin (1967).
94. Brenner, M. and Niederwieser, A., *Experientia* **17**, 237 (1961).
95. Davies, B. H., *J. Chromatog.* **10**, 518 (1963).
96. Hara, S., Takeuchi, M. and Matsumato, N., *Bunseki Kagaku* **13**, 359 (1964).
97. Jänchen, D., *J. Chromatog.* **14**, 261 (1964).
98. Reimers, F. and Thomsen, H., *Medd. Norsk. Farm. Selskap.* **28**, 178 (1966).
99. Reimers, F. and Thomsen, H., *Arch. Pharm. Chemi* **74**, 117 (1967).
100. Wassicky, R., *Naturwiss.* **50**, 569 (1963).
101. de Zeeuw, R. A., *Pharm. Weekblad.* **102**, 113 (1967).
102. de Zeeuw, R. A., *J. Chromatog.* **32**, 43 (1968).
103. de Zeeuw, R. A., *J. Chromatog.* **33**, 222 (1968).
104. de Zeeuw, R. A., *J. Chromatog.* **33**, 227 (1968).
105. de Zeeuw, R. A., *Anal. Chem.* **40**, 2134 (1968).
106. de Zeeuw, R. A., *Anal. Chem.* **50**, 915 (1968).
107. Reimers, F., *Arch. Pharm. Chem.* **74**, 531 (1967).
108. Thomsen, H. and Reimers, F., *Arch. Pharm. Chem.* **74**, 969 (1967).
109. Geiss, F. and Schlitt, H., *Chromatographia* **1**, 387 (1968).
110. Vanhaelen, M., *Ann. Pharm. Franc.* **26**, 565 (1968).
111. Badings, H. T., *J. Chromatog.* **14**, 265 (1964).
112. Geiss, F. and Schlitt, H., *Naturwiss.* **50**, 350 (1963).
113. Geiss, F., Schlitt, H., and Klose, A., *Z. Anal. Chem.* **213**, 321 (1965).
114. Kelemen, J., and Pataki, G., *Z. Anal. Chem.* **195**, 81 (1963).
115. Hesse, G., Engelhardt, H. and Kowallik, W., *Z. Anal. Chem.* **214**, 81 (1966).
116. Sandroni, S. and Geiss, F., *Chromatographia* **2**, 165 (1969).
117. Reichel, W. L., *J. Chromatog.* **26**, 304 (1967).
118. Geiss, F. and Schlitt, H., *J. Chromatog.* **33**, 208 (1968).
119. Niederwieser, A., *Chromatographia* **2**, 23 (1969).
120. Geiss, F. and Schlitt, H., *Chromatographia* **1**, 392 (1968).
121. Hanes, C. S., *Canad. J. Biochem. Physiol.* **39**, 119 (1961).
122. Hanes, C. S., Harris, C. K., Mascarello, M. A. and Tigane, D., *Canad. J. Biochem. Physiol.* **39**, 163 (1961).

123. Leemann, H. G. and Stich, K., *Helv. Chim. Acta* **45**, 1275 (1962).
124. Schroder, W. A., Shelton, J. R., Shelton, J. B., Cormick, J. and Jones, R. T., *Biochemistry* **2**, 992 (1963).
125. Tomisek, A. J. and Allan, P. W., *J. Chromatog.* **14**, 232 (1964).
126. Tomisek, A. J. and Allan, P. W., *J. Chromatog.* **33**, 35 (1968).
127. Wade, E. H. M., Matheson, A. T. and Hanes, C. S., *Canad. J. Biochem. Physiol.* **39**, 141 (1961).
128. Pataki, G. and Keller, M., *Helv. Chim. Acta* **46**, 1054 (1963).
129. Dallas, M. S. J., *J. Chromatog.* **33**, 193 (1968).
130. Jänchen, D., *J. Chromatog.* **33**, 195 (1968).
131. Stárka, L. and Hampl, R., *J. Chromatog.* **12**, 347 (1963).
132. Vaedtke, J., Gajewska, A. and Czarnocka, A., *J. Chromatog.* **12**, 208 (1963).
133. Honegger, C. G., *Helv. Chim. Acta* **46**, 1730 (1963).
134. Copius-Peereboom, J. W. and Beekes, H. W., *J. Chromatog.* **20**, 43 (1965).
135. Stewart, C. H. and Gierke, T. D., *J. Chromatog. Sci.* **8**, 129 (1970).
136. Gocan, S. and Liteanu, C., *Rev. Roumaine Chim.* **17**, 661 (1972).
137. Bate-Smith, E. C., *Partition Chromatography, Biochem. Soc. Symp.* **3**, 62 (1960).
138. Zimmermann, G., *Z. Anal. Chem.* **138**, 321 (1953).
139. Lacourt, A. and Heyndryckx, P., *Microchim. Acta* 1211 (**1956**).
140. Clayton, R. A., *Anal. Chem.* **14**, 209 (1964).
141. Grüne, A., *Chimia* **11**, 173 (1957).
142. Lacourt, A., *Mikrochim. Acta* 269 (**1957**).
143. Lacourt, A., *Mikrochim. Acta* 700 (**1957**).
144. Decker, P., *Naturwiss.* **45**, 464 (1958).
145. Hasegawa, H., *Yakugaku Zasshi* **80**, 1175 (1960), **81**, 655, 985, 994 (1961).
146. Racinski, V. V., *Gen. Probl. Paper Chromatog. Symp. Liblice Czech.* **1961** 49 (Publ. 1962).
147. Gouthier, H. and Mangeney, G., *J. Chromatog.* **14**, 209 (1964).
148. Matthias, W. and Wagner, J., *J. Chromatog.* **33**, 316 (1968).
149. Churáček, J., *J. Chromatog.* **33**, 45 (1968).
150. Soczewiński, E. and Mańko, R., *J. Chromatog.* **33**, 40 (1968).
151. Grüne, A., *J. Chromatog.* **33**, 28 (1968).
152. Fairbairn, J. W. and Relph, S. J., *Chromatographia* **2**, 204 (1969).
153. Finley, K. T., *J. Chromatog.* **19**, 443 (1965).
154. Pokorný, J. and Hladík, J., *J. Chromatog.* **33**, 267 (1968).
155. Tichý, J. and Dencker, S. J., *J. Chromatog.* **33**, 262 (1968).
156. Liteanu, C. and Gocan, S., *Stud. Univ. Babes-Bolyai, Chem.* **13**, 135 (1968).
157. Liteanu, C. and Gocan, S., *Stud. Univ. Babes-Bolyai, Chem.* **14**, 29 (1969).
158. Kirchner, J. C., Miller, J. M. and Keller, G. J. *Anal. Chem.* **23**, 420 (1951).
159. Miller, J. M. and Kirchner, J. C., *Anal. Chem.* **25**, 1107 (1953).
160. Stahl, E., Schroeter, G., Kraft, G. and Renz, R., *Pharmazie* **11**, 633 (1956).
161. Furukawa, T., *J. Sci. Hiroshima Univ. Ser. A* **21**, 285 (1958); C.A. **53**, 809 (1959).
162. Brenner, M. and Niederwieser, A., *Experientia* **16**, 378 (1960).
163. Černý, V., Joska, J. and Lábler, L., *Collection Czech. Chem. Commun.* **26**, 1658 (1961).
164. Birkofer, L., Kaiser, C., Meyer-Stoll, H. A. and Suppan, F., *Z. Naturforsch.* **17B**, 352 (1962).
165. Vaedtke, J., Gajewska, A. and Czarnocka, A., *J. Chromatog.* **12**, 208 (1963).

166. Brodasky, T. F., *Anal. Chem.* **36**, 996 (1964).
167. Pataki, G., *Helv. Chim. Acta* **47**, 784 (1964).
168. Versino, C., Fogliano, L., Giaretti, F., *Riv. Combust.* **20**, 527 (1966).
169. Geiss F. and van der Venna, M. Th., *Chromatog. Electrophor. Symp. Int.*, 4th 1966, p. 153 (Publ. 1968).
170. Geiss, F., *J. Chromatog.* **33**, 9 (1968).
171. Dallas, M. S. J., *J. Chromatog.* **33**, 58 (1968).
172. Pataki, G. and Zürcher, H., *J. Chromatog.* **33**, 103 (1968).
173. Jolliffe, G. H. and Shellard, E. J., *J. Chromatog.* **33**, 165 (1968).
174. Wohlleben, G., *Z. Anal. Chem.* **243**, 498 (1968).
175. Waksmundzki, A. and Różyło, J., *Wiad. Chem.* **23**, 1 (1969).
176. Ebing, W., *J. Chromatog.* **44**, 81 (1969).
177. Halpaap, H., *J. Chromatog.* **33**, 144 (1968).
178. Bark, L. S., Graham, R. J. T. and McCormick, D. M., *Talanta* **12**, 122 (1965).
179. Macek, K., *J. Chromatog.* **33**, 257 (1968).

CHAPTER 4

ESTIMATION OF SEPARATION EFFICIENCY

4.1 Separation Efficiency of a Closed Chromatographic Column

The separation efficiency of a chromatographic column is proportional to the number of theoretical plates in the column, given by the length of the column divided by the height equivalent to a theoretical plate (HETP). The notion of height equivalent to a theoretical plate was introduced by Martin and Synge [1] and developed later by Mayer and Tompkins [2] and Glueckauf [3]. The basis of every chromatographic process is the principle of repetition of equilibration between the phases. According to this principle, the height equivalent to a theoretical plate, H, represents the length of the column on which one equilibration of the solute between the mobile and the stationary phase is achieved. The concentration profile of the solute in the effluent from the column has the form of a Gaussian curve if the sorption isotherm is linear (Fig. 4.1).

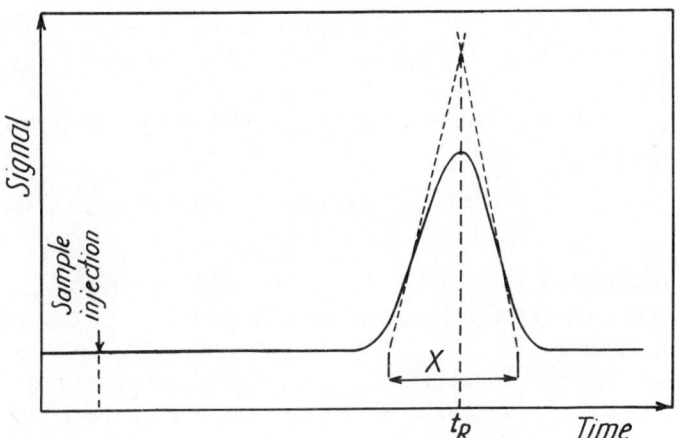

FIG. 4.1. Characteristics of a chromatogram and the shapes of recorded peaks at the outlet of the column.

In Fig. 4.1 t_R represents the retention time of the solute to be separated, that is to say the time elapsed between the insertion of the sample and the appearance of the maximum concentration in the detector (the maximum of the chromatographic peak), and X denotes the width of the base of the peak, as determined by tangents drawn through the inflexion points. The area under the peak is proportional to the quantity of solute.

Strictly speaking, the concentration profile is described by a Poisson distribution but if the number of equilibrations is large enough, this approximates to a Gaussian distribution.

According to the plate theory the number of theoretical plates (N) of a chromatographic column can be calculated from

$$N = 16\frac{t_R^2}{X^2} \tag{4.1}$$

The standard deviation σ of the Gaussian peak is approximately $X/4$, so Eq. (4.1) may be written as

$$N = \frac{t_R^2}{\sigma^2} \tag{4.2}$$

The length of the column is given by $L = R_f v t_R$, so L and t_R are proportional since R_f and v are constant for a given solute and chromatographic system. Thus, t_R and σ can be replaced by measures proportional to the length of the column, that for σ being $\sigma R_f v$. Consequently Eq. (4.2) may be written:

$$N = \frac{L^2}{R_f^2 v^2} \cdot \frac{R_f^2 v^2}{\sigma^2} = \frac{L^2}{\sigma^2} \tag{4.3}$$

where σ is now measured in terms of length, and the height of the theoretical plate will be defined thus:

$$H = \frac{L}{N} = \frac{\sigma^2}{L} \tag{4.4}$$

The height of the theoretical plate is generally accepted as a measure of the zone width and of the separation efficiency of the column for a particular solute.

The theoretical-plate theory does not relate the physical and thermodynamic characteristics of the phases and the internal geometry of the column to the parameter H. This is one of the greatest shortcomings of the theory. Another is that the plate theory tackles the chromatographic process somewhat statically and does not offer the possibility of an *a priori* calculation of the efficiency.

A step towards better understanding of the phenomena taking place in the column was taken by the kinetic theories [4]. However, it must be stressed that these theories do not contradict the plate theory; on the contrary, they complement it.

The kinetic theories succeed in expressing the dependence of the variance of the zone, σ^2, on a series of physiochemical and geometrical parameters of the chromatographic system. These expressions can sometimes have simplified forms, but are generally complicated, owing to the fact that a great number of factors influencing the value of H are taken into account. This arises from the complexity of the chromatographic processes. The more accurate the description is to be, the higher the number of parameters needed and the more complicated the mathematical expression.

Taking into account longitudinal diffusion [4–8], eddy diffusion [4, 9–14] and the local non-equilibrium [15–25], Giddings [4] shows that for most analytical systems, H may be expressed by the equations:

$$H = \frac{2\gamma D_M}{v} + \omega \frac{d_p^2 v}{D_M} + qR_f(1 - R_f)\frac{d^2 v}{D_S}, \qquad (4.5)$$

when the partition mechanism is dominant, and

$$H = \frac{2\gamma D_M}{v} + \omega \frac{d_p^2 v}{D_M} + 2R_f(1 - R_f)\bar{t}_d, \qquad (4.6)$$

when the adsorption mechanism is dominant. In these equations, γ is the tortuosity factor for diffusion through the support material, D_M and D_S are the diffusion coefficients for the mobile and stationary phases respectively, ω is the coefficient expressing the contribution of the mobile phase to the height of the plate, q is the flux of fluid per unit cross-sectional area of the column, d_p is the grain-diameter, d is the thickness of unit volume of the stationary phase, R_f is the relative migration rate of the zone, v is the migration rate of the eluent and $\bar{t}_d$ is the average desorption time; γ, ω and q are structural coefficients.

In Eqs. (4.5) and (4.6) the first term represents the contribution of molecular diffusion, the second eddy diffusion and the third local non-equilibrium. These relations have a particular importance in chromatographic practice, as they indicate precisely which factors must be adjusted to obtain the lowest possible value for H, that is to say a column with the highest possible efficiency.

Recently Huber [26] expressed the height of the theoretical plate as a sum of terms corresponding to four independent column processes: molecular diffusion, convection, and mass transfer in the mobile and stationary phases.

The stochastic approach [27–33] to chromatographic processes gives complicated expressions (for example for the probability density, etc.) of little practical interest. Oxtoby, nevertheless, arrived finally in a recent paper [34] at the following equation for the HETP:

$$H = \frac{2D_M}{v} + \frac{2\alpha}{\lambda(1+\alpha)^2}v, \tag{4.7}$$

in which D_M is the diffusion coefficient in the mobile phase, v the flow-rate of the eluent and $\alpha = \mu/\lambda$ is the partition coefficient ($\mu = p_s/\Delta t$ and $\lambda = p_M/\Delta t$, where p_s and p_M are the probabilities of the passage of a molecule from one phase to the other in the time interval Δt, p_s for stationary → mobile phase, p_M for mobile → stationary phase). This equation has a similar form to the van Deemter equation [9]:

$$H = A + \frac{B}{v} + Cv$$

The optimum flow-rate of the mobile phase, v_{opt}, is inferred by differentiating Eq. (4.7) with respect to v and equating it to zero:

$$v_{opt} = (1+\alpha)\sqrt{\frac{D_M \lambda}{\alpha}} \tag{4.8}$$

This must be taken into account when studying the optimization of the separating conditions.

The stochastic approach does not take into account all the parameters involved. By simple computing methods more complex expressions have been derived for H, namely Eqs. (4.5) and (4.6), which contain more of the factors involved in the chromatographic process. Although these solutions are ingenious and rigorous, the models chosen are still far from reality; for instance diffusion is considered as a simple process, local equilibrium is not taken into account, etc.

An important step forward in chromatographic process theory on the position, width and form of the chromatographic peak was taken by applying the method of statistical moments to the solving of the differential equations governing the process.

The statistical moment of first order μ' characterises the retention parameter, the central statistical moment of second order μ_2 is equal to the variance σ^2, and characterises the width of the peak. The third order central moment, μ_3 characterises the asymmetry of the peak.

In developing the transport theory for chromatographic systems with linear partition, Kragten [35] solved the differential equations governing the transport of matter, by the method of statistical moments. He finally obtained an expression for the second order central statistical moment μ_2, and hence the width of the chromatographic zone:

$$\mu_2 = \mu_{0_2} + 2\left[\left(\frac{1}{1+\alpha}\right)D_f + \left(\frac{\alpha}{1+\alpha}\right)D_b\right]t$$
$$+ 2\left[\left(\frac{1}{1+\alpha}\right)x_f^2 + \left(\frac{\alpha}{1+\alpha}\right)x_b^2\right]\frac{t}{\tau} \quad (4.9)$$

where μ_{0_2} represents the width of the concentration peak at $t = 0$, α is the partition coefficient, x_f and x_b are the co-ordinates of the mass centres in the free and bound states, τ is the relaxation time (the average time for a molecule to stay in the free or bound state), and D_f and D_b are the diffusion coefficients in the free and bound states.

Equation (4.9) shows that the peak broadens by diffusion, in proportion to the weighted mean of D_f and D_b. Besides diffusion there is another process of peak broadening, proportional to the relaxation time. This contribution is due to the kinetics of the separation processes and has its maximum value for $\alpha = 1$, and is zero for $\alpha = 0$ and $\alpha = \infty$. According to Eq. (4.4) $H = \mu_2/L$ and $N = L^2/\mu_2$ and thus the value of H rises with increase in μ_2 and that of N decreases, in other words the efficiency of the column decreases.

On the basis of a physical model, Vink [36–40] has worked out the theory of partition chromatography by solving the differential equations for the operational conditions on a chromatographic column, and also used the moment method. From his studies we may draw the following conclusions regarding the influence of lateral diffusion (local non-equilibrium) on the efficiency of chromatographic separations.

(a) It reduces the efficiency of separation, causing spreading of the zones (increase of μ_2) and decreasing N.

(b) It affects only slightly the rate of movement of the chromatographic peak (the retention volume).

(c) It is generally impossible to separate two solutes by column chromatography based only on the diffusion coefficients, but if one has a very low diffusion coefficient, the two solutes can easily be separated. Thus it can be concluded that in partition chromatography the separation is chiefly due to the difference between the partition coefficients.

(d) Increase in longitudinal diffusion in the mobile phase broadens the peak, and therefore decreases the efficiency.

Kučera [20] characterised the chromatographic concentration peak by means of the same statistical moments. As already said, the standard deviation $\sigma = \sqrt{\sigma^2} = \sqrt{\mu_2}$, characterises the width of the chroma-topographic peak. The value $2\sigma\sqrt{2}$ represents the width of the elution curve at a height $1/e$ of the maximum, and the ratio $2\sigma\sqrt{2}/\sqrt{\mu'}$ defining the relative width of the peak has an important significance in determining the separation efficiency of the column [41, 42].

The optimum velocity, v_{opt}, for which the relative width of the chromatographic peak has a minimal value, may be derived from the minimum condition:

$$\frac{d}{dv}(2\sigma\sqrt{2}/\sqrt{\mu'_1}) = 0$$

Kučera's model [20] takes into consideration the longitudinal diffusion in the mobile phase, as well as the radial diffusion in the grains of the stationary phase, the finite mass-transfer rate through the interfaces, and the sorption on the inner surface of the grains. The following expressions are obtained for the statistical moments μ'_1 and μ_2:

$$\mu'_1 = \left(\frac{L}{v} + \frac{2D_M}{v^2}\right)[1 + \varepsilon K_c(1 + K_n)] \quad (4.10)$$

$$\mu_2 = \left(\frac{2D_M L}{v^3} + \frac{8D_M^2}{v^4}\right)[1 + \varepsilon K_c(1 + K_n)]^2$$

$$+ \left(\frac{2L}{v} + \frac{4D_M}{v^2}\right)\varepsilon K_c\left[\frac{r_p^2}{D_S}\frac{(1 + K_n)^2}{v(v + 2)} + \frac{\varepsilon(1 + K_n)^2}{l_c} + \frac{K_n}{l_n}\right] \quad (4.11)$$

where L stands for the length of the column, v the flow-rate of the eluent, D_M the longitudinal diffusion coefficient in the mobile phase (molecular diffusion and contribution of turbulence), D_S the radial diffusion coefficient in the stationary phase, r_p the radius of a grain, K_c and K_n the equilibrium constants at transfer and sorption, l_c and l_n being the corresponding mass-transfer coefficients, $\varepsilon = \phi(1 - \psi)/\psi$, where ϕ is the inner and ψ the outer porosity), and v is a parameter connected with the shape of the particle ($v = 3$ for spherical particles, 2 for cylindrical particles and 1 for pellet-shaped particles). The equilibrium constants are related to the partition coefficient α by $\alpha = \varepsilon K_c(1 + K_n)$. D_M depends only on v, according to an approximate equation of the type:

$$D_M = D + B_1 v + B_2 v^2 \quad (4.12)$$

where D is the diffusion coefficient at $v = 0$, and B_1 and B_2 are empirical constants depending on the characteristics of the column and its packing.

Equation (4.11) shows that $\sigma^2 (\equiv \mu_2)$ increases proportionally to L, D_M, K_c and K_n, and r_p, and decreases with increase in D_S, in the mass-transfer coefficients and in v.

By substituting μ'_1 and μ_1 from Eqs. (4.10) and (4.11) in the expression $2\sqrt{2\mu_2}/\sqrt{\mu'_1}$ and taking into account Eq. (4.12) we obtain an expression which, after differentiation with respect to v and equating to zero, gives a fourth degree equation in v, which practically cannot be solved. For high values of L, the terms containing L in the denominator are neglected and the fourth degree equation is reduced to the form

$$(B_2 + \chi)v^2 - D = 0 \tag{4.13}$$

where

$$\chi = \frac{\dfrac{r_p^2(1 + K_n)^2}{D_S v(v + 2)} + \dfrac{\varepsilon(1 + K_n)}{l_c} + \dfrac{K_n}{l_n}}{[1 + \varepsilon K_c(1 + K_n)]^2} \tag{4.14}$$

The optimum velocity, v_{opt}, is obtained after solving equation (4.13), rewritten as

$$v_{opt} = \sqrt{\frac{D}{B_2 + \chi}} \tag{4.15}$$

This relationship gives us the parameters needed to obtain chromatographic peaks with the minimum relative width and hence to obtain the most efficient separations.

Kubín [43, 44] has given approximate methods for computing the elution curve, starting directly from the Laplace–Carson transform of the mass-transport differential equation, based on the relations inferred for the statistical moments of the elution curve. Considering only longitudinal diffusion we obtain for the second order central statistical moment the equation:

$$\mu_2 = 2D_M \frac{K}{v^3} x + \frac{t^2}{12} \tag{4.16}$$

where v is the linear flow-rate of the eluent, t_0 the time needed to introduce the sample, $K = 1 + k_1/a$, (where k_1 is the constant in the adsorption isotherm equation and a is the free-volume fraction of the column) and x is the distance covered.

According to Eq. (4.16) the second order central statistical moment μ_2 is proportional to t_0^2, which in turn is proportional to the volume of sample. It follows that increase in sample volume increases μ_2, and hence H, and the separation efficiency decreases.

If longitudinal diffusion is neglected and only inner diffusion considered [43], the elution curves for different values of r_p^2/D_S and constant values

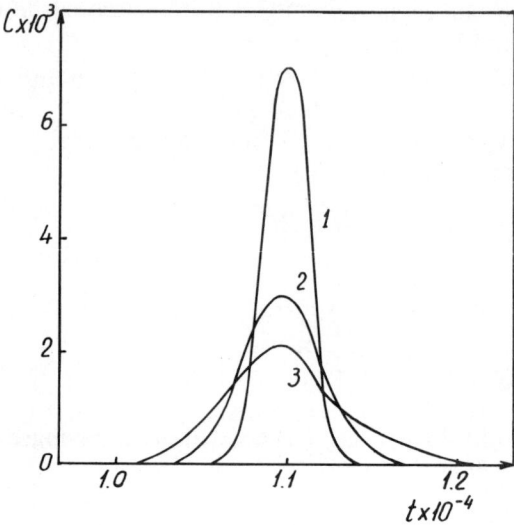

Fig. 4.2. Shapes of elution curves for three values of the ratio r_p^2/D_S: 1—$r_p^2/D_S = 100$ s, 2—$r_p^2/D_S = 500$ s, 3—$r_p^2/D_S = 1000$ s. ($x = 10^2$ cm, $v = 10^{-2}$ cm/s, $h = 0.1$, $t_0 \to 0$) [43].

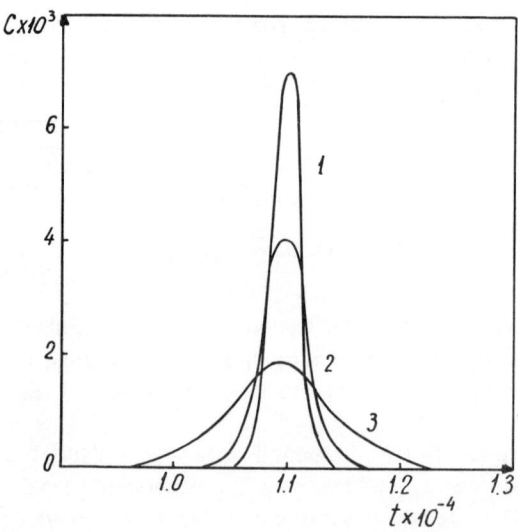

Fig. 4.3. Shapes of elution curves for three values of the longitudinal diffusion coefficient D_M: 1—$D_M = 10^{-5}$ cm^2/s, 2—$D_M = 10^{-4}$ cm^2/s, 3—$D_M = 10^{-3}$ cm^2/cm. ($r_p^2/D_S = 100$ s, $x = 10^2$ cm, $v = 10^{-2}$ cm/s, $h = 0.1$, $t_0 \to 0$) [43].

of x, v, t_0, c and h are as shown in Fig. 4.2. In this case the central statistical moment of second order is:

$$\mu_2 = 2\frac{hxr_p^2}{15vD_S} + \frac{t_0^2}{12} \tag{4.17}$$

where $h = \alpha(1 - a)/a$.

If both longitudinal and radial diffusion are considered, the second order statistical central moment will be:

$$\mu_2 = 2\left[\frac{hr_p^2}{15D_S} + D_M\left(\frac{1+h}{v}\right)^2\right]\frac{x}{v} + \frac{t_0^2}{12} \tag{4.18}$$

and Fig. 4.3 gives the corresponding elution curves for different values of D_M and constant values of r_p^2/D_S, x, v, h and t_0. As shown by Fig. 4.3 not only the width of the elution curves, but also the symmetry, increases when the value of D_M increases. This may be seen from the expression for the third order central statistical moment:

$$\mu_3 = 3\left[\delta_2 + 4D_M\delta_1 + \frac{1+h}{v^2} + 4D_M^2\frac{1+h}{v^3}\right]\frac{x}{v}, \tag{4.19}$$

where

$$\delta_1 = \frac{h}{15}\frac{r_p^2}{D_S} \quad \text{and} \quad \delta_2 = \frac{12}{945}\left(\frac{r_p^2}{D_S}\right)^2,$$

and is also confirmed by qualitative results obtained by means of an analogue computer, simulating the chromatographic process [43].

For ion-exchange chromatography, Dybczyński [45] showed that the number of theoretical plates in a column may be calculated on the Mayer and Tompkins [21] theory, starting from the formula [46]:

$$N = 8\left(\frac{K_d}{K_d + 1}\right)\frac{(V_R - V_M)^2}{w^2} \tag{4.20}$$

where K_d is the distribution coefficient (ratio of solute concentrations in the resin and solution), V_R the retention volume of the solute, V_M the interstitial volume of the column, and w the width of the peak at the concentration $C = C_{max}/e = 0.369\, C_{max}$.

The value $K_d/(K_d + 1)$ tends towards 1 when $K_d > 10$, and Eq. (4.20) becomes

$$N = 8\frac{(V_R - V_M)^2}{w^2} \tag{4.21}$$

This equation is identical to the equation deduced by Glueckauf [3] and shows the advantage of the continuous flow model against the discontinuous flow model used by Mayer and Tompkins [2].

Because the elution curves are Gaussian, w has the value $2\sigma\sqrt{2}$, where σ is the standard deviation of the peak. Substitution for w in Eq. (4.21) gives

$$N = \frac{(V_R - V_M)^2}{2\sigma^2} \tag{4.22}$$

and H is given by

$$H = \frac{L}{N} = \frac{L^2\sigma^2}{(V_R - V_M)^2} \tag{4.23}$$

The value of H in ion-exchange chromatography may also be deduced by means of the theoretical formula [47, 48]:

$$H = 1.64\, r_p + \frac{K_d}{(K_d + F_I)^2}\frac{0.142\, r_p^2 v}{D_S} + \frac{K_d^2\, 0.266\, r_p^2 v}{(K_d + F_I)^2 F_{II} D_M(1 + 70\, r_p v)}$$

$$+ \frac{D_M F_I \sqrt{2}}{v} \tag{4.24}$$

where r_p is the radius of the resin particles, v is the linear velocity of eluent, K_d the distribution coefficient, D_S the diffusion coefficient in the resin, D_M the longitudinal diffusion coefficient in solution, F_I the void fraction of column and $F_{II} = (1 - F_I)$ is the fraction of column occupied by the stationary phase.

As shown in Eq. (4.24), H contains four terms: the first represents the effect of the particle size, the second diffusion in the resin particles, the third diffusion in the liquid film surrounding the resin particles and having a thickness of approximately 10^{-3} cm, and the last longitudinal diffusion. The last term is negligible except at very low flow-rates.

Figure 4.4 shows the factors controlling the height of the theoretical plate. The plate height depends mainly on the linear flow-rate, but if diffusion in the particle is dominant it also depends on the distribution coefficient. In practice, diffusion in the particles is usually a controlling factor.

Diffusion in the liquid film is dominant if the distribution coefficient is high and the flow-rate is optimal. For high flow-rates the height of the theoretical plate becomes independent of the distribution coefficient, and, in this case the thickness of the liquid film becomes smaller with the increase in flow-rate and the contribution of diffusion in the film becomes unimportant.

Separation Efficiency of a Closed Chromatographic Column

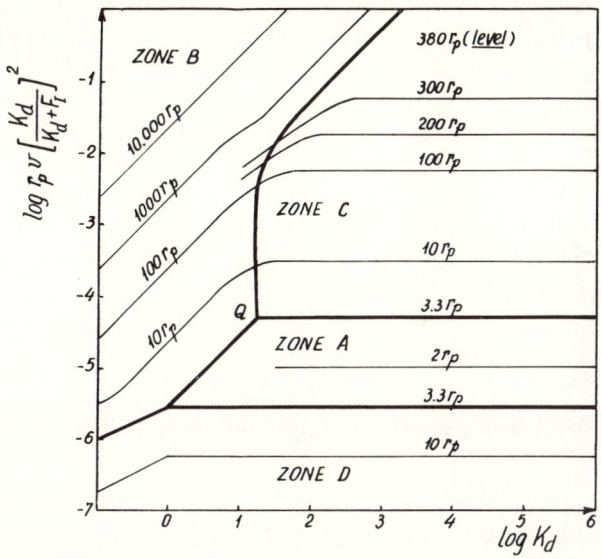

FIG. 4.4. Factors controlling theoretical plate height. Zone A, effective equilibrium; zone B, particle diffusion; zone C, film diffusion; zone D, longitudinal diffusion in column. Contour lines give effective plate height. $D_M = 10^{-5}$ cm^2/s; $D_S = 3 \times 10^{-7}$ cm^2/s; r_p = radius of resin beads [47].

The best operating conditions are found in the region of point Q. In these conditions all three terms contribute about equally to the value of H, which is then approximately 5 times the radius of the resin particles.

Let us consider an ideal chromatogram of two solutes A and B (Fig. 4.5). To obtain a clear separation, the following condition must be fulfilled [49]:

$$V_R^B - V_R^A \geqq \tfrac{1}{2}W_A + \tfrac{1}{2}W_B \qquad (4.25)$$

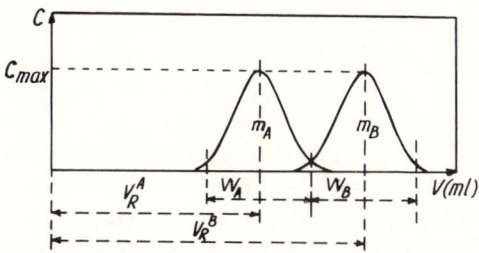

FIG. 4.5. Elution diagram of two components of similar behaviour [49].

where V_R is the retention volume for the species indicated and W is the peak-width. The fundamental equation deduced by Glueckauf [47] states that

$$C_{max} = \frac{m_i}{V_i}\sqrt{\frac{N}{2\pi}}$$

where N is the number of theoretical plates, and m_i the total number of moles of the ith solute, which has retention volume V_i and since

$$m_B \sim \tfrac{1}{2}W_B C^B_{max} \quad \text{and} \quad m_A \sim \tfrac{1}{2}W_A C^A_{max}$$

then

$$W_B \sim 2V^B_R\sqrt{\frac{2\pi}{N}} \quad \text{and} \quad W_A \simeq 2V^A_R\sqrt{\frac{2\pi}{N}}$$

The retention volume may be expressed by means of the distribution coefficient and the parameters of the column by

$$V_R = V(K_d + F_I)$$

where V is the total volume of the column.

Equation (4.25) may also be written:

$$\frac{V^B_R}{V^A_R} - 1 \geq \frac{\tfrac{1}{2}(W_A + W_B)}{V^A_R} \tag{4.26}$$

and after substitution gives

$$\frac{K^B_d + F_I}{K^A_d + F_I} - 1 \geq \frac{(K^A_d + F_I) + (K^B_d + F_I)}{(K^A_d + F_I)}\sqrt{\frac{2\pi}{N}}$$

hence:

$$N \geq 2\pi\left[\frac{K^B_d + F_I}{K^A_d + F_I} + 1\right]^2 \bigg/ \left[\frac{K^B_d + F_I}{K^A_d + F_I} - 1\right]^2 \tag{4.27}$$

By means of this relation, deduced by Inczédy [49], and the distribution coefficients of the two substances, the number of theoretical plates necessary for their separation can be calculated. The values obtained for different ratios of $(K^B_d + F_I)/(K^A_d + F_I)$ were calculated and compared with the data of Glueckauf's diagram [47] which is based on a rigorous mathematical treatment of the overlap of two chromatographic peaks. The values calculated by means of Eq. (4.27) correspond to separations giving products with <0·3 per cent contamination if the quantities of the two components are in 1:1 ratio, but the real curves always differ from the ideal form and therefore the calculations are only approximate and Eq. (4.27) may only be used for preliminary calculations when planning separations.

Experimental determination of the number of plates and the HETP. The number of plates N is calculated from Eq. (4.1) written in the form

$$N = 16\left(\frac{V'_R}{w}\right)^2 \tag{4.28}$$

where the corrected retention volume V'_R is the total volume of eluent necessary for elution of the band centre, less the hold-up volume in the tubing between column and detector cell, and w is the band width (equal to 4σ) expressed as a volume and determined in the same way as in Fig. 4.1.

The height equivalent to the theoretical plate, H, is calculated from $H = L/N$. Snyder [50] showed that $H_{real} = H_{observed} - H_{(extra\text{-}column)}$ in which $H_{(extra\text{-}column)}$ is determined practically and represents the contribution to H of the volumes of the connecting tube between column and detector, of the detector, and of the sample introduction unit.

4.2 Separation Efficiency of an Open Chromatographic Column

In a series of works [51–53] it was shown that the HETP concept may also be applied to chromatographic systems having an irregular eluent flow-rate. These studies are based on the well-known van Deemter equation [9]:

$$H = A + \frac{B}{v} + Cv \tag{4.29}$$

where A is the eddy-diffusion coefficient, B the molecular diffusion coefficient, C the non-equilibrium coefficient, and v the solvent velocity at the zone of solute. Thus H_{obs} is the average plate-height observed for a completed chromatogram, because v in Eq. (4.29) represents the average velocity of the eluent along the column [54].

De Ligny and van de Meent [55] showed that in paper or thin-layer chromatography the contribution of the macroscopic mobile phase velocity profile to longitudinal dispersion may be described by

$$H = \frac{\sigma_K^2}{l} = \frac{B}{v} + C_M v + C_S v + C_F(v)v + C_E(v)v \tag{4.30}$$

where σ_K is the standard deviation of the peak for the solute in the medium used for chromatography, l is the distance travelled by the solute and

v the mean flow-rate of the eluent. The coefficients in the various terms are concerned with longitudinal diffusion (B), resistance to attainment of the partition equilibrium in the mobile phase (C_M) and in the stationary phase (C_S), the macroscopic mobile phase velocity [$C_F(v)$] and eddy diffusion [$C_E(v)$]. The expressions for these terms for chromatography on paper and thin layers are given in the papers by de Ligny et al. [56–59]. To conclude, longitudinal dispersion depends on transverse (convective and diffusive) dispersion, the latter being decisive in paper as well as thin-layer chromatography.

De Ligny and Bax [52] showed that peak-broadening in paper chromatography at low eluent flow-rates (0·0005 cm/s) is exclusively caused by longitudinal diffusion, both in the mobile and the stationary phase. The contributions of eddy-diffusion and of resistance to mass-transfer to the broadening of the peak in paper chromatography are negligible. This broadening may be described by

$$\frac{\sigma^2}{l} = \left[1{\cdot}36\, D_M + 0{\cdot}12\, D_S \frac{(1 - R_f)}{R_f}\right] \cdot \frac{1}{v} \qquad (4.31)$$

For eluent velocities >0·002 cm/s de Ligny and Remijnse [60] also demonstrated the slow attainment of partition equilibrium. The slow mass-transfer between the mobile and the stationary phase appeared to be caused mainly by slowness of diffusion in the mobile phase [56]. Similar results were obtained for thin-layer chromatography [57].

Mallik and Giddings [51] have studied the role of non-equilibrium in the broadening of the zone, and related measurements of C (Eq. 4.29) to factors affecting mass-transfer in the stationary phase.

Stewart has given an ample account of the structural factors of the stationary phase in paper chromatography as well as of the problems connected with the migration and spreading of the zone. By analysing the diffusion processes referring to the penetration of eluent into horizontal chromatographic paper, and to the migration of the zone, he found the following equation for the height of the theoretical plate:

$$H_{obs} = \frac{\sigma^2}{l} = A + \frac{B[R_f^2(z_f - z_0)^2 - z_0^2]}{\theta k[R_f(z_f - z_0) - z_0]}$$
$$+ \frac{C\theta k R_f}{2[R_f(z_f - z_0) - z_0]} \ln \frac{R_f(z_f - z_0)}{z_0} \qquad (4.32)$$

As the distance from the level of the eluent in the tank to the front (z_f) of the eluent is far bigger than the distance to the initial zone of the solute (z_0)

Eq. (4.32) may be written:

$$H_{\text{obs}} = A + \frac{BR_f z_f}{\theta k} + \frac{C\theta k}{2z_f} \ln \frac{R_f z_f}{z_0} \qquad (4.33)$$

where $R_f = A_M/(A_M + \alpha A_S)$, that is the relative velocity of the zone, k is a 'flow parameter' such that $z_f^2 = kt$, and depends on the structure of the paper as well as on the characteristics of the eluent (surface tension, viscosity), and θ is the ratio of the solvent velocity at the level of the zone to that at the front and A, B and C are constants.

Equations (4.32) and (4.33) show that the observed height of the plate is the sum of three terms. The first is independent of the distance covered by the eluent front, the second increases with the distance covered by the zone, and the last decreases with the increase in the value of z_f. The first term represents eddy diffusion and seems negligible for paper. The second represents molecular diffusion and is dominant for high values of z_f whereas for low values the non-equilibrium term will be dominant. There is therefore a development distance for which H_{obs} will be minimal.

In a later paper Stewart [54] deals again with this problem and gives the following equation for local plate heights:

$$H = \left(\frac{1}{A} + \frac{1}{C_M v}\right)^{-1} + \frac{B}{v} + C_S v \qquad (4.34)$$

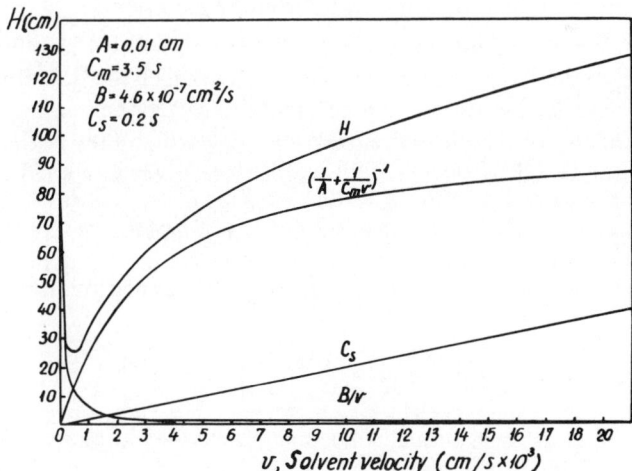

FIG. 4.6. The total local plate height as a function of solvent velocity is the sum of three terms: the coupled eddy diffusion, molecular diffusion, and non-equilibrium spread [54].

The contribution of each of these three terms to the total plate height is illustrated in Fig. 4.6 as a function of v. This figure shows that for low eluent velocities there is indeed a contribution from molecular diffusion (B/v), as observed by de Ligny. As can be seen in Eq. (4.29) the non-equilibrium coefficient C was split into the contributions from the mobile phase, C_M and of the stationary phase C_S ($C = C_M + C_S$). The term A in (4.34) is also interpreted not as eddy diffusion but as fluctuations depending on the R_f value, owing to the lack of homogeneity of the fibrous lattice.

The term $(1/A + 1/C_M v)^{-1}$, representing the so-called coupled eddy diffusion, has a strong variation with v at the beginning, but then tends towards linearity. The contribution of non-equilibrium ($C_S v$) to the plate height increases proportionally to the eluent velocity.

Stewart [61] showed that the mechanism of eluent migration in paper chromatography [53] may be extended to thin-layer chromatography and used as a basis for the interpretation of the zone-broadening mechanism, as shown by the height of a theoretical plate.

4.3 Chromatographic Selectivity

The partition coefficient α plays an important role in the separation of two components in a given chromatographic system. From the thermodynamic point of view, the value of α is given by

$$\alpha = \exp(-\Delta G^\circ / RT) \qquad (4.35)$$

where ΔG° is the change in standard free energy when the solute passes from the mobile to the stationary phase. The bigger the difference of the partition coefficients of the two solutes, the better their separation. An estimate of the possibility of separating any two solutes is the ratio of the partition coefficients, α_1/α_2, also called the selectivity coefficient β.

$$\beta = \frac{\alpha_1}{\alpha_2} = \exp[-(\Delta G_2^\circ - \Delta G_1^\circ)/RT] \qquad (4.36)$$

Another concept is the relative selectivity [4] defined by the ratio $\Delta\alpha/\alpha$ where

$$\Delta\alpha = (\alpha_1 - \alpha_2) = (\alpha_1 \alpha_2)^{1/2} \left[\left(\frac{\alpha_1}{\alpha_2}\right)^{1/2} - \left(\frac{\alpha_2}{\alpha_1}\right)^{1/2} \right]. \qquad (4.37)$$

Assuming that the partition coefficients of the two solutes are very similar ($\alpha_1 \alpha_1 \sim \alpha^2$), substitution of (4.36) in (4.37) gives

$$\frac{\Delta\alpha}{\alpha} = \exp(\Delta G_2^\circ - \Delta G_1^\circ)/2RT - \exp[-(\Delta G_2^\circ - \Delta G_1^\circ)/2RT] \qquad (4.38)$$

By means of hyperbolic functions, (4.38) may be written:

$$\frac{\Delta\alpha}{\alpha} = 2 \sinh\left(\frac{\Delta G_2^\circ - \Delta G_1^\circ}{2RT}\right) \qquad (4.39)$$

From the assumption about the partition coefficients it follows that ΔG_1° and ΔG_2° are also similar, and therefore (4.39) becomes

$$\frac{\Delta\alpha}{\alpha} = \frac{\Delta G_2^\circ - \Delta G_1^\circ}{RT} = -\frac{\Delta(\Delta G^\circ)}{RT}$$

This relation shows that $\Delta\alpha$ increases in direct proportion to the difference between the standard free energy changes for the two solutes. Consequently the possibility of separating the two solutes will be best when the difference is biggest. This relation also permits the study of the influence of temperature on the selectivity factor.

In ion-exchange chromatography, when the sample B + C is introduced into the column of resin (which is in form AR), the reactions (a) and (b) occur from left to right:

$$AR + B \rightleftharpoons BR + A \qquad (a)$$

$$AR + C \rightleftharpoons CR + A \qquad (b)$$

On introduction of an eluent which contains the electrolyte AX, reactions (a) and (b) are reversed and the equilibria can be written as

$$K_B^A = \frac{[A]_R[B]}{[A][B]_R} = \frac{[A]_R}{[A]} \bigg/ \frac{[B]_R}{[B]} = \frac{K_d^A}{K_d^B},$$

and

$$K_C^A = \frac{[A]_R[C]}{[A][C]_R} = \frac{[A]_R}{[A]} \bigg/ \frac{[C]_R}{[C]} = \frac{K_d^A}{K_d^C} \qquad (4.40)$$

where K_d is the distribution coefficient of the ions. The exchange constant K_B^A, or as it is also called, the separation factor [62], indicates the degree to which one ion is preferred to another in the exchange process.

The competition between B and C for retention by the resin during passage of the AX solution is described by the selectivity coefficient (β) defined as the ratio of the separation factors:

$$\beta = K_B^A/K_C^A = \frac{[B]_R}{[B]} \bigg/ \frac{[C]_R}{[C]} = \frac{K_d^B}{K_d^C} \qquad (4.41)$$

Data on ion-exchange selectivity can be found in several works [45, 63–66].

Ion-exchange selectivity has also been discussed from the thermodynamic point of view [67–71]. Myers and Boyd [72] show that the separation factor may be expressed as

$$\ln K_B^A = \ln \frac{\gamma_B}{\gamma_A} + \ln \frac{\bar{\gamma}_{AR}}{\bar{\gamma}_{BR}} + \frac{P}{RT}(\bar{V}_{AR} - \bar{V}_{BR}) \qquad (4.42)$$

where P is the swelling pressure of the resin, $\bar{V}_{AR}$ and $\bar{V}_{BR}$ are the partial molar volumes of the 'resinates', γ_A and γ_B are the activity coefficients of ions A and B in the solution, $\bar{\gamma}_{AR}$ and $\bar{\gamma}_{BR}$ are the activity coefficients of the resinates referred to a new standard state, a hypothetical infinitely dilute resin, T is the absolute temperature and R the gas constant.

The following conclusions can be drawn from Eq. (4.42). First, since the activity coefficients are unity for sufficiently dilute solutions, K_B^A will then be given by the ratio of the activity coefficients of the two ions in the resin and by the osmotic term. Secondly when the two ions are of similar dimensions, $\bar{V}_{AR} - \bar{V}_{BR} \approx 0$, and the separation factor for dilute solutions will depend only on the ratio of the activity coefficients of the ions in the resin. When $\bar{V}_{AR} \neq \bar{V}_{BR}$ the osmotic term becomes significant, and is determined by the structure of the resin, which means that the bigger the degree of reticulation, the bigger the osmotic pressure, and therefore the selectivity [73] will be better.

Eisenman [74], in explaining the selectivity of cation-exchange at a glass membrane, considers not only the hydration of ions, but also the electrostatic interaction, as the position of the ion-exchange equilibrium is controlled primarily by the intensity of the electric field of the macroanion of the glass.

Let us now express the selectivity coefficient as a function of the R_f values, and examine the practical consequences resulting for paper and thin-layer chromatography [75].

Let the R_f values be

$$R_{f_1} = \frac{1}{1 + \alpha_1 k} \quad \text{and} \quad R_{f_2} = \frac{1}{1 + \alpha_2 k} \qquad (4.43)$$

where $k = V_S/V_M$, that is the ratio between the volumes of the stationary and the mobile phase, and α is the partition coefficient.

Writing (4.43) as

$$\alpha_1 k = \frac{1}{R_{f_1}} - 1 \quad \text{and} \quad \alpha_2 k = \frac{1}{R_{f_2}} - 1 \qquad (4.44)$$

and taking the ratio, we obtain

$$\beta = \frac{\alpha_1}{\alpha_2} = \frac{R_{f_2} - R_{f_1} R_{f_2}}{R_{f_1} - R_{f_1} R_{f_2}} \qquad (4.45)$$

By substituting $R_{f_2} = R_{f_1} + \Delta R_f$ in (4.45) we obtain

$$\beta = \frac{(1 - R_{f_1})(R_{f_1} + \Delta R_f)}{(1 - R_{f_1} - \Delta R_f)R_{f_1}} \qquad (4.46)$$

For the study of Eq. (4.46) we shall consider ΔR_f to be constant. Differentiating with respect to R_{f_1} and equating to zero, we find a minimum for

$$R_{f_1} = \frac{1 - \Delta R_f}{2} = 0.50 - \frac{\Delta R_f}{2} \qquad (4.47)$$

The minimum value of β is obtained by substituting this value in (4.46):

$$\beta_{min} = \frac{\Delta R_f^2 + 2\Delta R_f + 1}{\Delta R_f^2 - 2\Delta R_f + 1} = \frac{(\Delta R_f + 1)^2}{(\Delta R_f - 1)^2} \qquad (4.48)$$

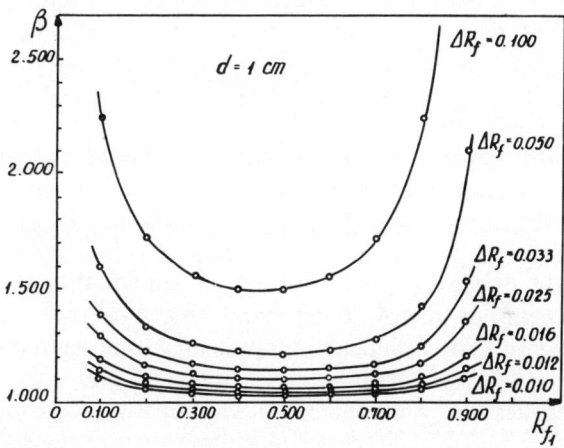

FIG. 4.7. Variation of β with R_f and ΔR_f, $d = 1$ cm (distance between the centres of zones) [75].

Let us examine the practical consequences of these relationships. It can be shown that if ΔR_f is kept constant, then β as a function of R_{f_1} has a minimum at $R_{f_1} = 0.5$ (see Fig. 4.7). That is to say, if the zones are to be a given distance apart, the necessary difference between the partition coefficients will be lowest if the conditions are arranged so that $R_{f_1} = 0.5$. It is also seen from Fig. 4.7 that the curves for $\beta = f(R_{f_1})$ for constant ΔR_f become flatter as the column length is increased, so that for columns more than 50 cm long there is practically no change in selectivity when R_{f_1} is varied from 0.3 to 0.7. It follows that it is advantageous to use fairly long columns and to arrange for the value of R_{f_1} to be 0.5 ± 0.2.

Martire and Locke [76] worked out the first theory of selectivity in liquid–liquid chromatography (LLC), as a result of a theoretical research [77] on chromatographic separations. The thermodynamic basis for selectivity in liquid–liquid chromatography was considered in terms of a new approach to solution free energies. Locke and Martire [78] had already shown that at moderate column pressures, the LLC solute specific retention volume, V_g, is given by

$$V_g = \frac{\gamma_M m_M}{\gamma_S m_S \rho_M} \qquad (4.49)$$

where γ_M and γ_S respectively are the infinite-dilution solute activity coefficients in the mobile and stationary phases, which have molecular weights m_M and m_S, and ρ_M is the density of the eluent at the temperature of the column.

The separation of the two solutes 1 and 2 can be characterised by the relative retention β_r:

$$\beta_r = \frac{(V_g)_1}{(V_g)_2} = \frac{\gamma_{M_1}\gamma_{S_2}}{\gamma_{M_2}\gamma_{S_1}} \qquad (4.50)$$

In two earlier papers [79, 80] Martire and Locke showed that the activity coefficient of a solute is additive:

$$\ln \gamma = \ln \gamma^{th} + \ln \gamma^{ath} \qquad (4.51)$$

where γ^{ath} is the configurational or athermal contribution to the activity coefficient arising from size differences between solute and solvent molecules, and γ^{th} is the thermal component which accounts for intermolecular forces between solute and solvent molecules: γ^{ath} is purely an entropy term, whereas γ^{th} includes both enthalpy and entropy contributions.

Based on these considerations β_r may be written:

$$\ln \beta_r = (\ln \beta_r)^{ath} + (\ln \beta_r)^{th} \qquad (4.52)$$

where the two activity coefficient contributions have been separated. The explicit expression of the athermal contribution to relative retention is then

$$(\ln \beta_r)^{ath} = \ln (\gamma_{M_1}/\gamma_{M_2})^{ath} - \ln (\gamma_{S_1}/\gamma_{S_2})^{ath} = (V_2^\circ - V_1^\circ)\left(\frac{1}{V_M^\circ} - \frac{1}{V_S^\circ}\right) \qquad (4.53)$$

where V° is the molar volume of the solute or phase indicated by subscript. Figure 4.8 shows the dependence of $\ln \beta_r^{ath}$ on $(1/V_M^\circ - 1/V_S^\circ)$ for different values of $(V_2^\circ - V_1^\circ)$ [76]. The greater the differences $(V_2^\circ - V_1^\circ)$

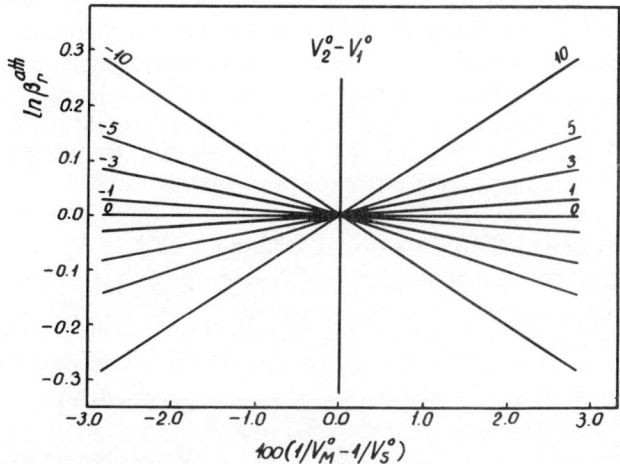

FIG. 4.8. Athermal contribution to relative retention with $(V_2^0 - V_1^0)$ as a parameter [76].

and $(1/V_M^0 - 1/V_S^0)$, the greater β_r and the better the separation. The eluent molecules in LLC are generally smaller than the molecules of the stationary phase $(V_M^0 < V_S^0)$. Evidently isomeric molecules that are identical from the energetic point of view can be separated by LLC if the molar volumes are sufficiently different $(V_1^0 \neq V_2^0)$.

The expression for the thermal contribution to relative retention was also given by Martire and Locke [76], starting from a given expression for this term [77] and substituting Hildebrand solubility parameters [81, 82] for cohesive energy densities:

$$(\ln \beta_r)^{th} = (\ln \gamma_{M_1}/\gamma_{M_2})^{th} - (\ln \gamma_{S_1}/\gamma_{S_2})^{th}$$
$$= \frac{V_r^0}{RT}(\delta_S^2 - \delta_M^2)\left(\frac{T_1^* - T_2^*}{T_r^*}\right) \quad (4.54)$$

where V_r^0 is the molar volume of a reference solute, T_1^*, T_2^* and T_r^* are the critical absolute temperatures of the solutes and of the reference solute, and δ_S and δ_M are solubility parameters, identical to $(\Delta E_i/V_i^0)^{1/2}$ in which ΔE_i represents the molar vaporization energy of component i.

In the case of solutes (1) and (2) having similar energetic properties to each other and to the reference solute (r) we can write the approximate equations

$$V_r^0 = V_1^0 \quad \text{and} \quad T_r^* = T_1^* \quad (4.55)$$

so that Eq. (4.54) becomes

$$(\ln \beta_r)^{th} = \frac{V_1^\circ}{RT}(\delta_S^2 - \delta_M^2)\left(\frac{T_1^* - T_2^*}{T_1^*}\right) \quad (4.56)$$

where we define $(T_1^* - T_2^*)/T_1^* = f(T)$ as the fractional difference between the critical temperatures of pure solutes 1 and 2.

Figure 4.9 shows the dependence of $(\ln \beta_2)^{th}$ on $V_1^\circ(\delta_S^2 - \delta_M^2)/RT$, for different values of the fractional difference $(T_1^* - T_2^*)/T_1^*$ [76]. The greater the difference between the critical temperatures of the two solutes and between the cohesive energy densities, the greater (β_r^{th}) and the better the separation.

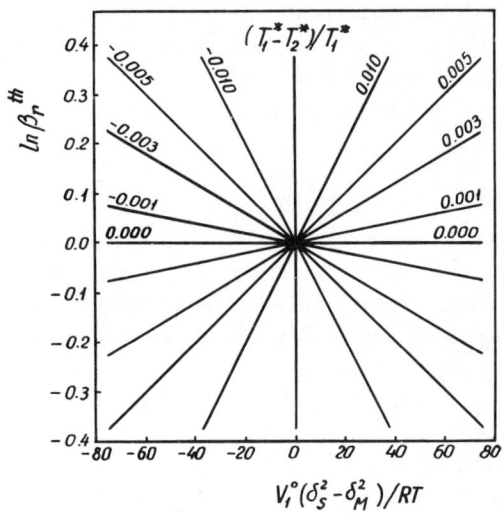

Fig. 4.9. Thermal contribution to relative retention with $(T_1^x - T_2^x)/T_1^x$ as a parameter [76].

Martire and Locke [76] listed a series of papers [80–93] giving values of critical temperatures and solubility parameters.

4.4 Resolution

4.4.1 Resolution on a Closed Chromatographic Column

Chromatographic analysis aims ultimately at the separation of the components in a given sample, but the separation is determined by the dynamics of the processes on the column (efficiency of the column) as well as the thermodynamic properties of the phases and solutes (separation selectivity). The degree of separation of two solutes is expressed by the resolution R_s and includes the efficiency of the column as well as the selectivity. The resolution is defined [94, 95] by

$$R_s = \frac{t_2 - t_1}{2(\sigma_2 + \sigma_1)} \qquad (4.57)$$

where t_1 and t_2 are the necessary time-intervals for the elution of solutes 1 and 2, and σ_1 and σ_2 are the standard deviations of the two chromatographic peaks. Equation (4.57) may be written:

$$R_s = 2\frac{\Delta L}{X_1 + X_2} \qquad (4.58)$$

where X_1 and X_2 are the widths of the peaks, measured at the base between the tangents to the inflexion points (Fig. 4.10) and ΔL is the distance between the peaks on the chromatogram.

If the two peaks are close together and have equal areas, expression (4.58) becomes:

$$R_s = \frac{\Delta L}{4\sigma} \qquad (4.59)$$

The effect of the thermodynamic properties of the chromatographic system is manifested through ΔL, and the efficiency through σ. If $\Delta L = 4\sigma$, $R_s = 1$ and corresponds to a 98 per cent separation of the two components.

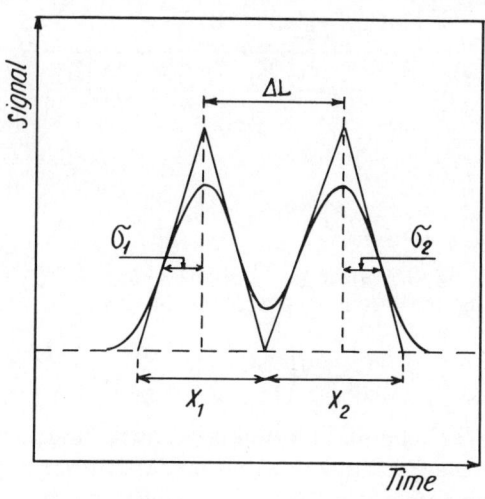

FIG. 4.10. Resolution parameters.

ΔL is the difference between the distances L_1 and L_2 traversed by the two zones along a uniform column. Thus $L_1 = tvR_{f_1}$ and $L_2 = tvR_{f_2}$ where v is the velocity of the mobile phase, and t the migration time, so

$$\Delta L = L_2 - L_1 = v(R_{f_2} - R_{f_2})t = tv\,\Delta R_f \tag{4.60}$$

When R_{f_1} and R_{f_2} are close enough for the peaks to overlap, we may write $R_{f_1} \sim R_{f_2} = R_f$, and $vt = L/R_f$ (L being the length of the column) and therefore Eq. (4.60) may be written as

$$\Delta L = L\,\Delta R_f/R_f \tag{4.61}$$

The standard deviation for each peak, σ, is obtained by means of Eq. (4.4) defining the HETP, $H = \sigma^2/L$, or

$$\sigma = \sqrt{LH} \tag{4.62}$$

Substitution of (4.61) and (4.62) into (4.59) gives

$$R_s = \frac{1}{4}\sqrt{\frac{L}{H}} \cdot \frac{\Delta R_f}{R_f} \tag{4.63}$$

and on further substitution by $\Delta R_f/R_f = V_s\,\Delta\alpha/(V_M + \alpha V_s)$ we obtain

$$R_s = \frac{1}{4}\sqrt{\frac{L}{H}} \cdot \frac{V_s\,\Delta\alpha}{V_M + \alpha V_S} \tag{4.64}$$

where α is the partition coefficient, and V_S and V_M are the volumes of the stationary and mobile phases. Further manipulation and substitution of $N = L/H$ gives a new expression for R_s:

$$\frac{V_S}{V_M + \alpha V_S} = \frac{\alpha V_S}{\alpha(V_M + \alpha V_S)} = \frac{1}{\alpha}\left[\frac{V_M + \alpha V_S}{V_M + \alpha V_S} - \frac{V_M}{V_M + \alpha V_S}\right] = \frac{1 - R_f}{\alpha} \tag{4.65}$$

$$R_s = \frac{1}{4}\sqrt{N}\frac{\Delta\alpha}{\alpha}(1 - R_f)$$

$\Delta\alpha/\alpha$ is the relative selectivity.

Equation (4.65), obtained by Giddings [4] for the case of liquid–solid chromatography is analogous to the equation given by Snyder [96–98]:

$$R_s = \frac{1}{4}\sqrt{N}\left(\frac{\alpha_1^*}{\alpha_2^*} - 1\right)\left(\frac{\bar{\alpha}^*}{1 + \bar{\alpha}^*}\right) \tag{4.66}$$

(where α_1^* and α_2^* are the distribution coefficients and $\bar{\alpha}^*$ their average. This equation indicates that the resolution is expressed by three independent and important factors (*a*) the efficiency of the column, (*b*) the relative selectivity and (*c*) the capacity.

To connect the resolution with the second order central statistical moment μ_2, we replace N by L^2/μ_2 in Eq. (4.65) to obtain

$$R_s = \frac{1}{4} \frac{L}{\sqrt{\mu_2}} \frac{\Delta\alpha}{\alpha}(1 - R_f) \qquad (4.67)$$

As shown above, H [see Eqs. (4.5)–(4.7) and (4.24)] and μ_2 [see Eqs. (4.9), (4.11), (4.16) and (4.17)] have complicated expressions, and therefore the expression for the resolution is also sometimes too inconvenient to be calculated. Research-workers must therefore choose the particular expression for the resolution which best satisfies the chosen chromatographic model, as well as the aim of the research.

Hamilton et al. [48] show that the separation of two solutes in the case of ion-exchange chromatography is practically complete when:

$$R_s = \frac{1}{\sqrt{2}}\left(\frac{L}{v}\right)^{1/2}\left[\frac{\Delta K_d}{\sqrt{X_1} + \sqrt{X_2}}\right] = 2 \qquad (4.68)$$

where $\Delta K_d = K_d^1 - K_d^2$ is the difference between the distribution coefficients of the two solutes, v is the linear velocity of the eluent, L the length of the column and $X_1 = \lambda d_p(K_d^1 + F_I)^2/v + d_p^2 K_d^1/60 D_s$ (X_2 is similarly expressed with K_d^2 for solute 2). In the equation for X, λ is dimensionless (more or less constant for a given column, with a value probably between 0·5 and 2) and related to the physical properties of fluid flow through the packed bed, d_p is the resin particle diameter, D_s is the coefficient of diffusion of solute into the resin particles, and F_I is the fraction of the column volume occupied by the fluid between the particles.

If the widths of the two peaks differ, their ratio being defined by $n = X_1/X_2$, the general expression for the resolution [100] is

$$R_s = \frac{1}{2(n+1)} \frac{\Delta R_f}{R_f}\sqrt{N} \qquad (4.69)$$

[in Eq. (4.63) $n = 1$].

It is important to note that the resolution of two peaks is the poorer, the higher the ratio of the two peak widths. This frequently happens in trace analysis where the peak of the main solute is followed by the peak of the 'trace' solute. In such cases, according to Eq. (4.69), for a fixed $\Delta R_f/R_f$ ratio a much larger number of theoretical plates is needed to yield the same value of R_s as that for equal amounts of the solutes.

Dybczyński [45, 101, 102], starting from $R_s = \Delta V_R/n(\sigma_1 + \sigma_2)$, (in which ΔV_R is the difference between the retention volumes of the two solutes, σ_1 and σ_2 are the respective standard deviations, and n is a positive number, so that for instance, if n is 3 cross-contamination is 0·14 per cent) reached the following equation for the resolution:

$$R_s = \frac{(\beta - 1)\sqrt{L}}{n(\beta + 1)\sqrt{\overline{H}}} = \frac{(\beta - 1)\sqrt{\overline{N}}}{n(\beta + 1)} \qquad (4.70)$$

where β is the selectivity coefficient, $\overline{H} = (H_1 + H_2)/2$ and $\overline{N} = L/\overline{H}$.

In good separations, $R_s \geq 1$ and if we know the height of a plate H (by measurement or calculation), the partition coefficients, the relative velocity difference $\Delta R_f/R_f$ and the volumes of the two phases V_M and V_S, then the length of the column necessary for separation with the desired resolution can be computed from

$$L = \frac{16 H R_s^2 R_f^2}{\Delta R_f^2} = \frac{16 H R_s^2 (V_M + \alpha V_S)^2}{(V_S \Delta \alpha)^2} = \frac{16 H R_s^2}{[\Delta \alpha (1 - R_f)/\alpha]^2} \qquad (4.71)$$

Glueckauf [3, 47] has also studied the relation of overlapping of two neighbouring bands to the degree of separation. The eluate is cut at the volume corresponding to the minimum between the two peaks. Writing ΔA_1 for the area of the peak corresponding to solute 1 that is associated with the peak of solute 2, and ΔA_2 for the area of peak 2 overlapping peak 1 (Fig. 4.11) the degree of impurity η is given by

$$\eta_1 = \frac{\Delta A_2}{A_1} \quad \text{and} \quad \eta_2 = \frac{\Delta A_1}{A_2}$$

where A_1 and A_2 are the total areas of the peaks. If $\Delta A_2 = \Delta A_1 = \Delta A$ and $A_1 = A_2 = A$ we have:

$$\eta = \frac{\Delta A}{A}$$

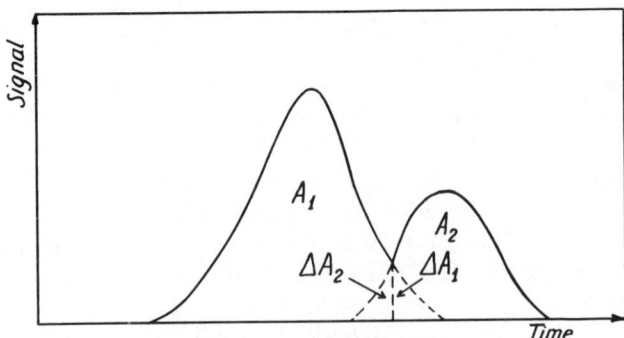

FIG. 4.11. Overlapping peaks.

If there is a normal distribution of the concentration for a certain value of η and of the separation coefficient $\beta_s = (\alpha_2 + F_1)/(\alpha_1 + F_1)$, the number of plates can be calculated [3, 103]:

$$N = \frac{z^2}{(\sqrt{\beta_s} - 1)^2} \qquad (4.72)$$

where z is the normalized deviation, its value depending on the value of the Laplace function $\Phi(z) = 0.5 - \eta$ which is tabulated [104]; $(\alpha + F_1)$ is the solute concentration per ml of solution at equilibrium and $\alpha = C_S/C_M$ is the partition coefficient.

As an example consider the separation of amino-acids [48] on a column of Dowex 50-XB, for which $r_p = 24\ \mu\text{m}$, $F_1 = 0.38$, $T = 54°C$, pH = 2.93 (citrate buffer). For a mixture of glycine ($K_d = 3.72$) and alanine ($K_d = 4.11$), we have $\beta_s = (4.11 + 0.38)/(3.72 + 0.38) = 4.49/4.10 = 1.093$.

For 0.1 per cent degree of impurity ($\eta = 10^{-3}$), $\Phi(z) = 0.5 - 10^{-3} = 0.499$ and $z = 3.09$, so $N = (3.09)^2/(\sqrt{1.095} - 1)^2 = 4435$. For $\eta = 10^{-2}$ (1 per cent impurity), $\Phi(z) = 0.5 - 10^{-2} = 0.490$, $z = 2.35$ and $N = 2522$.

It was shown earlier that in the region near point Q in Fig. 4.4 the height of a theoretical plate is approximately 5 times the radius of the resin particle, i.e. for optimum performance $H = 5r_p = 5 \times 24\ \mu\text{m} = 120 \times 10^{-4}$ cm and the length of the column will be $L = HN = 53$ cm for $\eta = 10^{-3}$, or $L = 30$ cm for $\eta = 10^{-2}$.

A very useful diagram for the practical determination of the number of plates N for two solutes giving unequal peaks is presented in Fig. 4.12 [3]. The normalized areas A'_1 and A'_2 of the two peaks are calculated:

$$A'_1 = \frac{A_1}{A_1 + A_2} \quad \text{and} \quad A'_2 = \frac{A_2}{A_1 + A_2}$$

and η is multiplied by the factor

$$\frac{A'^2_1 + A'^2_2}{2A'_1 A'_2}$$

to allow for the inequality of the areas.

Suppose that the normalized areas are $A'_1 = 0.6$ and $A'_2 = 0.4$. For two solutes having a separation coefficient $\beta_s = 1.3$ and degree of impurity of 0.1 per cent ($\eta = 10^{-3}$) we obtain:

$$\eta \frac{(A'^2_1 + A'^2_2)}{2A'_1 A'_2} = 1.08 \times 10^{-3}$$

From Fig. 4.12, for the intersection of the diagonal line $\beta_s = 1.3$ with the vertical line $\eta(A'^2_1 + A'^2_2)/2A'_1 A'_2 = 1.08 \times 10^{-3}$ we find $N \sim 600$.

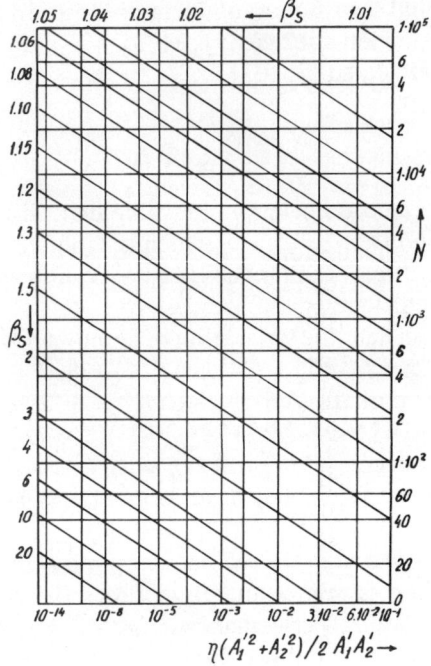

FIG. 4.12. Purity of separated products as function of separation coefficient β_s, number of theoretical plates N and mass ratio η [3].

Suppose the height of a theoretical plate is 0·25 cm in the case of a column filled with silica gel containing 4 per cent H_2O. To separate the two solutes on such a column the length would need to be $L = HN = 0.25 \times 600 = 150$ cm.

Another measure of the separability of two substances is the separation function [105, 106]:

$$F = \frac{(\Delta L)^2}{8(\sigma_1^2 + \sigma_2^2)} = \frac{(\Delta L)^2}{16\sigma^2} = R_s^2 \qquad (4.73)$$

which is ≥ 1 for a satisfactory separation. The resolution and the separation function are equally satisfactory as criteria of separability.

To characterise the separation of two solutes, Dixon [107] proposed the resolution power P of a chromatographic column, expressed by

$$P = \frac{\sqrt{N}}{k}\sqrt{1 - R_f} \qquad (4.74)$$

where k is a constant. For $k = 4$ the maximum mutual contamination of the two solutes as 2·3 per cent.

4.4.2 Resolution on Open Chromatographic Columns

In the case of an open column (paper and thin-layer chromatography) a resolution relationship has been established [108] which contains the values R_f and H for the two solutes, as well as the distances travelled by the zone and the eluent front.

Consider two chromatographic spots on a paper strip or a chromatographic plate, and their concentration profile (Fig. 4.13). The distance

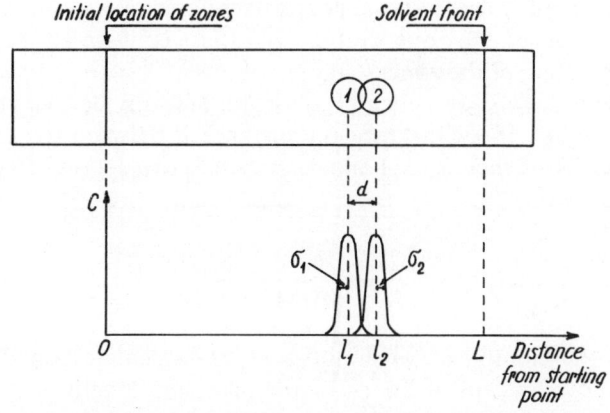

Fig. 4.13. Chromatogram and densitogram of two zones.

between the centres of the zones, $d = l_2 - l_1 = R_{f_2}L - R_{f_1}L = L\Delta R_f = \Delta z$, and the distance measured at the base of the peak between the tangents passing through the inflexion points of the concentration curve is equal to $X_1 = 4\sigma_1 = 4\sqrt{l_1 H_1}$ and $X_2 = 4\sigma_2 = 4\sqrt{l_2 H_2}$. Substitution in Eq. (4.58) gives

$$R_s = \frac{L \Delta R_f}{2(\sqrt{l_1 H_1} + \sqrt{l_2 H_2})} \quad (4.75)$$

Replacing l_1 and l_2 by $R_{f_1}L$ and $R_{f_2}L$ yields

$$R_s = \frac{\sqrt{L} \Delta R_f}{2(\sqrt{R_{f_1} H_1} + \sqrt{R_{f_2} H_2})} \quad (4.76)$$

Equation (4.76) shows how the resolution depends on the parameters determining the efficiency of the column, namely H_1 and H_2. It also shows that the resolution is proportional to $\sqrt{L}$ and that therefore substances having very similar R_f values can be separated if paper strips or thin layers of appropriate length are used. This also results from the fact that the distance between the centres of the zones, or of the peaks on the profile diagram is proportional to L, that is, to the distance covered by the eluent front, while the breadth of the peak grows in proportion to $\sqrt{l}$, the square root of the distance by which the zone has migrated.

It is evidently preferable to use the longest possible columns for separating substances with similar properties and to choose the mobile and stationary phases so as to make the R_f values between 0·3 and 0·7, to obtain the best possible resolution. It is also advisable to add the sample on as small an area as possible and to choose the working conditions so that diffusion of the zone during migration is as small as possible, i.e. to avoid tailing of the zones.

By means of Eq. (4.76) the length of column for any desired resolution can be computed. If for instance the distance d between the centres of the two peaks is 4σ, then $R_s = 1$ and corresponds to a degree of 98 per cent of separation, and

$$L = \frac{4R_s^2(\sqrt{R_{f_1}H_1} + \sqrt{R_{f_2}H_2})^2}{(\Delta R_f)^2}. \qquad (4.77)$$

Consider the example of two amino-acids and calculate from de Ligny's data [59] and by means of Eq. (4.77) the necessary length of column for their separation on Whatman No. 1 paper with butanol–acetic acid–water (4:1:5) as eluent, and descendent chromatography. For L-glutamic acid, $R_{f_2} = 0.21$ and $H_2 = 0.34$ mm, and for L-threonine, $R_{f_1} = 0.20$ and $H_1 = 0.24$ mm. To make this separation possible in isothermal conditions, a column having a length of 9·45 m is necessary, which is rather difficult to achieve.

Now take another pair of amino-acids: L-α,γ-diaminobutyric acid, $R_{f_2} = 0.08$, $H_1 = 0.31$ mm, and L-glutamic acid, $R_{f_2} = 0.21$ and $H_2 = 0.34$ mm. In this case a 4·25-cm column is needed.

Consider another example from the same work [59] but this time on a thin-layer of cellulose MN 300 by ascendent chromatography. For L-glutamic acid, $R_{f_2} = 0.38$ and $H_2 = 0.04$ mm, and for L-threonine, $R_{f_1} = 0.36$ and $H_1 = 0.05$ mm. In this case a 61-cm column is needed.

Snyder and Saunders [109], using the general expression for the resolution of a column

$$R_s = \frac{(l_2 - l_1)}{2(\sigma_1 + \sigma_2)}$$

(l_1 and l_2 being the distances travelled from the start by the two solutes) have deduced another expression for this parameter for two narrow, closely adjacent bands, assuming that $\sigma_1 = \sigma_2$ and $\alpha_1 \sim \alpha_2$. With these approximations it can easily be shown [96] that:

$$R_s = \frac{1}{4}\left(\frac{\alpha_1}{\alpha_2} - 1\right)\sqrt{N'}\left(\frac{\alpha_1}{1 + \alpha_2}\right) \qquad (4.78)$$

where N' is the number of theoretical plates in the bed length that has been traversed by solute 1 and 2 by the end of the separation. For elution from a column, N' is equal to the total number of plates in the column. For separation by thin-layer chromatography N' is equal to $R_f N$, where R_f is the average distance migrated by the solutes relative to the solvent front, and N is the number of plates in the adsorbent bed behind the solvent front.

De Ligny et al. [59, 110–112] introduced the *efficiency* E for the characterisation of the separation of two solutes. The separation of two substances by paper or thin-layer chromatography is defined by means of the concentration profile (Fig. 4.13, p. 83). The degree of separation of two solutes R (called the peak resolution by de Ligny et al.) is given by

$$R = \frac{l_2 - l_1}{\sigma_2 + \sigma_1} \qquad (4.79)$$

where l is the distance from the starting point to the point of maximum concentration on the profile, σ is the standard deviation. [Note that this definition of R is just twice that for R_s in Eq. (4.57).]

Even in the case of slightly asymmetric peaks, frequently obtained in adsorption chromatography, Eq. (4.79) may be applied [110]. By approximation of the sum of standard deviations $(\sigma_2 + \sigma_1)$ by their geometric mean, Eq. (4.79) becomes

$$R = \frac{l_2 - l_1}{2\sqrt{\sigma_2 \sigma_1}} \qquad (4.80)$$

or:

$$R^2 = \frac{1}{4}\frac{(l_2 - l_1)}{\sigma_2}\frac{(l_2 - l_1)}{\sigma_1} = \frac{1}{4}\frac{l_2 l_1}{\sigma_2 \sigma_1}\left(1 - \frac{l_1}{l_2}\right)\left(\frac{l_2}{l_1} - 1\right). \qquad (4.81)$$

The distance l measured on the profile is evidently proportional to the distance l' measured on the paper strip or thin layer. Moreover, according to the principle of independent additivity of distributions, it follows that σ^2 (and hence R^2) is also proportional to l. Thus a criterion for efficiency is obtained, independent of the distance measured on the band, so that R^2 may be divided by $\sqrt{l_2 l_1}$ to give the following expression for efficiency [111]:

$$E = \frac{R^2}{\sqrt{l_2 l_1}} = \frac{1}{4} \frac{l_2}{\sigma^2 \sqrt{l_2}} \frac{l_1}{\sigma_1 \sqrt{l_1}} \left(1 - \frac{l_1}{l_2}\right)\left(\frac{l_2}{l_1} - 1\right) \quad (4.82)$$

and as $H = \sigma^2/l$ it follows that:

$$E = \frac{1}{4} \frac{1}{\sqrt{H_2 H_1}} \left(1 - \frac{l_1}{l_2}\right)\left(\frac{l_2}{l_1} - 1\right). \quad (4.83)$$

The fraction of solute in the mobile phase being $1/(1 + \alpha)$, it can easily be shown that $R_f = 1/(1 + \alpha)$ and that $l_1/l_2 = (1 + \alpha_2)/(1 + \alpha_1)$, so that:

$$1 - \frac{l_1}{l_2} = (1 - R_{f_1})\left(1 - \frac{1}{\beta}\right) \quad \text{and} \quad \frac{l_2}{l_1} - 1 = (1 - R_{f_2})(\beta - 1)$$

where α is the partition coefficient, and $\beta = \alpha_1/\alpha_2$.

By substituting these expressions in (4.83) we obtain:

$$E = \frac{(\beta - 1)^2}{4\beta}(1 - R_{f_2})(1 - R_{f_1})\frac{1}{\sqrt{H_2 H_1}} \quad (4.84)$$

The efficiency of a chromatographic procedure for separation of two solutes depends on three factors:
(a) the nature of the solutes to be separated
(b) the R_f values of the two solutes
(c) the widths of the peaks, which depend, for a two-solute system, on the characteristics of the stationary and the mobile phase, as well as on the texture of the support [see Eqs. (4.31)–(4.33)].

The concept of efficiency, introduced by de Ligny, is really the separation function referred to unit length, as can easily be demonstrated. As already shown [Eq. (4.73)], $F = R_s^2$, and after substitution in Eq. (4.82) we obtain $E = F/\sqrt{l_1 l_2} \sim F/l$.

Duncan [113] also gave a formula for thin-layer chromatography and showed that two substances may be separated if the value of r in Eq. (4.85) is >1:

$$r = \frac{R_{f_2}}{R_{f_1} + 0.1 R_{f_2}}; \quad (R_{f_2} > R_{f_1}) \quad (4.85)$$

As shown [75] the condition for r to be >1 is not sufficient to characterise the conditions for separation of two substances.

Thoma and Perisho [114] suggest that evaluation of the resolution should be based on an *index of resolution*, I_R, defined as the distance between the centres of the spots $L\,\Delta R_f$, less the sum of the semimajor axes of the two spots:

$$I_R = L\,\Delta R_f - (a_1 + a_2), \tag{4.86}$$

where a is the semiaxis of an elliptical spot in the z-direction.

Obviously, I_R has positive values when the two spots are separated, and negative values when they overlap.

REFERENCES

1. Martin, A. J. P. and Synge, R. L. M., *Biochem. J.* **35**, 1358 (1941).
2. Mayer, S. W. and Tompkins, E. R., *J. Am. Chem. Soc.* **69**, 2866 (1947).
3. Glueckauf, E., *Trans. Faraday Soc.* **51**, 34 (1955).
4. Giddings, J. C., *Dynamics of Chromatography, Part I, Principles and Theory*, Dekker, New York (1965).
5. Bogue, D. C., *Anal. Chem.* **32**, 1777 (1960).
6. Kaminskii, V. A., Kononenko, L. E. and Timashev, S. F., *Zh. Fiz. Khim.* **41**, 2760 (1967).
7. Dvorkin, L. B., *Zh. Fiz. Khim.* **42**, 438 (1968).
8. Dvorkin, L. B., *Zh. Fiz. Khim.* **42**, 1452 (1968).
9. Van Deemter, J. J., Zuiderweg, F. J. and Klinkenberg, A., *Chem. Eng. Sci.* **5**, 271 (1956).
10. Beran, M. J., *J. Chem. Phys.* **27**, 270 (1957).
11. Giddings, J. C., *Nature* **184**, 357 (1959).
12. Giddings, J. C., *Nature* **187**, 1023 (1960).
13. Giddings, J. C., *Anal. Chem.* **35**, 1338 (1963).
14. Knox, J. H., *Anal. Chem.* **38**, 253 (1966).
15. Giddings, J. C., *J. Chem. Educ.* **35**, 588 (1958).
16. Giddings, J. C., *J. Chromatog.* **2**, 44 (1959).
17. Goldstein, S. and Murray, J. D., *Proc. Roy. Soc. London* **A 252**, 334, 348, 360 (1959).
18. Giddings, J. C., *J. Chromatog.* **3**, 445 (1960).
19. Giddings, J. C., *Anal. Chem.* **35**, 439 (1963).
20. Kučera, E., *J. Chromatog.* **19**, 237 (1965).
21. Schettler, P. D., Russell, C. A. and Giddings, J. C., *Anal. Chem.* **37**, 609 (1965).
22. Giddings, J. C., *Anal. Chem.* **37**, 1580 (1965).
23. Kočiřík, M., *J. Chromatog.* **30**, 459 (1967).
24. Schettler, P. D., Jr. and Giddings, J. C., *J. Phys. Chem.* **73**, 2582 (1969).
25. Giddings, J. C. and Schettler, P. D., Jr., *J. Phys. Chem.* **73**, 2577 (1969).
26. Huber, J. F. K., *J. Chromatog. Sci.* **7**, 85 (1969).
27. Giddings, J. C. and Eyring, H., *J. Phys. Chem.* **59**, 416 (1955).
28. Giddings, J. C., *J. Chem. Phys.* **26**, 169 (1957).
29. Giddings, J. C., *J. Chem. Phys.* **31**, 1462 (1959).
30. Giddings, J. C., *J. Chromatog.* **3**, 443 (1960).
31. Beynon, J. H., Clough, S., Crooks, D. A. and Lester, C. R., *Trans. Faraday Soc.* **54**, 705 (1958).
32. McQuarrie, D. A., *J. Chem. Phys.* **38**, 437 (1963).
33. Weiss, G. H., *Sepn. Sci.* **5**, 51 (1970).
34. Oxtoby, J. C., *J. Chem. Phys.* **51**, 3886 (1969).
35. Kragten, J., *J. Chromatog.* **37**, 373 (1968).
36. Vink, H., *J. Chromatog.* **15**, 488 (1964).

37. Vink, H., *J. Chromatog.* **18**, 25 (1965).
38. Vink, H., *J. Chromatog.* **20**, 305 (1965).
39. Vink, H., *J. Chromatog.* **25**, 71 (1966).
40. Vink, H., *J. Chromatog.* **36**, 237 (1968).
41. Zhukhovitskii, A. A. and Turkeltaub, N. M., *Gazovaia chromatographia*, p. 12, Gostopechizdat, Moscow (1962).
42. Purnell, R., *Gas Chromatography*, p. 104, Wiley, New York (1962).
43. Kubín, M., *Collection Czech. Chem. Commun.* **30**, 1104 (1965).
44. Kubín, M., *Collection Czech. Chem. Commun.* **30**, 2904 (1965).
45. Dybczyński, R., *J. Chromatog.* **14**, 79 (1964).
46. Minczewski, J. and Dybczyński, R., *Chem. Analit. (Warsaw)* **6**, 725 (1961).
47. Glueckauf, E., *Ion Exchange and its Applications*, p. 34, Society of Chemical Industry, London (1955).
48. Hamilton, P. S., Bogue, D. C. and Anderson, R. A., *Anal. Chem.* **32**, 1782 (1960).
49. Inczédy, J., *J. Chromatog.* **50**, 112 (1970).
50. Snyder, L. R., *Anal. Chem.* **39**, 698 (1966).
51. Mallik, K. L. and Giddings, J. C., *Anal. Chem.* **34**, 760 (1962).
52. De Ligny, C. L. and Bax, D., *Z. Anal. Chem.* **205**, 333 (1964).
53. Stewart, G. H. in *Advances in Chromatography*, Vol. 1, p. 93, Giddings, J. C. and Keller, R. A. (eds.), Arnold, London (1965).
54. Stewart, G. H., *Sepn. Sci.* **1**, 135 (1966).
55. De Ligny, C. L. and van de Meent, W., *J. Chromatog.* **53**, 469 (1970).
56. De Ligny, C. L. and Remijnse, A. G., *J. Chromatog.* **35**, 257 (1968).
57. De Ligny, C. L. and Remijnse, A. G., *J. Chromatog.* **33**, 242 (1968).
58. De Ligny, C. L., *J. Chromatog.* **49**, 393 (1970).
59. De Ligny, C. L. and Remijnse, A. G., *Rec. Trav. Chim.* **86**, 421 (1967).
60. De Ligny, C. L. and Remijnse, A. G., *Rec. Trav. Chim.* **86**, 410 (1967).
61. Stewart, G. H., *Sepn. Sci.* **1**, 747 (1966).
62. Helfferich, F., *Ionenaustauscher*. Bd. I, p. 150, Verlag Chemie, Weinheim/Bergstr. (1959).
63. Bonner, O. D. and Smith, L. L., *J. Phys. Chem.* **61**, 326 (1957).
64. Gregor, H. P., Belle, J. and Marcus, R. A., *J. Am. Chem. Soc.* **77**, 2731 (1955).
65. Aveston, J., Everest, D. A. and Wells, R. A., *J. Chem. Soc.* 231 (1958).
66. Dybczyński, R., *J. Chromatog.* **50**, 487 (1970).
67. Dybczyński, R., *Roczniki Chem.* **37**, 1411 (1963).
68. Gregor, H. P., *J. Am. Chem. Soc.* **75**, 642 (1951).
69. Boyd, G. B. and Soldano, B. A., *Z. Elektrochem.* **57**, 162 (1953).
70. Soldano, B. and Chesnut, D., *J. Am. Chem. Soc.* **77**, 1334 (1955).
71. Soldano, B., Larson, Q. V. and Myers, G. E., *J. Am. Chem. Soc.* **77**, 1339 (1955).
72. Myers, G. E. and Boyd, G. E., *J. Phys. Chem.* **60**, 521 (1956).
73. Senyavin, M. M., Kolosova, G. M. and Nikasyna, V. A., *Zh. Neorgan. Khim.* **3**, 104 (1958).
74. Eisenman, G., *Biophys. J. Suppl.* **2**, Part 2, 259 (1962).
75. Liteanu, C. and Gocan, S., *Talanta* **17**, 1115 (1970).
76. Martire, D. E. and Locke, D. C., *Anal. Chem.* **43**, 68 (1971).
77. Luckhurst, G. R. and Martire, D. E., *Trans. Faraday Soc.* **65**, 1248 (1968).
78. Locke, D. C. and Martire, D. E., *Anal. Chem.* **39**, 921 (1967).
79. Ashworth, A.-J. and Everett, D. H., *Trans. Faraday Soc.* **56**, 1609 (1960).

80. Martire, D. E. in *Gas Chromatography 1966*, p. 21, Littlewood, A. B. (ed.), Elsevier, Amsterdam (1967).
81. Hildebrand, J. H. and Scott, R. L., *Regular Solutions*, Prentice-Hall, Englewood Cliffs, N.J. (1962).
82. Hildebrand, J. H. and Scott, R. L., *Solubility of Nonelectrolytes*, 3rd Ed., Reinhold, New York (1950).
83. Burrel, H., *Interchem. Rev.* **14** (1), 3 (2), 31 (1955).
84. Lieberman, E. P., *Off. Dig. Fed. Soc. Paint Technol.* **34**, 30 (1962).
85. Gordon, J. L., *J. Paint Technol.* **38**, 43 (1966).
86. Crowley, J. D., Teague, G. S. and Lowe, J. W., Jr., *J. Paint Technol.* **38**, 269 (1966).
87. Morrison, G. H. and Freiser, H., *Solvent Extraction in Analytical Chemistry*, Wiley, New York (1957).
88. Polák, J., *Collection Czech. Chem. Commun.* **31**, 1483 (1966).
89. Small, P. A., *J. Appl. Chem.* **3**, 71 (1953).
90. Dreisbach, R. R., *Advan. Chem. Ser.* **15** (1953), **22** (1959), **29** (1961).
91. Timmermans, J., *Physicochemical Constants of Pure Organic Compounds*, Elsevier, New York, Vols. 1 and 2, 1950 and 1965.
92. Stull, D. R. and Sinke, G. C., *Advan. Chem. Ser.* **18**, (1956).
93. Kudchadker, A. P., Alani, G. H. and Zwolinski, B. J., *Chem. Rev.* **68**, 659 (1968).
94. Nomenclature Committee, in *Gas Chromatography Amsterdam 1958*, Desty, D. H. (ed.), Butterworths, London (1958).
95. Jones, W. L. and Kieselbach, R., *Anal. Chem.* **30**, 1590 (1958).
96. Snyder, L. R., *Anal. Chem.* **39**, 705 (1967).
97. Snyder, L. R. and Saunders, D. L., *J. Chromatog. Sci.* **7**, 195 (1969).
98. Snyder, L. R., *J. Chromatog.* **7**, 352 (1969).
99. Hamilton, P. B. in *Advances in Chromatography*, Vol. 2, p. 3, J. C. Giddings and Keller, R. A. (eds.), Arnold, London (1966).
100. Karger, B. L., *J. Gas Chromatog.* **5**, 161 (1967).
101. Dybczyński, R., *J. Chromatog.* **31**, 155 (1967).
102. Dybczyński, R., *Proceedings Analytical Chemical Conference III*, Vol. 1, p. 35, Budapest (1970).
103. Heftmann, E., *Chromatography*, 2nd Ed., p. 316, Reinhold, New York (1967).
104. Lark, P. D., Craven, B. R. and Bosworth, R. C. L., *The Handling of Chemical Data*, p. 335, Pergamon, Oxford (1969).
105. Giddings, J. C., *Anal. Chem.* **32**, 1707 (1960).
106. Svensson, H., *J. Chromatog.* **25**, 266 (1966).
107. Dixon, H. B. F., *J. Chromatog.* **7**, 467 (1962).
108. Gocan, S. and Liteanu, C., *Rev. Roumaine Chim.* **17**, 661 (1972).
109. Snyder, L. R. and Saunders, D. L., *J. Chromatog.* **44**, 1 (1969).
110. De Ligny, C. L., Schmidt, H. M. and de Vries, W., *Rec. Trav. Chim.* **82**, 1061 (1963).
111. De Ligny, C. L., Schmidt, H. M. and de Vries, W., *Rec. Trav. Chim.* **82**, 1071 (1963).
112. De Ligny, C. L., Schmidt, H. M. and de Vries, W., *Rec. Trav. Chim.* **82**, 1075 (1963).
113. Duncan, C. R., *J. Chromatog.* **8**, 37 (1962).
114. Thoma, J. A. and Perisho, C. R., *Anal. Chem.* **39**, 745 (1967).

CHAPTER 5

OPTIMISATION OF THE CHROMATOGRAPHIC PROCESS

As shown in Chapter 4, the resolution of two components depends on a series of factors. The optimisation of the resolution does not entirely solve the problem of optimisation of the separation. The time needed for the analysis must also be taken into account, besides other parameters which may influence the separation process and are not included in the equation for the resolution.

Optimisation of the process is the key to chromatography. All the theoretical and experimental studies converge towards the problem of choosing the best conditions for separating a mixture as quickly and completely as possible.

This chapter will deal with the conventional classical optimisation of the chromatographic process, i.e. by systematic variation of one parameter at a time, and selection of the best values for these parameters.

5.1 Adsorption and Partition Chromatography on Closed Columns

5.1.1 DIMENSIONS OF THE COLUMN

The ratio between the length of the column and the particle size. For optimal separation ($R_s \geq 1$), the ratio between the length of the column and the diameter of the particles can be calculated [1], starting from Eq. (4.65) for the resolution:

$$R_s = \frac{1}{4}\sqrt{\frac{L}{H}}\frac{\Delta\alpha}{\alpha}(1 - R_f) \tag{5.1}$$

or

$$\frac{L}{H} = \frac{16}{\left[\frac{\Delta\alpha}{\alpha}(1 - R_f)\right]^2} \tag{5.2}$$

which for $H > 2d_p$ becomes

$$\frac{L}{d_p} > \frac{32}{\left[\frac{\Delta\alpha}{\alpha}(1 - R_f)\right]^2} \tag{5.3}$$

The magnitude of $(1 - R_f)$ can never exceed 1, so

$$\frac{L}{d_p} > 32\left(\frac{\alpha}{\Delta\alpha}\right)^2 \tag{5.4}$$

or, if the values of R_f are more accessible than those of α,

$$\frac{L}{d_p} > 32\left(\frac{R_f}{\Delta R_f}\right)^2 \tag{5.5}$$

Equations (5.4) and (5.5) allow the calculation of the optimum length of the column according to the particle diameter d_p, which is known, and the ratio $\alpha/\Delta\alpha$ or $R_f/\Delta R_f$.

The ratio between the quantity of packing and of sample. In preparative separations by adsorption chromatography, this ratio varies from 20:1 to 100:1, and in partition chromatography from 50:1 to 500:1 [2–4]. In the case of analytical separations the ratio may be increased by a factor varying from 10 to 50.

The ratio between the length and breadth of the column. In early column chromatography this ratio varied from 5:1 to 10:1 [5–7], and later from 10:1 to 100:1 [2, 3, 8]. In high efficiency columns (with more than 2000 theoretical plates) [9], a ratio between 100:1 and 1000:1 is used [10, 11]. The efficiency of the separation evidently increases with increase in this ratio (i.e. increase of column length), but this is offset by the longer time of analysis and difficulties experienced in the filling the column uniformly.

Knox and Parcher [12] have deduced the relationship $L/d_c = 0.4 d_c/d_p$ which expresses the relation between the column length L, the column diameter d_c and the particle diameter d_p. This relationship helps in choosing these parameters.

Diameter of the column. Stewart et al. [13] show that columns with a 4-mm diameter can be packed well. Columns of 6·5-mm diameter give zone breadths about 4 times those obtained on 4-mm diameter columns of the same length. Columns with a 2-mm diameter are less efficient [14], but it was proved that in special conditions high efficiencies may be obtained with columns having a diameter up to 11 mm [15, 16].

The effect of the column diameter on efficiency was studied by Snyder [17], who showed that concurrently with the increase of the column diameter, the equivalent height of a theoretical plate (H) also increases, and therefore the efficiency of the column decreases.

King and Chambers [18] have found that in adsorption chromatography the efficiency improves with increase in column diameter in the range 1·0–2·3 cm. Giddings [19] has also discussed the contribution of the column diameter to H, adding to the van Deemter equation a term proportional to d_c^2. This term will make a serious contribution to H for $d_c > 2.5$ cm.

These observations are in agreement with the finding that the value of H is approximately independent of the column diameter for $d_c < 1$ cm. For columns broader than 2 cm, the value of H increases with increase in d_c. It was also shown [20] that the value of H is maximal for a d_c/d_p ratio between 10 and 30, and Smuts and Pretorius [21] have suggested that the optimal value is approximately 75.

Nowadays, columns most frequently have a diameter of 3–4 mm, similar to that used in gas chromatography [22].

5.1.2 SAMPLE SIZE, PARTICLE SIZE OF THE PACKING, TYPE AND ACTIVITY OF THE ADSORBENT

Dimensions of the particles in the column. The column efficiency increases with decreasing particle size [10, 13, 17, 23–26].

Snyder [27] finds empirically that $H = Dv^{0.4}$, which approximates well enough the coupling of the eddy-diffusion equation with that for mass transfer in the mobile phase, assumed by Giddings [1]. In this equation v is the linear flow-rate of the eluent, and D is a constant for a given column and experimental conditions. Let D be D° for $d_p = 1$; then, for a regular packing $D = D^\circ d_p^{1.4}$, and substitution in the equation gives $H = D^\circ d_p^{1.4} v^{0.4}$. This equation shows that the height of a plate depends on the diameter of the particles as well as on the mode of packing, and on the eluent flow-rate.

In the case of a column with $L = 75$ cm, $d_c = 3.5$ mm, packed with Davison 952 particles having a diameter from 47 to 53 μm, $H = 1.0$ mm for a flow-rate of 1 cm/s, and 0.4 mm at a flow-rate of 0.3 cm/s [22].

The particle size used in adsorption and partition chromatography on columns is usually 100–200 mesh.

Quantity of sample placed on the column. It was found [17] that for lower concentrations of the sample put on the column, H is independent of sample size, but increases rapidly above a certain concentration limit. This was also observed in other liquid–solid chromatographic systems [2, 4, 5, 28] where it is shown that high resolution is obtained for small quantities of sample. For instance, Kirkland [29] has shown that in a column with $d_c = 2.1$ mm a maximum sample volume of only 50 μl may be used without seriously affecting the efficiency. On the other hand, columns with a large diameter have a higher capacity. De Stefano and Beachell [16] showed that in a column with $d_c = 10.9$ mm, a sample volume of 0.9 ml may be used without a decrease in the resolution.

Type and activity of the adsorbent. These factors affect both the separation selectivity and the capacity, and hence implicitly the efficiency of the column.

Snyder [17] studied the effect of the activity of silica gel adsorbent (0.4, 10 and 20 per cent water) on the parameters A and C of the van Deemter equation ($H = A + B/v + Cv$) and on k (the proportionality coefficient of Darcy's equation $v = kp/L$, where v is the linear flow-rate of eluent and p the pressure drop along the column of length L) and found that the smallest values for A, C and k (and of course, for H) were obtained when the material containing 4 per cent water was used. Similar work with alumina showed that Al_2O_3–4 per cent H_2O gave values for k about 20 per cent higher than those for SiO_2–4 per cent H_2O. The values of A for alumina were comparable to those for silica gel, and the values of C were higher.

The role of the adsorbent and its activity in separations by column chromatography have been further discussed in the literature [25, 30].

5.1.3 Flow-rate of the Eluent

The resolution depends on the flow-rate of the eluent, and the choice of the optimal rate was dealt with by Wilson [31] as early as 1940.

As in the van Deemter equation (see Chapter 4) the term A for the contribution of eddy diffusion may be considered to a first approximation as independent of the flow-rate, the van Deemter equation may be written:

$$H = \frac{B}{v} + Cv \qquad (5.6)$$

The minimal value of H is obtained by differentiating and equating to zero, yielding:

$$v_{opt} = \sqrt{\frac{B}{C}} \qquad (5.7)$$

and after substitution in Eq. (5.6):

$$H_{min} = 2\sqrt{BC} \qquad (5.8)$$

Giddings [1] showed that the flow-rate can be varied around the optimal value without changing the resolution very much.

Different aspects of the effect of flow-rate on H have been extensively studied [13, 17, 26, 27, 32–37]. From a practical point of view the optimal rate is in the 5–50 mm/min range or 0.3–3 column-volumes per hour, for a column of 100 cm length [5].

5.1.4 VISCOSITY OF THE ELUENT

The viscosity of the eluent does not directly affect the separation selectivity or the capacity factor in the resolution equation, but these terms depend very much on the type of solvent used. It is generally possible to adjust the viscosity to give nearly maximal efficiency with an eluent chosen to give optimal values for the separation selectivity and the capacity factor [32].

Studies [17, 38] on different types of solvents showed that C in the van Deemter equation generally decreases with the decrease in the viscosity, but the value of k (in Darcy's equation) increases. The value of A remains practically constant. The increase in C may be explained by diffusion in the liquid phase and will become slower as the viscosity increases.

These results show the importance of using solvents with the lowest possible viscosity in high-efficiency separations [22, 32].

5.1.5 STRUCTURE OF THE BED

At first the structure of the chromatographic bed was considered to be very important, and still is for broad and medium diameter columns, but chromatography has developed very much towards narrow columns, where the structure loses some of its importance.

The structure of the bed is determined by various factors such as the material used, the average particle diameter d_p, the column geometry and the method used to fill the column. The contributions of the first factors have already been dealt with so here we shall deal with the methods used for filling.

Several procedures are available [17]. Stewart et al. [13] have studied the influence of some of them on the values of H and have found that the most efficient packing method consists in gradually adding enough adsorbent to fill 10 cm of the column, followed by light tamping with a rod (glass or metal) of diameter near to that of the column, followed by vertical shaking accompanied by light tapping on a table, the rod being left over the adsorbent in the column.

There is an interdependence between H, d_p and the structure of the bed. Theoretically, for a bed with a fixed structure (all the particles of the bed keep the same position and relative configuration), this interdependence should be described by a single curve for h ($h = H/d_p$, the reduced height of the plate) plotted against v ($v = vd_p/D_M$, the reduced flow-rate of eluent, where D_M is the diffusion coefficient of the component in the mobile phase) for a chromatographic system for which the particle size is varied [1, 27]. In practice this seems fairly true if d_p is about 100 μm or more. For smaller values of d_p a family of curves is obtained [39, 40]

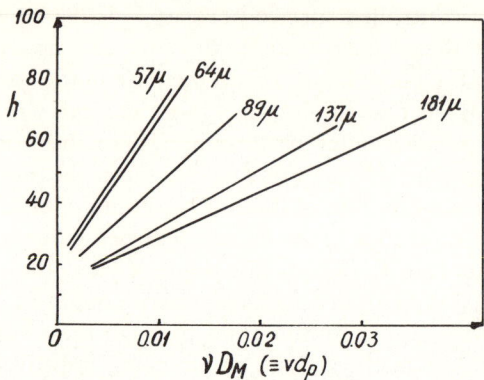

FIG. 5.1. Reduced plate height h vs. reduced solvent velocity in LSCC [27].

for which D_M is constant. When d_p is decreased, the value of h increases. These facts are illustrated by Fig. 5.1 [27] based on results obtained earlier by Snyder [17].

The most acceptable theoretical explanation of the disagreement between theory and practice (the fact that a family of lines is obtained instead of a single line) is that it is easier to obtain beds of similar structure with large particles ($d_p \geq 100\ \mu m$) of different diameters than with fine particles. Giddings [1] has stated that in the case of particles with small d_p, large aggregates are formed through agglomeration or aggregation, and the channels between them are of much larger diameter than they would be if agglomeration had not occurred. This leads to less compact packing with small particles than with big ones.

A recent study [41] has shown that H differs by a factor of up to 10 for beds of silica gel, alumina and Sil-O-Cel, for columns filled with dry adsorbents having the same particle size.

Huber [26] has achieved efficient columns filled with small diameter particles ($d_p \simeq 20\ \mu m$) by consecutively adding small quantities (1–2 cm of column length) of adsorbent, and tapping the column between additions. Joynes [42] similarly found that more efficient columns are obtained by adding the dry packing material to the charge.

In the case of the polystyrene resins used in ion-exchange chromatography, columns of high efficiency can be obtained by filling with slurry at high pressure. The resin is suspended in a solvent with the same density, and then pumped into the column at a pressure of 45–120 atm [27].

Adsorbents of another class, nowadays of increasing importance, are the particles of porous glass, or particles of glass with a thin porous coating [14, 29, 36, 39, 43–45]. Such adsorbents, used by Kirkland to obtain beds with controlled surface porosity [14, 29, 44, 45], present the advantage of a reduced mass-transfer distance and hence the reduction of the term C_S in Eq. (4.34) for H, i.e. an increase in the column efficiency. Another advantage is that the filling is more regular. Controlled surface porosity supports gave better performance than the usual supports in liquid chromatography [16, 46]. For such a column, the results showed [45] that a value of H about a fifth of that for a column of an efficient diatomaceous earth, the difference in the values of H being due to the change in resistance to mass transfer. From this point of view, the columns with a filling of controlled surface porosity have superior qualities. The asymptotic variation of H with v, at high flow-rates through these supports, may be due to mass transfer in the mobile phase being improved by turbulent diffusion [47]. The turbulence effect inside the bed must also be taken into account [48, 49]. The final result is some efficient separations by liquid chromatography at high flow-rates of solvent.

The maximum efficiency of the bed, in liquid chromatography, was discussed theoretically by Giddings [50]. His conclusions were confirmed by the experimental results of Snyder [32]. Giddings [50] showed that by using small particles of adsorbent, high numbers of theoretical plates are obtained but this requires the use of high pressures. Very difficult separations requiring $N > 10^5$ may be done by liquid chromatography but the necessary time is greater than that required for the same separation by gas chromatography.

Knox and Saleem [51] also made a comparative study of efficiency of liquid and gas chromatography, using different supports.

Kirkland [52] compared the performances of columns filled with commercial support-materials, and this study underlined the fact that in liquid partition chromatography the efficiency is strongly influenced by the nature of the support used.

At present research is being devoted to use of some adsorbent beds with small dimension particles, operated at very high pressures. For instance, a reduction of the particle size by 50 per cent will double the column efficiency, if the column length is kept constant and the pressure increased fourfold [22].

5.1.6 TEMPERATURE

The temperature is an important parameter in liquid–solid chromatography, a fact noted in a great many papers. In 1948 Le Rosen and Rivet [53] studied the change in rate of zone transfer in liquid–solid chromatography, when the temperature of the column was varied from 10 to 70°, and drew the conclusion that for the system studied, the effect is not very marked over the range 20–35°. A little later, Chang [54] studied the dependence of the retention of a component on temperature, using silica gel as adsorbent. He found that the retention could be made to increase, decrease, or stay approximately constant by variation of the temperature in the column. Recently, Lake and Martire [55] have also studied the effect of temperature on performance of the column in liquid–liquid chromatography. The thermodynamics of the processes taking place in liquid–liquid chromatography were studied by Locke [56].

Giddings [1] showed that temperature influences the resolution in a complex way, affecting all the terms of the resolution equation [5.1)], except the term containing the length L of the column (if the expansion of the column is neglected).

Hesse and Engelhardt [57] have studied the dependence of the retention volume on temperature, for different substances (Sudan II, Sudan III and p-aminoazobenzene, using activated basic-form aluminium oxide as adsorbent). The retention volume decreases with increase in temperature. The highest variation was obtained with p-aminoazobenzene. From the elution curves [57], the number of the plates was calculated by means of

$$N = 8 \ln 2 \left(\frac{t_R}{b_{1/2}} \right)^2, \tag{5.9}$$

where t_R is the retention time, and $b_{1/2}$ is the breadth of the chromatographic peak at half-height. The results are given in Fig. 5.2, which shows that N, and hence the efficiency, increases with temperature.

Snyder [17] showed that the improvement of the separation efficiency at higher temperatures is probably due to reduction in viscosity of the solvent, when linear separation isotherms are involved.

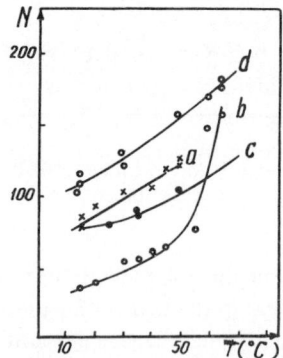

FIG. 5.2. Plot of the number of plates N vs. temperature for the different test-dyes; eluent xylene, $L = 45$ cm. Adsorbent: Al_2O_3 (basic) activity II: (*a*) Sudan II, (*b*) Sudan III, (*c*) p-aminoazobenzene; Al_2O_3 (basic) activity I, (*d*) azobenzene [57].

An ample theoretical and experimental study of influence of temperature and concentration of the moderator (a polar solvent added in small quantities to a non-polar eluent) on the retention parameters was made by Maggs and Young [59]. Recently, Maggs [60] has published new experimental results on the role of temperature and of the moderator in liquid–solid column chromatography. It was shown that the asymmetry of the chromatographic peak increases with increase in column temperature and decreases with the concentration of the moderator. An important role is played by the polarity of the mobile phase in the dependence of retention volume on temperature. When a less polar mobile phase is used, the separation is poorer. In this latter case, there is a bigger difference between the partition coefficients, which contributes to a better separation.

Scott and Lawrence [61] have studied the effect of temperature on the retention volume, the efficiency and the resolution. In the study of resolution they used a mixture of squalane, methyl palmitate, dinonyl phthalate, and dimethyl phthalate as components to be separated, and as mobile phase n-heptane, with isopropyl alcohol as moderator. The results for variation of resolution with temperature and concentration of moderator are shown in Fig. 5.3, which shows that in this case too, the resolution increases with temperature.

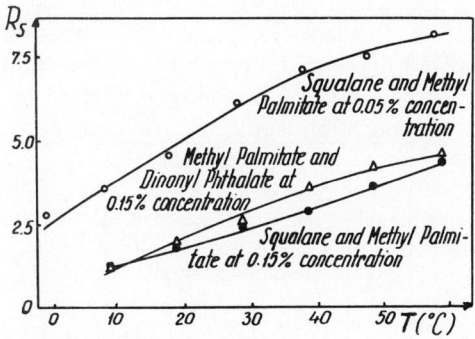

FIG. 5.3. Effect of temperature and moderator concentration on resolution at constant mobile-phase velocity (2 mm/s); $L = 100$ cm, $d_c = 4$ mm, adsorbent silica gel 200–250 mesh [61].

In the search for a new and more efficient separation procedures, a promising new technique is that in which the working conditions (temperature, pressure, etc.) are chosen so as to have values near those for the critical point of the mobile phase [62–67]. This technique is known as supercritical-fluid chromatography and theoretically can offer the following advantages: easier migration of high-boiling components and analysis of thermally labile components at lower temperatures, with efficiency and time of analysis comparable to those of conventional gas chromatography, owing to the reduced viscosity and relatively high diffusion rate. Novotný et al. [67] showed that both pressure and inverse temperature programming may be used to alter the retention parameters. The selectivity of separation can also be controlled by means of a moderator. The technique may be used for analytical purposes, and there is still a lot of theoretical work to be done on it.

5.1.7 Pressure

Modern liquid chromatography uses pumps creating a pressure of 34–340 atm to transfer eluent through the column [52]. This allows the use of short columns packed with finely granulated materials. Thus separations 100–1000 times more rapid than in conventional liquid chromatography, and 10–100 more rapid than in thin-layer chromatography, are rendered possible.

The effect of pressure on separation efficiency was the subject of several papers [14, 17, 26, 43, 44, 62-72] in which special attention was given to the use of higher pressures with column fillings with very small particle dimensions (approximately 10 μm), bringing about a marked increase in the column efficiency, explained by the differential changes in the partition coefficients with increase of pressure [71] (Fig. 5.4).

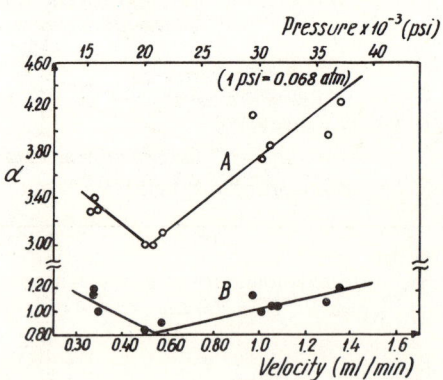

Fig. 5.4. Pressure effect on the partition coefficient. A—Ethyl Orange, B—Methyl Orange [71].

The partition coefficient in liquid chromatography is calculated from

$$V_R = V_m + \alpha V_s, \qquad (5.10)$$

where V_R is the retention volume of the component, V_m the interstitial volume of the column, and V_s the volume of the stationary phase.

The dependence of the specific retention volume V_g ($V_g = V_n/W_s$, where V_n is the retention volume, and W_s the weight of the stationary phase in the column) on the average pressure of the column $\bar{P} = (P_i + P_0)/2$ was given by Locke [73] for liquid–liquid partition chromatography, by the equation:

$$\left(\frac{\partial V_g}{\partial P}\right)_T = \frac{\Delta \bar{v}_2^\infty}{RT} \qquad (5.11)$$

where $\Delta \bar{v}_2^\infty = \bar{v}_2^{m,\infty} - \bar{v}_2^{s,\infty}$ is the difference between the partial molar volumes for infinite dilution in the mobile and stationary phases. In ordinary conditions the effect of the pressure on V_g may be neglected. The difference between the partial molar volumes of the component in the two phases is generally rather small, rarely larger than 10 cm^3/mole, so that the magnitude of the pressure term is of the order of 10^{-4}–10^{-2} in the equation:

$$\ln V_g = \ln \frac{\gamma_2^{m,\infty}(T,1)M_m}{\gamma_2^{s,\infty}(T,1)M_s\rho_m(T)} + \frac{\bar{P}-1}{RT}(\bar{v}_2^{m,\infty} - \bar{v}_2^{s,\infty}) \qquad (5.12)$$

where M_m and M_s are the molecular weights of the mobile and stationary phases, $\rho_m(T)$ is the density of the eluent at the temperature of the system, and $\gamma_2^{m,\infty}(T,1)$ and $\gamma_2^{s,\infty}(T,1)$ are the activity coefficients of the components for infinite dilution in the mobile and stationary phases, at temperature T and pressure $P = 1$ atm.

The effect of the pressure can be observed only at pressures of 10^2–10^3 atm and by precise measurements. For pressures higher than 10^3 atm Eq. (5.12) loses its validity.

Novotný et al. [67] have also shown that the migration of the components in supercritical-fluid chromatography depends very much on pressure and temperature, and these two parameters can hence be used to obtain optimal separations.

5.1.8 Time of Analysis

In chromatographic practice, a resolution R_s higher than that necessary for a good separation ($R_{s\,opt} = 1$) is sometimes obtained. In these cases, the difference in resolution $R_s - R_{s\,opt}$ may be utilised by increasing the flow-rate of the eluent, and implicitly shortening the analysis.

Giddings [1] discussed this problem theoretically. Guiochon [74, 75] has also analysed the normalised chromatography time from a theoretical point of view, finding a series of rules that the research-worker should follow to obtain better separations. The same problem was also studied in other papers [76–78], especially the minimal time of analysis [34, 37, 79–81].

5.2 Ion-exchange Chromatography on Closed Columns

The resolution in ion-exchange chromatography, as shown by Eq. (4.70) (p. 80) depends on the plate height H and on the ratio of the distribution coefficients β (the selectivity coefficient). But H, defined by Eq. (4.24) (p. 64) depends on a series of factors such as the radius of the resin granules, the distribution coefficients, the diffusion coefficients, the flow-rate of the eluent, etc., which in turn depend on the temperature, the ionic strength, the pH, the degree of cross-linking of the resin, and so on. This dependence is therefore very complicated and it is hence very difficult to observe the simultaneous influence of all these factors in the optimization process. In a series of researches, satisfactory results were obtained by studying the influence of a certain parameter, while maintaining the other working conditions as constant as possible. We shall try to classify below the results of these studies in accordance with the chosen parameter.

5.2.1 Size and Degree of Cross-linking of the Particles, and the Quantity of Resin

The size of the resin particle. The resolution of two components is improved as the standard deviations of the two chromatographic peaks decrease. Sharp peaks presuppose the rapid establishment of equilibrium between the mobile and stationary phases. In such conditions, the contribution of diffusion and the resistance to mass-transfer to the plate-height is small. Mayer and Tompkins [82] have shown that in the case of resins with small particles, the equilibrium is more quickly established, and the flow-rate of the eluent can be increased, without altering the separation of the components.

To define the optimum operational conditions for an ion-exchange column, it is necessary to know very precisely the dimensions of the resin particles, and the particle size distribution should be as narrow as possible. Means of obtaining resins with particles of known dimensions have been described [83, 84].

By supposing all parameters of the chromatographic system to be constant, except the radius r_p of the resin granules, the equation for the plate height [(4.24)] may be written:

$$H = C_0 + C_i r_p + C_2 r_p^2 \tag{5.13}$$

It is hence evident that the smaller the radius of the particles, the more H will decrease and the resolution increase, as may be observed in Fig. 5.5, plotted for Hamilton's data [85].

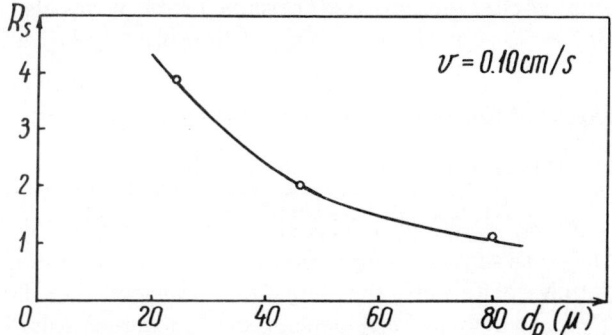

Fig. 5.5. Resolution *vs.* resin-grain diameter (Dowex 50-X8) for the mixture glycine–alanine, $L = 108$, $d_c = 6\cdot30$ mm, data from [85].

The results reproduced in this figure are in agreement with the theoretical considerations on the role of the grain-size on separation.

The influence of the grain size on the form of the elution curves, the breadth of the chromatographic peak, the plate-height and the resolution was shown in a series of papers [86–90]. The smaller the particles, the narrower the peaks, the smaller the plate-heights and the better the resolution.

The degree of cross-linking. Dybczyński [91] showed in a recent article that the influence of the degree of cross-linking of the resin on separations by ion-exchange has not been thoroughly investigated. Generally it is supposed [85, 92] that increased cross-linking should result in an increase of separation selectivity. Though such an increase has been reported several times [92–97], there exist other papers [98–100] showing that this situation is not by any means general. Moreover, the degree of separation depends on the dynamics of the processes in the column, which in turn will change with the degree of cross-linking of the resin. The influence of cross-linking on the separation of rare-earth complexes with EDTA on strongly basic anion-exchange resins [91] is reproduced in Fig. 5.6.

As is known [101, 102], for a high degree of cross-linking the diffusion coefficients decrease, and the plate-height will therefore increase. The diffusion coefficients will increase with decrease in degree of cross-linking, and the value of H will again rise, after passing through a minimal value. Good separations are generally obtained on resins with 3–4 per cent DVB cross-linking.

Quantity of resin. A suitable quantity of resin is needed to obtain good resolution for a given set of conditions (size of resin granules, degree of cross-linking, flow-rate, pH, temperature, etc.). A larger amount of resin than necessary increases the time for analysis, and a smaller quantity

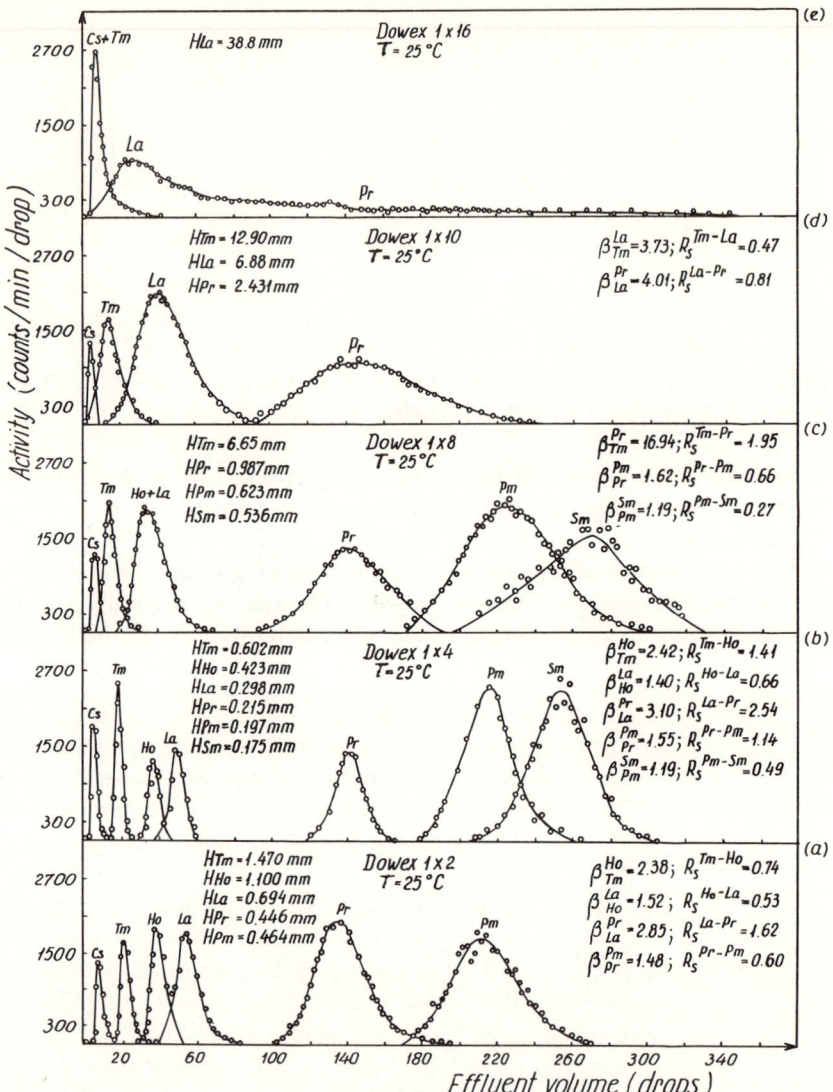

FIG. 5.6. Effect of resin cross-linking on the separation of Tm, Ho, La, Pr and Sm ethylenediaminetetra-acetates at 25°C. (a) Dowex 1-X2 [H_2Y^{2-}](6 μm $\leq d_p \leq$ 38 μm); column 5.30 cm × 0.0310 cm^2; eluent 0.011M Na$_2$H$_2$Y; flow-rate 0.91 cm/min. (b) Dowex 1-X4 [H_2Y^{2-}](11 μm $\leq d_p \leq$ 42 μm); column 5.30 cm × 0.0310 cm^2; eluent 0.036M Na$_2$H$_2$Y; flow-rate 1.19 cm/min. (c) Dowex 1-X8 [H_2Y^{2-}](17 μm $\leq d_p \leq$ 44 μm); column 5.30 cm × 0.0310 cm^2; eluent 0.078M Na$_2$H$_2$Y; flow-rate 1.22 cm/min. (d) Dowex 1-X10 [N_2Y^{2-}](17 μm $\leq d_p \leq$ 46 μm); column 5.30 cm × 0.0310 cm^2; eluent 0.096M Na$_2$H$_2$Y; flow-rate 1.17 cm/min. (e) Dowex 1-X16 [H_2Y^{2-}](20 μm $\leq d_p \leq$ 46 μm); column 5.30 cm × 0.0310 cm^2; eluent 0.029M Na$_2$H$_2$Y; flow-rate 1.21 cm/min [91].

leads to poorer resolution. The optimum quantity is that for which a good resolution is obtained, as well as adequate precision and accuracy of the chromatogram [103].

5.2.2 Dimensions of the Column

Hamilton et al. [104], analysing the diffusion mechanism (mass transfer), in the case of ion-exchange chromatography of amino-acids, deduced the following equation for the variance of the chromatographic peak:

$$\sigma^2 = 2A^2L\left[d_p\lambda(K_d + F_I)^2 + \frac{K_d d_p^2 \bar{v}_0}{60\,D_S} + \frac{K_d^2 d_p^2 v_0}{C_1 F_{II} D_M(1 + \Phi(N_{Re}N_{Sc}))}\right], \quad (5.14)$$

where A is the cross-sectional area of the column, L the column length, d_p the diameter of the resin particles, K_d the distribution coefficient, v_0 the linear flow-rate, D_S and D_M are the diffusion coefficients in the stationary and mobile phases, F_I is the void fraction of the column, $F_{II} = 1 - F_I$ is the fraction of the column occupied by the solid, $\lambda = 1/N_{Pe}$ where N_{Pe} is Peclet's number which can take the theoretical limiting value of 2, C_1 is a constant, and $\Phi(N_{Re}, N_{Sc})$ a function depending on the Reynolds and Schmidt numbers.

For optimal resolution the dimensions of the column, i.e. length and cross-section must be taken into account. If all other dimensions are considered constant, then from (5.14) the variance and the standard deviation are proportional to the length of the column and to the cross-sectional area respectively:

$$\sigma^2 = C_1 L \qquad \sigma = C_2 A \qquad (5.15)$$

The breadth of the peak will evidently be proportional to L and to A [89].

Equation (5.15) was experimentally verified by Hamilton et al. [104] and good agreement with theory obtained. Fritz et al. [87] showed the influence of the column length on the elution curves for zinc.

Mondiano [103] has dealt with the problem of optimising the column dimensions in automatic ion-exchange chromatography of amino-acids.

From Eq. (5.15) very small columns should be used to obtain the narrowest possible peaks, but the resolution would be very poor, and no separation could be achieved. The length of the column must therefore be chosen so as to give also a good resolution. Nevertheless, good resolution can be obtained with shorter columns by using a sufficiently small grain size. It has been shown [105] that the best separations are obtained for length/diameter ratio between 10:1 and 20:1.

5.2.3 Pressure and Flow-rate of the Eluent

Nowadays, very finely granulated resins are used to increase the separation efficiency of the column, which brings about an important decrease in the eluent flow-rate. Pumps are therefore used to increase the flow-rate. High efficiency columns have thus been created, without increasing the time needed for analysis.

In ion-exchange chromatography of amino-acids, the flow-rates used for analytical separations are between 0·00174 and 0·0131 cm/s, at a pressure up to 4·5 atm [83, 86, 106]. The flow-rate has been increased to 0·022 cm/s at a pressure of 14 atm [107]. Hamilton, in another paper [85], has achieved flow-rates of up to 0·17 cm/s at pressures of 41 atm, using a column packed with spherical Dowex 50-XB particles having diameters of 2·4, 4·6 or 8·0 × 10^{-3} cm. The best resolution is theoretically obtained with particles of the smallest possible diameter if all the other variables are constant, but necessitates increasing the pressure to give an acceptable flow. These statements are illustrated in Figs. 5.7 and 5.8.

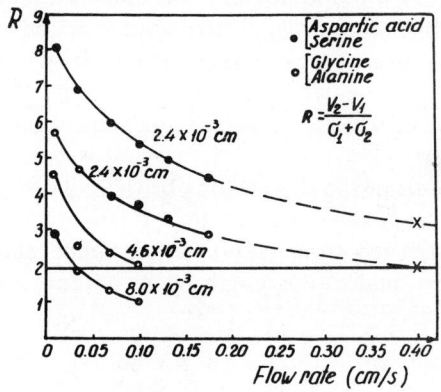

FIG. 5.7. Resolution of amino-acid pairs as a function of linear flow-rate through the column [85].

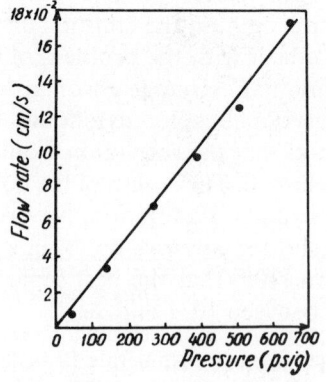

FIG. 5.8. Flow-rate as function of pressure at top of column (1 psig = 0·068 atm) [85].

Hamilton [108] in 1955 drew attention to the fact that the ion-exchange chromatography of amino-acids, and liquid chromatography generally, could be expected to undergo a new development, namely use of small diameter columns packed with very fine particles, probably smaller than 1 μm in diameter, working at flow-rates 10–100 times those of conventional chromatography. The pressures needed would have to be from 5000 to at least 10^5 psig (340–6800 atm). In this case the analysis time could probably be reduced from 4–6 hr to a few seconds.

5.2.4 TEMPERATURE

The effect of temperature on ion-exchange chromatography has lately been increasingly studied. Papers on work in which the temperature is used as a fundamental parameter appear more frequently than before. The effect of temperature was studied for the ion-exchange separation of amino-acids [109], some fatty acids [110], alkali metals and rare earths [111–113], and Zn^{2+} and Co^{2+} [114].

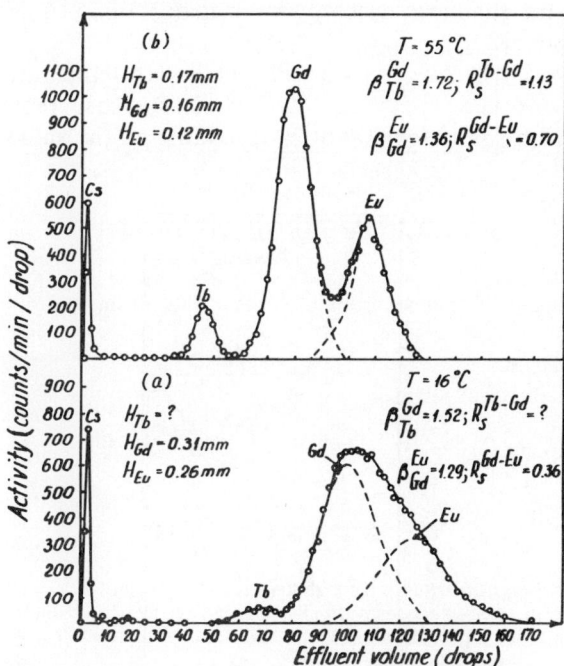

FIG. 5.9. Influence of temperature on the chromatographic separations of Tb, Gd and Eu ethylenediaminetetra-acetates. Amberlite IRA-400 $[H_2Y^{2-}]$ (10 μm $\leq d_p \leq$ 35 μm); column 2·50 cm × 0·0360 cm^2, eluent 0·0769M Na$_2$H$_2$Y; flow-rate 1·97–2·01 × 10^{-2} mole/cm^2/s [111].

The effect of temperature on the shapes of elution curves, the plate-height and the resolution is shown in Fig. 5.9.

The resolution, as shown by Eq. (4.70) depends on the selectivity coefficient β, the length of the column and the plate-height. Except for the length of the column (if expansion is neglected), all dimensions vary with the temperature.

It is well known that the ion-exchange reaction is a diffusion-controlled process and change in temperature will evidently influence its kinetics [115]. On the other hand, the relative affinities may change very much with the temperature. This is a consequence of the fact that the enthalpies of the ion-exchange reactions may differ considerably, even if the ions concerned have similar chemical features [112, 116–119].

The selectivity coefficient β is the ratio of the distribution coefficients K_d, which depend in turn on the temperature, through the equation

$$\Delta H = -2 \cdot 303 \, R \frac{\mathrm{d} \log K_d}{\mathrm{d}(1/T)}, \qquad (5.16)$$

where ΔH is the change in enthalpy for exchange of an equivalent of one ion for another [120].

The change in the selectivity coefficient β with temperature was also pointed out in other papers [121–124]. It can increase, remain constant or decrease with rise in temperature, depending on the value of ΔH.

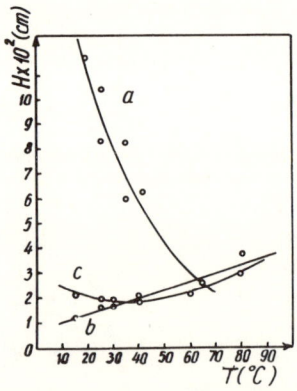

FIG. 5.10. Plate height normalised for distribution coefficient $K_d = 10$, as a function of temperature in various systems. (a) Rare earth ethylenediaminetetra-acetates in the system Amberlite IRA-400 [H_2Y^{2-}]–$Na_2H_2Y_{aq}$. Resin particle size 10 μm $\leq d_p \leq$ 35 μm; eluent 0·008–0·009M Na_2H_2Y; pH 4·6–4·7; flow-rate 1·08–1·25 cm/min. (b) Alkali metals in the system Amberlite IR-120 [H^+]–HCl_{aq}. Resin particle size 17 μm $\leq d_p \leq$ 51 μm; eluent 0·5199M HCl; flow-rate 0·47–0·51 cm/min. (c) Alkali metals in the system cation-exchanger MK$_3$ [H^+]–HCl_{aq}. Resin particle size 11 μm $\leq d_p \leq$ 31 μm; eluent 0·0712M HCl; flow-rate 0·84–0·91 cm/min [112].

Figure 5.10 shows the variation of the normalised plate-height for three systems, with $K_d = 10$.

The variation of H is explained by the changes in K_d, in the diffusion coefficients, and in the eluent's hydrodynamic properties, caused by changing the temperature.

The diffusion coefficient D depends on the temperature according to the equation [1]

$$D = D_0 \exp(-E/RT) \tag{5.17}$$

where E is the activation energy for transfer of a molecule from its initial position to a neighbouring one, and D_0 is a constant.

Hamilton et al. [104] have determined the diffusion coefficient in the solid phase (resin)(D_S) by means of the equation [(5.14)] for variance of the chromatographic peak, in which the term for the diffusion in the liquid phase is ignored. The variation of the diffusion coefficient D_S with $1/T$ for some amino-acids, is given in Fig. 5.11.

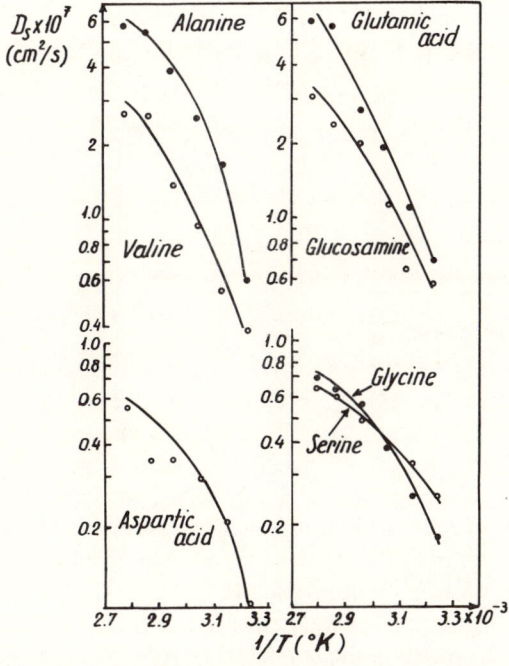

FIG. 5.11. Plot of diffusion coefficient D_S vs. inverse of temperature, $1/T$ [104].

The increase of D_M with temperature will decrease H for the whole range of D, if the last two terms of Eq. (5.18) can be neglected, but if not, H has a minimum [91].

$$H = 1.64\, r_p + \frac{K_d\, 0.142\, r_p^2 v}{(K_d + F_I)^2 D_S} + \frac{K_d^2\, 0.266\, r_p v}{(K_d + F_I)^2 (1 - F_I) D_M (1 + 70\, r_p v)}$$

$$+ \frac{D_M F_I \sqrt{2}}{v} + \frac{2\gamma D_S K_d}{v F_I} \qquad (5.18)$$

where γ is the tortuosity factor.

The variation of H with temperature is obviously very complex, and three cases may arise: a decrease, an increase, or no change with change in temperature. The first case occurs often enough, as shown in a series of papers [111, 114, 125–128].

The influence of temperature on the resolution must be interpreted (as was shown above) by means of the selectivity coefficient β and of H. All possibilities are schematically summarised [112] in Table 5.1.

TABLE 5.1

	$\beta \uparrow$	β constant	$\beta \downarrow$
$\bar{H} \uparrow$	$R_s \uparrow \downarrow$ or constant	$R_s \downarrow$	$R_s \downarrow$
$\bar{H}$ constant	$R_s \uparrow$	R_s constant	$R_s \downarrow$
$\bar{H} \downarrow$	$R_s \uparrow$	$R_s \uparrow$	$R_s \uparrow \downarrow$ or constant

The temperature is a parameter which, judiciously used, may increase the resolution and shorten the time of analysis.

5.2.5 Other Factors

The composition of the eluent is an important factor in obtaining satisfactory resolution. Changes in the eluent composition (pH, ionic strength, dielectric constant, etc.) will lead to a change in the distribution coefficient and the selectivity coefficient, and hence implicitly of the resolution.

The influence of pH and ionic strength is discussed by Hamilton [109] for amino-acids. It is found necessary to elute at different pH values to separate a great number of amino-acids.

When solvent mixtures of water and an organic solvent, generally an alcohol, are used, changing the concentration of the organic solvent alters the resolution [90, 129–133]. In partition chromatography of some sugars on an ion-exchange column [90, 134] it was noted that the separation was improved by increasing the concentration of ethyl alcohol in the eluent. Dixon [135] studied the reduction of tailing of chromatographic peaks and increase in their symmetry through changing the eluent composition.

To increase the separation efficiency in ion-exchange chromatography, complexing agents are often used, including sulphosalicyclic acid [136], potassium thiocyanate [137, 138], sodium carbonate [139], sodium thiosulphate [140, 141], ammonium citrate, ammonium lactate, EDTA, ammonium salt of α-hydroxyisobutyric acid, and nitrilotriacetic, glycollic, malonic, acetic, tartaric acids, glycerine, ammonia etc. [91, 111, 112, 143, 156].

5.3 Partition and Adsorption Chromatography on Open Columns

5.3.1 CHOICE OF THE MOBILE AND STATIONARY PHASES

The success of the separation of a complex mixture by paper and thin-layer chromatography greatly depends on the choice of the mobile and stationary phases. The mobile phase offers the wider choice of variation.

A series of research-workers [157–159] have established eluotropic series of the common solvents, setting them in order of increase of polarity. It is recommended to use the following series for practical choice of solvents:

Solvent	Dielectric constant
n-Hexane	1·9
Cyclohexane	2·1
Carbon tetrachloride	2·2
Benzene	2·3
Toluene	2·4
Dioxan	3·0
Trichloroethylene	3·4
Diethyl ether	4·3
Chloroform	4·8
Ethyl acetate	6·0
Methyl ethyl ketone	18·0
n-Butanol	19·0
Acetone	21·0
n-Propyl alcohol	22·0
Ethanol	24·0
Isopropyl alcohol	26·0
Methanol	31·0
Water	81·0

The elution power increases with increase in the dielectric constant of the solvent.

In practice, the elution system is chosen either by trial and error, running chromatograms with different solvents, or by the method given by Stahl [160–162], which consists of putting sample at several places on the plate, at a certain distance from one another. Different pure solvents are then applied to the centres of the spots. After the development, the solvent that has displaced the sample to approximately half the distance between the origin and the outer edge of the solvent zone is chosen. If no such solvent exists, a mixture of two solvents is made, one that displaces the sample too much, and the other too little. A polar solvent will displace a polar component more than a non-polar solvent does.

Thoma [163] has suggested two complementary quantitative indexes of the resolving power of a chromatographic system, namely the ratio between the chemical potentials of transfer of the components examined, and the ratio between the cross-sectional areas of the mobile and stationary phases (A_M/A_S). For substances with very similar chemical transfer potentials, changes in the mobile phase will not significantly improve the resolution. Use of thin-layer instead of paper chromatography increases the A_M/A_S ratio about 20-fold and increases the efficiency. Thoma showed that the separation of two components is best when the zones have migrated for approximately a quarter of the support length, and hence the solvent should be chosen to give an average R_f value around 0·25.

Knowledge of the basic mechanism of the chromatographic process is used in the choice of solvent, development condition and support. Thus, for the separation of lipophilic components in a medium of low polarity, the use of activated silica gel and alumina layers is recommended, and a large range of solvents may be used as mobile phase. For the separation of inorganic compounds soluble in water and polar organic compounds, cellulose or unactivated silica gel can be used as support with very good results and organic solvents saturated with water or buffer solutions as the mobile phase. For the separation of lipophilic compounds having very similar properties the support should be a sorbent (cellulose or silica gel) containing a very slightly polar organic liquid, and very polar solvents should be used for the mobile phase.

Stahl [164] has given a scheme (Fig. 5.12) allowing choice of the support and the mobile phase according to the nature of the mixture to be separated in adsorption chromatography.

The triangle is arranged with one peak pointing towards the arc representing the mixture to be separated. The positions of the other two peaks give the conditions to be fulfilled by the stationary and the mobile phases.

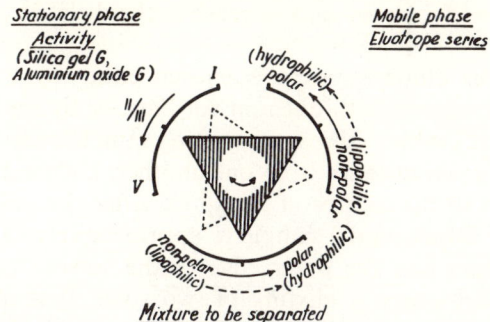

FIG. 5.12. The three fundamental variable parameters in adsorption chromatography (the figures refer to the Brockmann degree of activity for the stationary phase) [164].

The problem of choice of solvent system for different classes of combinations has already been dealt with in a series of papers [165–173]. Kananen et al. [174] made a comparative study of the adsorption capacity of a series of adsorbents used as supports in thin-layer chromatography. Studies were also undertaken to find some selective developers, which should allow the separation of a single ion from the rest [175].

5.3.2 Choice of Optimum pH

For organic electrolytes which have high R_f values when in non-ionised form, improved separations can often be achieved by modifying the pH of the aqueous phase [176], as the apparent partition coefficient depends on the pH. Figure 5.13 gives the curves for R_f and ΔR_f (corresponding to different pairs of substances) vs. pH [177]. As seen, the $\Delta R_f = f(\text{pH})$ curves generally have a maximum. The pH at which this maximum is obtained may correspond to the conditions of maximum resolution (when the difference between the R_f values of the components is the highest possible) for a pair of components, as results from Eq. (4.76). This optimum pH may be calculated [177] by equating to zero the derivative of Waksmundzki and Soczewiński's equation [178–180]:

$$\frac{d\Delta R_f}{d\text{pH}} = \frac{d}{d\text{pH}} \left[\frac{\alpha^{(1)} k / K_A^{(1)}}{(\alpha^{(1)} k + 1)/K_A^{(1)} + 10\,\text{pH}} - \frac{\alpha^{(2)} k / K_A^{(2)}}{(\alpha^{(2)} k + 1)/K_A^{(2)} + 10\,\text{pH}} \right] = 0$$

(5.19)

where α is the partition coefficient of the non-ionised substance, k is the volume ratio of the mobile and stationary phases ($k = V_M/V_S$), and K_A

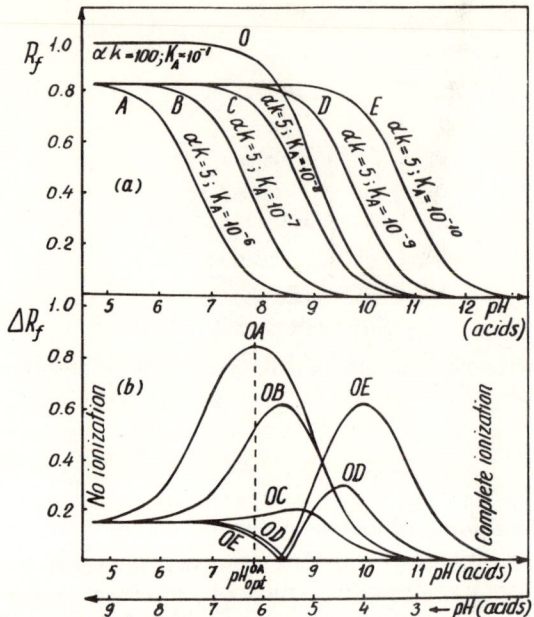

Fig. 5.13. Plot of R_f and ΔR_f as a function of pH [177].

the ionisation constant of the acid. The indexes (1) and (2) refer to the two components.

The solution of equation (5.15) is

$$\text{pH}_{\text{opt}} = \log \frac{(\alpha^{(1)}k + 1)/K_A^{(1)}\sqrt{\alpha^{(2)}k/K_A^{(2)}} - (\alpha^{(2)}k + 1)/K_A^{(2)}\sqrt{\alpha^{(1)}k/K_A^{(1)}}}{\sqrt{\alpha^{(1)}k/K_A^{(1)}} - \sqrt{\alpha^{(2)}k/K_A^{(2)}}}$$

(5.20)

By introducing the value of the optimum pH into the expression for ΔR_f, given in (5.19), and omitting the units, the maximum difference of the R_f values for the two components may be calculated:

$$\Delta R_{f\,\text{max}} = \frac{\sqrt{\beta'} - 1}{\sqrt{\beta'} + 1} \qquad (5.21)$$

where $\beta' = K_A^{(2)}\alpha^{(1)}/K_A^{(1)}\alpha^{(2)}$.

The $\Delta R_{f\,\text{max}}$ values given by Eq. (5.21) are plotted in Fig. 5.14 vs. the values of β' [177]. This figure shows that the maximum difference between the R_f values can be obtained (at pH$_{\text{opt}}$) for a certain value of the coefficient.

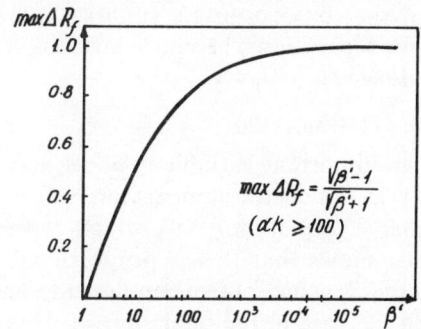

FIG. 5.14. Plot of $\Delta R_{f_{max}}$ as a function of the coefficient β' [177].

5.3.3 INFLUENCE OF HUMIDITY IN THIN-LAYER CHROMATOGRAPHY

Humidity plays an important role in the partition of the chromatographic zones between start and solvent front in a thin-layer chromatogram and can even lead to the reversion of the order of migration [181].

The influence of humidity on chromatographic separation is illustrated [182] in Fig. 5.15, which shows that the distribution of the zones on the chromatogram is much modified by variation of the humidity. The way in which humidity acts on the R_f values was discussed in Chapter 3. This problem was dealt with by Geiss et al. [183] and Sandroni et al. [184].

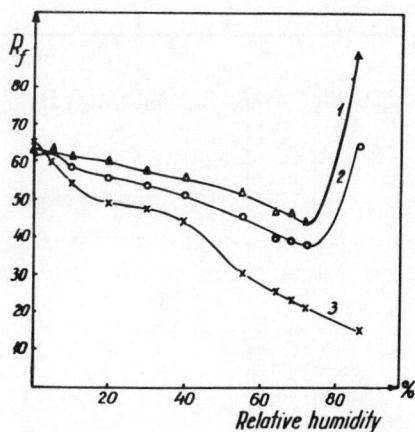

FIG. 5.15. Plot of R_f vs. relative humidity of the stationary phase: Al_2O_3. (D): mobile phase: benzene–methanol (9:1 v/v). 1. dioctylsulphoxide, 2. dibutylsulphoxide, 3. dimethylsulphoxide [182].

These studies show that by choosing a suitable humidity for the room, we can obtain better separations, i.e. more efficient columns, and consequently better resolution.

5.3.4 Influence of Temperature

Temperature is an important variable in paper and thin-layer chromatography. The way in which the temperature affects the R_f value was discussed in Chapter 3. Equation (4.69), which expresses the resolution of two components, shows that this is proportional to ΔR_f. Often the difference between the R_f values of two components, and hence the resolution, increases with increase in the temperature. This is illustrated [185] by the separation of Pb^{2+} and Co^{2+} on Whatman No. 2 paper, with n-butanol + 20 per cent $3N$ hydrochloric acid as eluent. The variation of ΔR_f with temperature was linear. For the same system, the duration of development decreases with increase in temperature, for instance solvent travel through 10 cm took 130 min, but only 90 min at 60°.

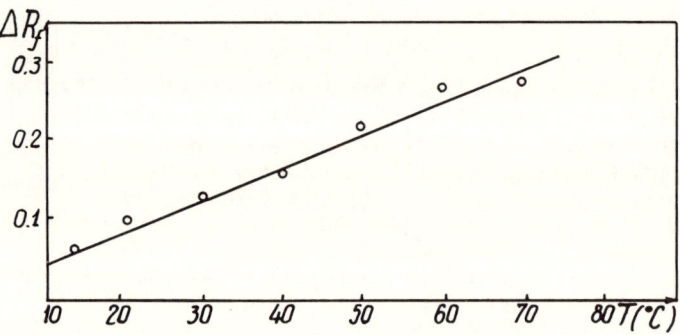

Fig. 5.16. Plot of ΔR_f vs. temperature [185].

REFERENCES

1. Giddings, J. C., *Dynamics of Chromatography, Part I, Principles and Theory*, Dekker, New York (1965).
2. Bush, I. E., *Chromatography of Steroids*, Pergamon, London (1961).
3. Macek, K., in *Chromatography*, Chap. 7, E. Heftman (ed.), 1st Ed., p. 1750, Reinhold, New York (1961).
4. Snyder, L. R., *Advan. Anal. Chem. Instr.* **3**, 251 (1964).
5. Snyder, L. R., in *Chromatography*, Chap. 5, E. Heftman (ed.), 2nd Ed., Reinhold, New York (1967).
6. Strain, H. H., *Chromatographic Adsorption Analysis*, Interscience, New York (1957).
7. Zechmeister, L. and Cholnoky, L., *Principles and Practice of Chromatography*, Wiley, New York (1943).
8. Mair, B. J., in *Treatise on Analytical Chemistry*, Part I, Vol. III, Chap. 34, I. M. Kolthoff and P. J. Elving (eds.), Interscience, New York (1957).
9. Lambert, S. M., *Anal. Chem.* **37**, 959 (1965).
10. Karr, C., Jr., Childers, E. E., Warner, W. C. and Estep, P. E., *Anal. Chem.* **36**, 2105 (1964).
11. Lambert, S. M. and Porter, P. E., *Anal. Chem.* **36**, 99 (1964).
12. Knox, J. H. and Parcher, J. F., *Anal. Chem.* **41**, 1599 (1969).
13. Stewart, H. N. M., Amos, R. and Perry, S. G., *J. Chromatog.* **38**, 209 (1968).
14. Kirkland, J. J., *J. Chromatog. Sci.* **7**, 7 (1969).
15. Sie, S. T. and van den Hoed, N., *Advances in Chromatography*, 1969, p. 318, A. Zlatkis (ed.), Preston Technical Abstracts, Evanston (1969).
16. De Stefano, J. J. and Beachell, H. C., *J. Chromatog. Sci.* **8**, 434 (1970).
17. Snyder, L. R., *Anal. Chem.* **39**, 698 (1967).
18. King, P. J. and Chambers, L. E., *J. Inst. Petrol.* **51**, 401 (1965).
19. Giddings, J. C., *Anal. Chem.* **34**, 37 (1962).
20. Horne, D. S., Knox, J. H. and McLaren, L., *Sepn. Sci.* **1**, 531 (1966).
21. Smuts, T. W. and Pretorius, V., *J. Gas Chromatog.* **4**, 404 (1966).
22. Heinekey, D. M., *Proc. Soc. Anal. Chem.* **9**, 11 (1972).
23. Green, F. C. and Kay, L. M. *Anal. Chem.* **24**, 726 (1952).
24. Alderweireldt, F., *J. Chromatog.* **5**, 98 (1961).
25. Cahnmann, H. J., *Anal. Chem.* **29**, 1307 (1957).
26. Huber, J. F. K., *J. Chromatog. Sci.* **7**, 85 (1969).
27. Snyder, L. R., *J. Chromatog. Sci.* **7**, 352 (1969).
28. Llemm, L. H. and Reed, D., *J. Chromatog.* **3**, 364 (1960).
29. Kirkland, J. J., *J. Chromatog. Sci.* **7**, 361 (1969).
30. Klein, P. D., *Anal. Chem.* **34**, 733 (1962).
31. Wilson, J. N., *J. Am. Chem. Soc.* **62**, 1583 (1940).
32. Snyder, L. R., *Anal. Chem.* **39**, 705 (1967).

33. Klemm, L. H. and Reed, D., *J. Chromatog.* **3**, 364 (1960).
34. Waters, J. L., Little, J. N. and Horgan, D. F. J., *J. Chromatog. Sci.* **7**, 293 (1969).
35. Simpson, D. W. and Wheaton, R. M., *Chem. Eng. Progr.* **50**, 45 (1954).
36. Halasz, I. and Walking, P., *J. Chromatog. Sci.* **7**, 129 (1969).
37. Knox, J. H. and Saleem, M., *J. Chromatog. Sci.* **7**, 614 (1969).
38. Snyder, L. R., *J. Chromatog.* **8**, 178 (1962).
39. Lepage, M., Beau, R. and de Vries, A. J., *J. Polymer Sci.* **21(C)**, 119 (1968).
40. Smuts, T. W., van Niekerk, F. A. and Pretorius, V., *J. Gas Chromatog.* **5**, 190 (1967).
41. Sie, S. T. and van den Hoed, N., *J. Chromatog. Sci.* **7**, 257 (1969).
42. Joynes, P. L., *Column* **3**, 9 (1969).
43. Horváth, C. and Lipsky, S. P., *J. Chromatog. Sci.* **7**, 109 (1969).
44. Kirkland, J. J., *Anal. Chem.* **40**, 391 (1968).
45. Kirkland, J. J., *Anal. Chem.* **41**, 218 (1969).
46. Little, J. N., Horgan, D. F. and Bombaugh, K. J., *J. Chromatog. Sci.* **8**, 625 (1970).
47. Giddings, J. C., *J. Chromatog.* **5**, 61 (1961).
48. Knox, J. H., *Anal. Chem.* **38**, 253 (1966).
49. Pretorius, V. and Smuts, T. W., *Anal. Chem.* **38**, 274 (1966).
50. Giddings, J. C., *Anal. Chem.* **36**, 1890 (1964).
51. Knox, J. H. and Saleem, M., *J. Chromatog. Sci.* **7**, 745 (1969).
52. Kirkland, J. J., *Anal. Chem.* **43**, (12) 36 A (1971).
53. LeRosen, A. L. and Rivet, C. A., Jr., *Anal. Chem.* **20**, 1093 (1948).
54. Chang, L. T., *Anal. Chem.* **25**, 1235 (1953).
55. Locke, D. C. and Martire, D. E., *Anal. Chem.* **39**, 921 (1967).
56. Locke, D. C., *J. Gas Chromatog.* **5**, 202 (1967).
57. Hesse, G. and Engelhardt, H., *J. Chromatog.* **21**, 228 (1966).
58. Kaiser, R., *Chromatographie in der Gasphase*, I, Band I, p. 50, Bibliographisches Institut, Mannheim (1960).
59. Maggs, R. J. and Young, T. E., *Gas Chromatography 1968*, p. 217, C.L.A. Harbourn and R. Stock (eds.), Elsevier, Amsterdam (1969).
60. Maggs, R. J., *J. Chromatog. Sci.* **7**, 145 (1969).
61. Scott, R. P. and Lawrence, J. G., *J. Chromatog. Sci.* **7**, 65 (1969).
62. Sie, S. T., van Beersum, W. and Rijnders, G. W. A., *Sepn. Sci.* **1**, 459 (1966).
63. Myers, M. N. and Giddings, J. C., *Sepn. Sci.* **1**, 761 (1966).
64. Sie, S. T. and Rijnders, G. W. A., *Sepn. Sci.* **2**, 729, 755 (1967).
65. McLaren, L., Myers, M. N. and Giddings, J. C., *Science* **159**, 197 (1968).
66. Giddings, J. C., Myers, M. N. and King, J. W., *J. Chromatog. Sci.* **7**, 276 (1969).
67. Novotný, M., Bertsch, W. and Zlatkis, A., *J. Chromatog.* **61**, 17 (1971).
68. Smith, T. W., van Niekerk, F. A. and Pretorius, V., *J. Gas Chromatog.* **5**, 190 (1967).
69. Horváth, C. G., Preiss, B. A. and Lipsky, S. P., *Anal. Chem.* **39**, 1422 (1967).
70. Felton, H., *J. Chromatog. Sci.* **7**, 13 (1969).
71. Bidlingmeyer, B. A., Hooker, R. P., Lochmüller, C. H. and Roger, L. B., *Sepn. Sci.* **4**, 439 (1969).
72. Henry, R. A., Schmit, J. A., Dieckman, J. F. and Murphey, F. J., *Anal. Chem.* **43**, 1053 (1971).
73. Locke, D. C., in *Advances in Chromatography*, Vol. **8**, p. 47, J. C. Giddings and R. A. Keller (eds.), Dekker, New York (1969).

REFERENCES

74. Guiochon, G., *Anal. Chem.* **38**, 1020 (1966).
75. Guiochon, G., in *Advances in Chromatography*, Vol. 8, p. 179, J. C. Giddings and R. A. Keller (eds.), Dekker, New York (1969).
76. Karger, B. L. and Cooke, W. D., *Anal. Chem.* **36**, 985 (1964).
77. Grushka, E., *Anal. Chem.* **42**, 1142 (1970).
78. Grushka, E., *Anal. Chem.* **43**, 766 (1971).
79. Horváth, C. G. and Lipsky, S. R., *Anal. Chem.* **39**, 1893 (1967).
80. Hawkes, S., *J. Chromatog. Sci.* **7**, 526 (1969).
81. Snyder, L. R. and Saunders, D. L., *J. Chromatog. Sci.* **1**, 195 (1969).
82. Mayer, S. W. and Tompkins, E. R., *J. Am. Chem. Soc.* **69**, 2866 (1947).
83. Moore, S. and Stein, W. H., *J. Biol. Chem.* **192**, 663 (1951).
84. Moore, S. and Stein, W. H., *J. Biol. Chem.* **211**, 893 (1954).
85. Hamilton, P. B., *Anal. Chem.* **32**, 1779 (1960).
86. Hamilton, P. B., *Anal. Chem.* **30**, 914 (1958).
87. Fritz, J. F. and Karraker, S. K., *Anal. Chem.* **31**, 921 (1959).
88. Anbonin, G. and Laverlochere, J., *J. Radioanal. Chem.* **1**, 123 (1968).
89. Hamilton, P. B., *Anal. Chem.* **35**, 2055 (1963).
90. Dahlberg, J. and Samuelson, O., *Acta Chem. Scand.* **17**, 2136 (1963).
91. Dybczyński, R., *J. Chromatog.* **50**, 487 (1970).
92. Trémillon, B., *Les séparations par resines échengeuses d'ions*, Gauthier-Villars, Paris (1965).
93. Herbert, R. H., Tongue, K. and Irvine, J. W., Jr., *J. Am. Chem. Soc.* **17**, 5840 (1955).
94. Alexa, J., *Collection Czech. Chem. Commun.* **33**, 1933 (1968).
95. Hulet, E. K., Gutmacher, R. G. and Coops, M. S., *J. Inorg. Nucl. Chem.* **17**, 350 (1961).
96. Kolosova, G. M. and Senyavin, M. M., *Zh. Fiz. Khim.* **41**, 1597 (1967).
97. Speecke, A. and Hoste, J., *Talanta* **2**, 332 (1959).
98. Bonner, O. D. and Smith, L. L., *J. Phys. Chem.* **61**, 326 (1957).
99. Talášek, V., *Collection Czech. Chem. Commun.* **33**, 35 (1968).
100. Diamond, R. M., *J. Am. Chem. Soc.* **77**, 2978 (1955).
101. Boyd, G. E. and Soldano, B. A., *J. Am. Chem. Soc.* **75**, 6091 (1953).
102. Soldano, B. A. and Boyd, G. E., *J. Am. Chem. Soc.* **75**, 6099 (1953).
103. Mondino, A., *J. Chromatog.* **50**, 260 (1970).
104. Hamilton, P. B., Bogue, D. C. and Anderson, R. A., *Anal. Chem.* **32**, 1782 (1960).
105. Sabău, C., *Schimbul ionic*, p. 41, Ed. Acad. RSR, Bucuresti (1967).
106. Speckman, D. H., Moore, S. and Stein, W. H., *Anal. Chem.* **30**, 1190 (1958).
107. Hamilton, P. B. and Anderson, R. A., *Anal. Chem.* **31**, 1504 (1959).
108. Hamilton, P. B., *Technicon Symposia, Automation in Analytical Chemistry* 1966, Vol. I, p. 447, Mediad, New York (1967).
109. Hamilton, P. B., in *Advances in Chromatography*, Vol. 2, p. 3, J. C. Giddings and R. A. Keller (eds.), Arnold, London (1965).
110. Larsson, U.-B., Norstedt, I. and Samuelson, O., *J. Chromatog.* **22**, 102 (1966).
111. Dybczyński, R., *J. Chromatog.* **14**, 79 (1964).
112. Dybczyński, R., *J. Chromatog.* **31**, 155 (1967).
113. Dybczyński, R. and Wódkiewicz, L., *J. Inorg. Nucl. Chem.* **31**, 1495 (1969).
114. Liteanu, C., Gocan, S., Hodişan, T., Naşcu, H. and Muroţoiu, C., *Rev. Roumaine Chim.* **17**, 497 (1972).
115. Helfferich, F., *Ion Exchange*, McGraw-Hill, New York (1962).
116. Kraus, K. A. and Raridon, R. J., *J. Phys. Chem.* **63**, 1901 (1959).

117. Baetsle, L., *J. Inorg. Nucl. Chem.* **25**, 271 (1963).
118. Dybczyński, R., *Roczniki Chem.* **37**, 1411 (1963).
119. Brunisholtz, G. and Roulet, R., *Chimia* **21**, 188 (1967).
120. Boner, O. D. and Pruett, R. R., *J. Phys. Chem.* **63**, 1417 (1959).
121. Holm, L., Choppin, G. R. and Moy, S., *J. Inorg. Nucl. Chem.* **19**, 251 (1961).
122. Koprada, V. and Fojtik, M., *Chem. Zvesti* **19**, 294 (1965).
123. Hulet, E. K., Gutmacher, R. G. and Coops, M. S., *J. Inorg. Nucl. Chem.* **17**, 350 (1961).
124. Surls, J. P. and Choppin, G. R., *J. Inorg. Nucl. Chem.* **4**, 62 (1957).
125. Thompson, S. G. Cunningham, B. B. and Seaborg, G. T., *J. Am. Chem. Soc.* **72**, 2798 (1950).
126. Cornish, F. W., Phillips, G. and Thomas, A., *Canad. J. Chem.* **34**, 1471 (1956).
127. Nelson, F. and Michelson, D. C., *J. Chromatog.* **25**, 414 (1966).
128. Rexen, B. and Christensen, B., *Technicon Symposia, Automation in Analytical Chemistry* (1966), Vol. II, p. 435, Mediad, New York (1967).
129. Samuelson, O. and Swenson, B., *Acta Chem. Scand.* **18**, 2056 (1962).
130. Korkisch, J. and Hanuaer, G. E., *Talanta* **9**, 957 (1962).
131. Edge, R. A., *J. Chromatog.* **6**, 452 (1961).
132. Edge, R. A., *J. Chromatog.* **8**, 419 (1962).
133. Samuelson, O., *Anal. Chim. Acta* **38**, 163 (1967).
134. Samuelson, O. and Swenson, E., *Acta Chim. Scand.* **16**, 2056 (1962).
135. Dixon, H. B. F., *J. Chromatog.* **24**, 199 (1966).
136. Fritz, J. S. and Palmer, T. A., *Talanta* **9**, 393 (1962).
137. Turner, J. B., Phil, R. H. and Day, R. A., Jr., *Anal. Chim. Acta* **26**, 99 (1962).
138. Hamaguchi, H., Kuroda, R., Aoki, K., Sugisita, R. and Onuma, N., *Talanta* **10**, 153 (1963).
139. Taketatsu, T., *Talanta* **10**, 1077 (1963).
140. Katsura, T., *Bunseki Kagaku* **10**, 366, 1207, 1211 (1961).
141. Katsura, T., *Bunseki Kagaku* **11**, 34 (1962).
142. Bandi, W. R., Buyok, E. G., Lewis, L. L. and Melnick, L. M., *Anal. Chem.* **33**, 1275 (1961).
143. Vickery, R. C., *J. Chem. Soc.* 1181 (1954).
144. Preobrayenski, B. K., Kaliamin, A. V. and Lilova, O. M., *Radiokhimiya* **2**, 239 (1960).
145. Choppin, G. and Silvan, R., *J. Inorg. Nucl. Chem.* **3**, 153 (1956).
146. Smith, L. and Hoffman, D., *J. Inorg. Nucl. Chem.* **3**, 245 (1956).
147. Starý, J., *Talanta* **13**, 421 (1966).
148. Choppin, G., Harvey, B. and Thompson, S., *J. Inorg. Nucl. Chem.* **2**, 66 (1956).
149. Preobreyenski, B. K., *Khim. Nauk i Prom.* **4**, 521 (1959).
150. Ketelle, B. H. and Boyd, G. E., *J. Am. Chem. Soc.* **69**, 2800 (1947).
151. Ketelle, B. H. and Boyd, G. E., *J. Am. Chem. Soc.* **73**, 1862 (1954).
152. Mayer, S. W. and Freiling, E. C., *J. Am. Chem. Soc.* **75**, 5647 (1953).
153. Inczédy, J., Klatsmanyi-Gábor, P. and Erdey, L., *Acta Chim. Acad. Sci. Hung.* **61**, 261 (1969).
154. Indzédy, J., Klatsmanyi-Gábor, P. and Erdey, L., *Acta Chim. Acad. Sci. Hung.* **62**, 1 (1969).
155. Inczédy, J., *Acta Chim. Acad. Sci. Hung.* **62**, 131 (1969).
156. Vanderleelen, J., *Anal. Chim. Acta* **49**, 361 (1970).
157. Trappe, W., *Biochem. Z.* **305**, 150 (1940).
158. Strain, H. H., *Chromatographic Adsorption Analysis*, Interscience, New York (1942).

159. Knight, H. S. and Groennings, S., *Anal. Chem.* **26**, 1549 (1954).
160. Stahl, E., *Chem. Ztg.* **82**, 323 (1958).
161. Stahl, E., *Arch. Pharm.* **292**, 411 (1959).
162. Stahl, E., *Pharm. Rundschau* **1**, 1 (1959).
163. Thoma, J. A., *Anal. Chem.* **37**, 500 (1965).
164. Stahl, E., *Angew. Chem.* **73**, 646 (1961).
165. Hasegawa, H., *Yakugaku Zasshi* **80**, 1175 (1960).
166. Bulenkov, T. J., *Zh. Analit. Khim.* **23**, 348 (1968).
167. Folk, E., Buchtela, K. and Gross, T., *Atomkernenergie* **15**, 297 (1970).
168. Bulenkov, T. J., *Zavodsk. Lab.* **33**, 418 (1967).
169. Soczewiński, E., *Anal. Chem.* **41**, 179 (1969); *Ann. Univ. Marie Curie-Sklodowska, Sect. D* **24**, 25 (1969).
170. Erhard, R., *Mitt. Deut. Pharm. Ges.* **40**, 176 (1970).
171. Procházka, Ž., *Chemie (Prague)* **9**, 736 (1957).
172. Pazdera, H. J., *Encycl. Ind. Chem. Anal.* **3**, 1 (1966).
173. Pie, A. and Giner, A., *Nature* **212**, 402 (1966).
174. Kananen, G., Sunshine, I. and Monforte, J., *J. Chromatog.* **52**, 291 (1970).
175. Născuţiu, T. and Oltenaşu, M., *Chim. Analit. (Bucharest)* **1**, 51 (1971).
176. Golumbic, C., Orchin, M. and Weller, S., *J. Am. Chem. Soc.* **71**, 2624 (1949).
177. Soczewiński, E., *Nature* **188**, 391 (1960).
178. Waksmundzki, A. and Soczewiński, R., *Roczniki Chem.* **32**, 863 (1958).
179. Waksmundzki, A. and Soczewiński, R., *Nature* **184**, 977 (1959).
180. Waksmundzki, A. and Soczewiński, R., *Roczniki Chem.* **33**, 1423 (1959).
181. Geiss, F. and Schlitt, H., *Chromatographia* **1**, 392 (1968).
182. Prinzler, H. W. and Tauchmann, H., *J. Chromatog.* **29**, 142 (1967).
183. Geiss, F., Sandroni, S. and Schlitt, H., *J. Chromatog.* **44**, 290 (1969).
184. Sandroni, S. and Schlitt, H., *J. Chromatog.* **52**, 169 (1970).
185. Liteanu, C. and Gocan, S., *Rev. Chim. Acad. RPR* **7**, 1041 (1962).

PART II
THE USE OF GRADIENTS

CHAPTER 6

INTRODUCTION TO USE OF GRADIENTS

6.1 History of Gradient Chromatography

Liquid phase chromatography, worked out by Tswett [1] in 1906, was practically neglected till 1938 when Reichstein [2] introduced the general notion of chromatographic elution. The decisive step in the development of chromatography was taken in 1941 by Martin and Synge [3] in developing partition chromatography. In 1948 Stein and Moore [4] made use of ion-exchange chromatography to separate aminoacids. These were the most important stages in the early development of chromatography.

Since the very beginning of chromatography, research-workers have realized the possibilities offered by the choice of certain working conditions in the separation of certain mixtures. The early appearance of works using gradients to improve chromatographic separations is a case in point. Gradient chromatography has developed in the general context of evolution of chromatography and is in full progress owing to the possibility of automation in the programming of certain parameters.

The greatest possibilities of achieving gradients are offered by the mobile phase, as it is much easier to obtain a continuous or stepwise variation of a certain property of a moving liquid (the eluent) than of the stationary phase or of the medium, in the case of open columns.

The idea of using the gradient, as shown by Synge [5], is ascribed to Tiselius. Concomitantly with him, Mitchell *et al.* [6] made use of concentration and pH gradients for the separation of some enzymes by paper chromatography. An important contribution to the progress of gradient chromatography was due to Tiselius's co-workers, Alm [7, 8], Williams [9, 10] and Hagdahl [11] as much for the development of certain theoretical considerations as for some practical uses. Simultaneously with them, Donaldson *et al.* [12] used the principle of the elution gradient, i.e. a polarity gradient, to separate organic acids, but without making use of the term 'gradient'. Strain [13] used a sorption gradient. Buch *et al.* [14–16] as well as Mader [17] also used gradient elution for the separation of some organic acids. As a consequence of the successes obtained, the field of application of gradient techniques broadened continuously. Thus Lederer [18] applied the elution gradient to cation separation, and Grande and

Beukenkamp [19] to the separation of inorganic anions. Desreux [20] applied the technique of gradient elution to the fractionation of polymers. The technique of gradient elution was also successfully used for the separation of some mixtures of organic and biological substances: sugars and oligosaccharides [21, 22], amino-acids and peptides [23, 24], steroids [25], enzymes [6, 26], proteins [27–29] and nucleic acids [30–32].

In 1962 the first works of Wieland and Determann [33] and Rybicka [34] appeared, showing the possibility of applying elution gradients to thin-layer chromatography. Later, Niederwieser and Honegger [35, 36] worked intensively to improve this technique.

Chromatography with stationary-phase gradients made slower progress, evidently owing to the difficulties with technique and apparatus. Composition gradients [39–42], impregnation gradients [39–40], and activity gradients [43] were achieved.

The temperature gradient, in the case of open column chromatography, was introduced in 1961 by Liteanu and Gocan [44]. In 1966, Drapron and Guilbot [45] used an evaporation gradient of the mobile phase in paper chromatography, and in 1968, Turina et al. [46] used an adapter for evaporation of the solvent during development in thin-layer chromatography.

Temperature gradients were used in 1956 by Baker and Williams [47] for polymer fractionation, and more recently for some ion-exchange separations by Liteanu and Gocan [48]. Hesse and Engelhardt [49] as well as Maggs and Young [50, 51] used temperature programming in liquid–solid column chromatography.

The possibility of using flow programming in column chromatography to reduce the time of analysis without worsening the resolution, was shown in 1969 by Scott and Lawrence [52]. A certain control of the flow-rate may be obtained in thin-layer chromatography by means of a grain-size gradient [39].

In 1965 Geiss and Schlitt [54, 55], and de Zeeuw [56], developed a working technique with a programmed vapour phase.

The use of gradients (in the mobile and stationary phase, and the environment) represents the only sensitive general method of differentiating the displacement rates of the different components of a mixture in the column. Gradient chromatography is in full progress, offering a new dimension to this separation method.

6.2 Classification of Gradients

The chromatographic method is based on the different rate of displacement of all the components of a mixture along an open or closed column.

The displacement rate will certainly be different to some extent, no matter how close the chemical characteristics of the substances in a mixture may be, and therefore, after a long enough time-interval, the distances covered may be sufficiently different to let separation take place.

The displacement of a substance along a column is based on its repeated partition between the two phases (the mobile and the stationary) forming the column. In a very general way, the displacement rate along the chromatographic column may be expressed as a function having the form

$$v = f(\alpha, k) \tag{6.1}$$

where α is the partition coefficient of the substance between the two phases (independent of the nature of the process on which the partition is based: dissolution, absorption, ion-exchange), and k is the volume ratio of the two phases ($k = V_S/V_M$).

The partition coefficient α may be defined as the ratio of the activities of the substance in the stationary and mobile phases, a_S and a_M respectively:

$$\alpha = \frac{a_S}{a_M} = \exp(\mu_{0M} - \mu_{0S})/RT \tag{6.2}$$

where μ_{0S} and μ_{0M} are the standard chemical potentials of the substance in the two phases. As the standard chemical potentials of the substances in a mixture are different: $\mu_{01} \neq \mu_{02}$, etc., it results that $\alpha_1 \neq \alpha_2$ etc., and consequently substances having equal displacement rates along a chromatographic column cannot exist.

Of course, as α_1/α_2 tends towards 1, the difference between the displacement rates of the two substances becomes so small as to cause very long columns to be needed for their separation, which evidently is a major practical inconvenience.

It is therefore evident that the more different and numerous the factors acting on the displacement rates, (or on the conditions of the partition equilibrium) the easier the separation of the components, i.e. shorter columns can be used.

As the partition equilibrium between the two phases depends on a series of parameters (temperature, pH, ionic strength, polarity etc.) we can write

$$\alpha = f(a, b, c, \ldots) \tag{6.3}$$

and

$$k = f(a', b', c', \ldots). \tag{6.4}$$

Therefore, evidently

$$v = f(a, a', b, b', \ldots). \tag{6.5}$$

Considering the parameters conditioning the displacement along the column, we can have: (1) constant condition chromatography and (2) variable-condition chromatography. In the latter case, one or several parameters conditioning the partition between the two phases, will be changed as a function of time:

$$a = f(t), \qquad b = f(t), \ldots$$

or

$$\frac{da}{dt} \neq 0 \text{ (gradient of parameter } a\text{)}$$

$$\frac{db}{dt} \neq 0 \text{ (gradient of parameter } b\text{),}$$

$$\frac{da}{dt} = k \text{ (constant gradient of parameter } a\text{)}$$

$$\frac{da}{dt} = f(t) \text{ (variable gradient of parameter } a\text{)}$$

The continuous variation of the chromatographic parameters may be linear, concave, convex, or compound. The discontinuous variation of a parameter (discontinuous gradient) may be approximated by a continuous curve. We speak of gradient elution and stepwise elution [57]. These types of gradient are illustrated in Fig. 6.1.

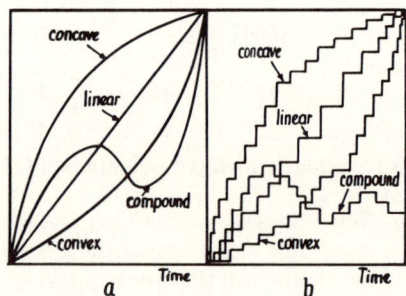

FIG. 6.1. The classification of gradients according to their shape (from data in [57]).

The use of gradients represents a basic means of optimizing the process of chromatographic separation, the major problem in chromatography, but in this case the optimisation actually takes place during the separation process, unlike conventional chromatography in which optimisation is achieved by choosing optimal parameters and keeping them constant during the whole chromatographic process.

It is because conventional chromatography does not always give a satisfactory resolution of some complex mixtures that gradient chromatography has developed of necessity, according to the requirements of practical needs.

6.3 Types of Gradient

Taking into account the principle worked out by Stahl [39] we can classify chromatographic gradient techniques according to the phases of the chromatographic process in which the change of certain properties or of certain combinations of them takes place.

Mobile phase gradients—concentration
—polarity
—pH
—ionic strength

Stationary phase gradients—composition
—impregnation
—activity

Environmental or medium gradients—temperature
—vapour pressure
—flow-rate
—grain size
—cross-sectional area
—thickness of the layer

Combined gradients—vapour composition–concentration
—concentration–layer thickness
—temperature–pH

6.4 Nomenclature in Gradient Chromatography

Paper and thin-layer chromatography are open column systems offering the possibility of applying a gradient in a direction other than the direction of flow of the eluent, or the application of certain combined gradients in different directions.

In the case of closed columns, the gradient must always coincide in direction with the flow of the eluent, but it can have a different sense. Niederwieser [58] therefore suggests a nomenclature referring to the gradient direction *vs.* the direction of flow of the eluent, as seen in Fig. 6.2.

Niederwieser [59] also shows the possibilities existing for combining two gradients, for instance in the mobile and stationary phases, in thin-layer

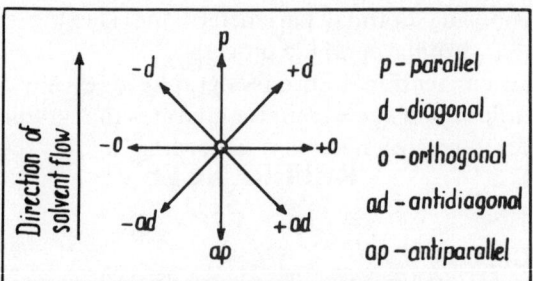

FIG. 6.2. The classification of gradients according to direction [58].

chromatography (Fig. 6.3). Other combinations of gradients may also exist, for instance medium gradients with mobile or stationary gradients, etc.

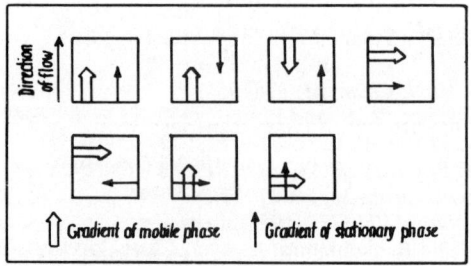

FIG. 6.3. The possibilities of combination of two gradients [59].

REFERENCES

1. Tswett, M., *Ber. Dtsch. Botan. Ges.* **24**, 316, 384 (1906).
2. Reichstein, T. and van Euw, J., *Helv. Chim. Acta* **21**, 1197 (1968).
3. Martin, A. J. P. and Synge, R. L. M., *Biochem. J.* **35**, 1358 (1941).
4. Moore, S. and Stein, W. H., *Ann. N.Y. Acad. Sci.* **49** (1948).
5. Synge, R. L. M., *Disc. Faraday Soc.* **7**, 164 (1949).
6. Mitchell, H. K., Gordon, M. and Haskins, F. A., *J. Biol. Chem.* **180**, 1071 (1949).
7. Alm, R. S., Williams, R. J. P. and Tiselius, A., *Acta Chem. Scand.* **6**, 826 (1952).
8. Alm, R. S., *Acta Chem. Scand.* **6**, 1186 (1952).
9. Williams, R. J. P., *Analyst* **77**, 905 (1952).
10. Williams, R. J. P., Hagdahl, C. and Tiselius, A., *Arkiv. Kemi* **7**, 4 (1954).
11. Hagdahl, L., Williams, R. J. P. and Tiselius, A., *Arkiv. Kemi* **4**, 193 (1952).
12. Donaldson, K. O., Tulane, V. J. and Marshall, L. M., *Anal. Chem.* **24**, 185 (1952).
13. Strain, H. H., *Anal. Chem.* **23**, 25 (1951).
14. Busch, H., Hurlbert, R. B. and Potter, V. R., *J. Biol. Chem.* **196**, 717 (1952).
15. Busch, H. and Potter, V. R., *Cancer Res.* **12**, 660 (1952).
16. Busch, H. and Potter, V. R., *Cancer Res.* **13**, 168 (1953).
17. Mader, C., *Anal. Chem.* **26**, 566 (1954).
18. Lederer, M., *Nature* **172**, 727 (1953).
19. Grande, J. A. and Beukenkamp, J., *Anal. Chem.* **28**, 1497 (1956).
20. Desreux, V., *Rec. Trav. Chim.* **68**, 789 (1949).
21. Alm, R. S., *Acta Chem. Scand.* **6**, 1186 (1952).
22. Parr, C. W., *Biochem. J.* **56**, XXVII (1954).
23. Hirs, C. H. E., Stein, W. H. and Moore, S., *J. Biol. Chem.* **211**, 941 (1954).
24. Hirs, C. H. E., Stein, W. H. and Moore, S., *J. Biol. Chem.* **221**, 151 (1956).
25. Lakshmanan, T. K. and Lieberman, S., *Arch. Biochem. Biophys.* **45**, 235 (1953).
26. Boman, H. G., *Nature* **173**, 447 (1954).
27. Peterson, E. A. and Sober, H. A., *Federation Proc.* **13**, 273 (1954).
28. Clauser, H. and Li, C. H., *J. Am. Chem. Soc.* **76**, 4337 (1954).
29. Morrison, M. and Cook, J. L., *Science* **122**, 920 (1955).
30. Hurlbert, R. B., Schmitz, H., Brumm, A. F. and Potter, V. R., *J. Biol. Chem.* **209**, 3 (1954).
31. Bendich, A., Fresco, J. R., Rosenkranz, H. S. and Beiser, S. M., *J. Am. Chem. Soc.* **77**, 3671 (1955).
32. Jackson, K., *Am. J. Physiol.* **189**, 315 (1957).
33. Wieland, T. and Determann, H., *Experientia* **18**, 431 (1962).
34. Rybicka, S. M., *Chem. Ind.* (*London*) 1947 (1962).
35. Niederwieser, A. and Honegger, C. G., *Helv. Chim. Acta* **48**, 893 (1965).
36. Niederwieser, A., *J. Chromatog.* **21**, 326 (1966).
37. Stahl, E., *German. Pat.* No. 1175912 (1964), (Appl. 1961).
38. Stahl, E., *Angew. Chem. Intern. Ed. Engl.* **3**, 784 (1964).

39. Stahl, E., *Z. Anal. Chem.* **221**, 3 (1966).
40. Stahl, E. and Vollmann, H., *Talanta* **12**, 525 (1965).
41. Berger, J. A., Meyniel, G., Petit, J. and Blanquet, P., *Bull. Soc. Chim. France*, 2662 (1963).
42. Stickland, R. G., *Anal. Biochem.* **10**, 108 (1965).
43. Honegger, C. G., *Helv. Chim. Acta* **47**, 2384 (1964).
44. Liteanu, C. and Gocan, S., *Studia Univ. Babeş-Bolyai, Chem.* **6**, 99 (1961).
45. Drapron, R. and Guilbot, A., *Chromatographia et Méthodes des Séparation Immédiates*, Vol. II, p. 27, Parissakis, G. (ed.), Publication de l'Union des Chimistes Hellénes, Athens (1966).
46. Turina, S., Marjanović-Krajovan, V. and Šoljić, Z. *Anal. Chem.* **40**, 471 (1968).
47. Baker, C. A. and Williams, R. J. P., *J. Chem. Soc.* 2352 (1956).
48. Liteanu, C., Gocan, S., Hodişan, T., Naşcu, H. and Măruţoiu, C., *Rev. Roumaine Chim.* **17**, 497 (1972).
49. Hesse, G. and Engelhardt, H., *J. Chromatog.* **21**, 228 (1966).
50. Maggs, R. J. and Young, T. E., *Gas Chromatography 1968 Copenhagen*, p. 217, Herbourn, C. L. A. (ed.), Butterworths, London (1969).
51. Maggs, R. J., *J. Chromatog. Sci.* **7**, 145 (1969).
52. Scott, R. P. W. and Lawrence, J. G., *J. Chromatog. Sci.* **7**, 65 (1969).
54. Geiss, F., Schlitt, H. and Klose, A., *Z. Anal. Chem.* **213**, 321, 331 (1965).
55. Geiss, F. and Schlitt, H., *Chromatographia* **1**, 392 (1968).
56. de Zeeuw, R. A., *Anal. Chem.* **40**, 2134 (1968).
57. Snyder, L. R., *Chromatog. Rev.* **7**, 1 (1965).
58. Niederwieser, A., *Chromatographia* **2**, 23 (1969).
59. Niederwieser, A. and Honegger, C. G., in *Advances in Chromatography*, Vol. 2, Chap. 4, p. 123, Giddings, J. C. and Keller, R. A. (eds.), Arnold, London (1966).

CHAPTER 7

MOBILE-PHASE GRADIENTS

The first part of this chapter deals with the problem of achieving a mobile-phase gradient, starting from the simplest devices and finally reaching the completely automated appliances now on the market. This will facilitate the understanding of the theoretical problems of elution gradient chromatography, which will be examined in the second part of the chapter. In the third part, the performance of elution gradient chromatography will be compared with that of conventional chromatography.

7.1 Devices for Achieving a Mobile-phase Gradient

7.1.1 CLOSED COLUMNS

By mobile-phase gradient we mean any change in the composition of the eluent, which may be achieved either discontinuously (i.e. stepwise) or continuously. The problem of designing apparatus for producing the desired type of gradient was examined early on, and a few review papers [1–5] are to be found in the literature.

7.1.1.1 *Discontinuous Elution Gradients*

For the separation of certain mixtures, several eluents must be used in sequence. To replace the manual change of eluent, Roberts and Mason [6] designed an appliance by means of which the polarity of the solvent may be increased automatically. The apparatus is shown schematically in Fig. 7.1. It consists of a glass tube (64 cm long, 51 mm in diameter) divided into compartments which communicate through 1-mm diameter apertures. Each compartment has a side-arm for filling (total volume of section and side-arm is 100 ml) and a stopcock at the bottom, which is closed during filling. The bottom compartment is filled with non-polar solvent, and its side-arm closed by clamping on a spherical stopper, then the other compartments are filled in turn with mixtures of non-polar solvent with increasing amounts of polar solvent. When stopcock (1) is opened, the contents of the first compartment will flow through the column, followed by the others in order of increasing polarity (Fig. 7.2).

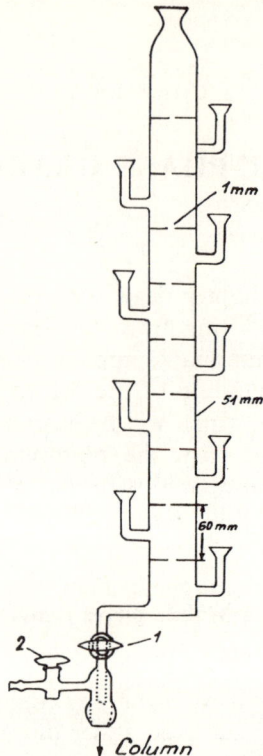

Fig. 7.1. Apparatus for automatically increasing the solvent polarity stepwise in partition chromatography [6].

Pressure may be applied at the top to increase the flow-rate, and a side-arm carrying a stopcock (2) permits application of pressure to the column alone.

The device also permits the use of water-saturated solvents. Usually the solvent mixtures (e.g. butanol and chloroform) decrease in density as they increase in polarity. With solvent mixtures such as butanol and hexane, however, the polarity and density increase simultaneously, and because the density is highest at the top of the set of compartments, diffusion will occur as the mixtures run from higher into lower compartments. The resulting increase in solvent polarity is logarithmic (see Fig. 7.2).

The apparatus has the drawback that only one programme at a time is possible, and any change involves cleaning and recharging the equipment. Harpur [7] has therefore built a device which can be dismantled

Devices for Achieving a Mobile-phase Gradient

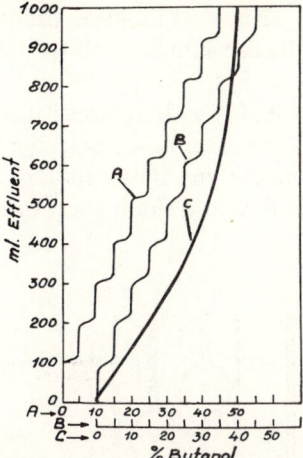

Fig. 7.2. Comparison of techniques to increase solvent polarity. *A*—solvent (butanol–chloroform mixture) changed manually; *B*—apparatus of Fig. 7.1 with butanol–chloroform mixture; *C*—apparatus of Fig. 7.1 with butanol–hexane mixture [6].

and arranged for a new programme in a few minutes. The principle of the appliance is the same, but the glass diaphragms are replaced by a series of spacer discs and diaphragms, the latter having four notches for passage of the eluent. The discs and diaphragms are set on a rod (A, see Fig. 7.3). The spacer discs permit the choice of compartment volumes of 10,

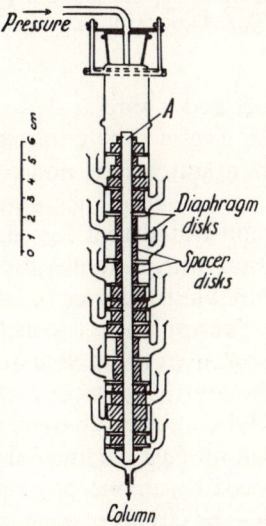

Fig. 7.3. Apparatus for obtaining a stepwise variable solvent gradient [7].

15, 20, 22·5, 25, 27·5, 30, 32·5 or 35 ml, thus adding more flexibility to the programming. The results are similar to those obtained with the Roberts and Mason apparatus.

Brusca and Gavienowski [8] built an apparatus for the stepwise elution of lipids, consisting of 5 identical flasks placed one above the other (Fig. 7.4) and connected to the column by a capillary and stopcock manifold. The lowest flask is filled first and the highest last. This device gives very good reproducibility.

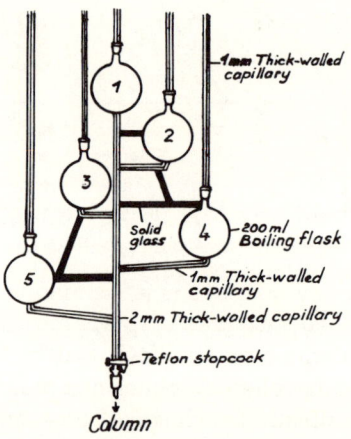

FIG. 7.4. Apparatus for making a stepwise elution gradient. The order of phase delivery is from reservoirs 1, 2, 3, 4 and 5 respectively [8].

Anderson et al. [9] described a similar device, made more flexible by connecting the flasks with Tygon tubing instead of glass, thus allowing arrangements of the flasks at any height, and continuous gradients to be obtained.

Teekell et al. [10] demonstrated that for separation of proteins and nucleotides stepwise elution with different buffers is necessary. As the elution takes 12–14 hr, they built a device to carry out these operations according to programme. The apparatus consists of ten 500-ml separatory funnels connected by Tygon tubes to the side-arm (f) of another flask (a) (Fig. 7.5). The tubes are closed by clamps (g), each connected to a solenoid by a stainless-steel wire. When current passes, the clamp opens and the buffer from that funnel will fill flask (a), in which there is a glass float (b) attached to an arm balanced by a counter-weight (c). When receiver (a) is empty, contact (d) closes the circuit of a small electric motor, which in turn moves the circuit selector (e), causing the solenoid to close and the next buffer to flow into (*a*) until the float causes (*d*) to stop the motor and the flow into (*a*). An elution in 11 stages can thus be carried out, including the initial solution in (*a*).

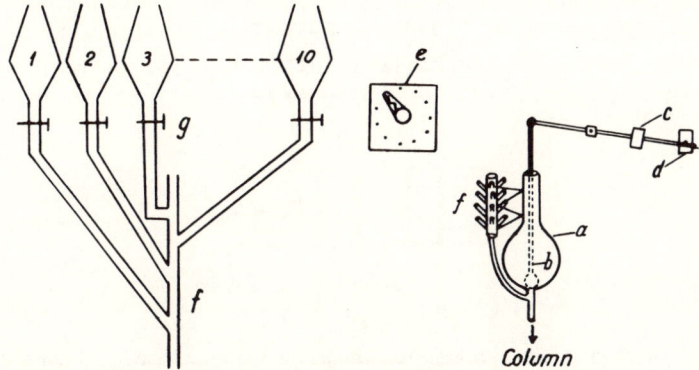

FIG. 7.5. Float valve and reservoir system used on automatic stepwise elution dispenser for column chromatography [10].

Another eluent programmer was described by Lerner [11]. Here the eluent distribution is controlled by means of a punched-card programme, the exact quantity of eluent at each stage being specified by the card.

Rombauts and Raftery [12] have also achieved a pH-gradient programmer that can be used for stepwise or continuous gradients. The device is equipped with a time programmer for the solenoids which operate on the clamps that open and close the storage containers for eluents of different pH.

7.1.1.2 *Concave and Convex Exponential Gradients*

An exponential elution gradient is theoretically achieved by means of a device formed of a mixing flask (M) and a storage vessel (R) placed above it (Fig. 7.6a). There also exist other types made of three flasks, two for mixing (M_1 and M_2) and one for storage (Fig. 7.6b). Solutions of concentrations C_1, C_2, (C_3) are placed in the two (three) flasks. The problem is to find the form of the function $C = f(V)$, i.e. the way in which the concentration of the solution in the mixing flask varies with the volume V flowing from the flask into the column.

In the case of the device in Fig. 7.6a, the concentration C_1 in the mixing flask will tend towards the concentration C_2 in the storage vessel if the volume in the latter is greater than that in the mixing vessel (V_m). When a volume of eluent dV flows into the column, the same volume will enter

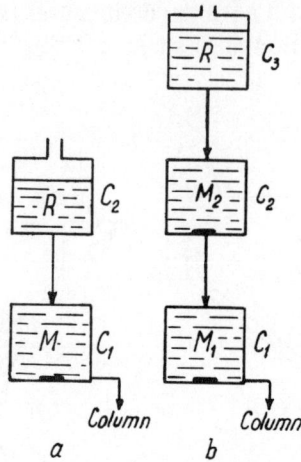

Fig. 7.6. Block diagram of apparatus for producing an exponential concentration gradient; *a*—with one mixing vessel, *b*—with two mixing vessels.

M from R, so the change dC in the concentration C in the mixing flask will be

$$dC = \frac{(C_2 - C)\,dV}{V_m} \qquad (7.1)$$

By separating the variables and integrating, we obtain

$$-\ln(C_2 - C) = \frac{V}{V_m} + K \qquad (7.2)$$

or

$$C_2 - C = K'\exp(-V/V_m)$$

To determine the integration constant we use the limit conditions $V = 0$, $\exp(-V/V_m) = 1$, and $C = C_1$, thus $K' = C_2 - C_1$, and on substitution in (7.2) this gives

$$C = C_2 - (C_2 - C_1)\exp(-V/V_m) \qquad (7.3)$$

i.e. the concentration C in the mixing flask is an exponential function of the volume V flowing into the column (Fig. 7.7). Equation (7.3) has been derived in several papers [13–23]. For this type of device Cherkin *et al.* [18] have also given the formula

$$C/C_2 = [\exp(V/V_m) - 1]/\exp(V/V_m) \qquad (7.4)$$

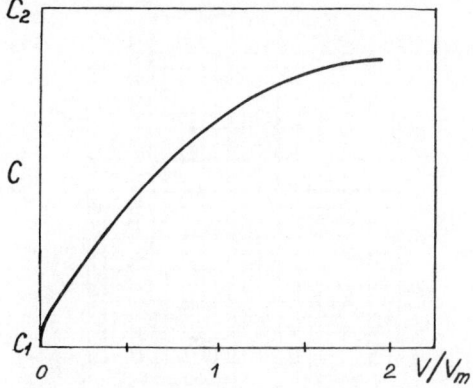

FIG. 7.7. Graph of Eq. (7.3).

Snyder [4] gives Eq. (7.4) in the form

$$V_B = [\exp(V/V_m) - 1]/\exp(V/V_m) \tag{7.5}$$

where V_B is the volume of component B which is initially present in R (there is no B in M initially).

From a graph of Eq. (7.4) we see that up to a value of $V/V_m = 1$, a linear gradient is obtained, and above this value the gradient becomes convex. The gradient is increased if the volume of the mixing flask is decreased.

Svensson [24] showed that in the case of a storage vessel and two mixing chambers of equal volumes V, coupled in series by capillary tubes, the following equation gives the concentration at the outlet:

$$C = C_3 - (C_3 - C_2)\exp(-V/V_m) - (C_3 - C_1(V/V_m)\exp(-V/V_m) \tag{7.6}$$

where C_1, C_2 and C_3 are the initial concentrations (see Fig. 7.6). This equation was also established by Drake [15].

For three mixing chambers and a reservoir:

$$C = C_4 - (C_4 - C_3)\exp(-V/V_m) - (C_4 - C_2)(V/V_m)\exp(-V/V_m)$$
$$- (C_4 - C_1)(V^2/2V_m^2)\exp(-V/V_m) \tag{7.7}$$

and in the general case for n mixing chambers:

$$C = C_{n+1} - \exp(-V/V_m) \sum_{i=1}^{n} (C_{n+1} - C_i)\frac{(V/V_m)^{n-i}}{(n-i)!} \tag{7.8}$$

This equation has also been derived by Hadwiger and Glur [25] as well as by Sorin and Vargues [17]. The concentration C given by Eq. (7.3) is

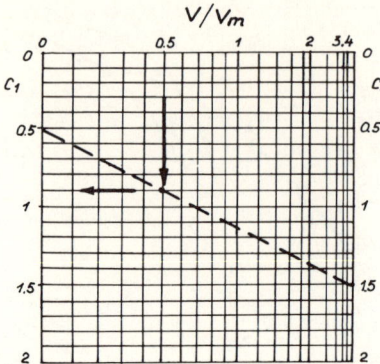

FIG. 7.8. Nomogram of Eq. (7.3) for a closed mixing chamber [26].

easier to obtain by means of the monogram in Fig. 7.8, due to Kocent [26]. For instance, suppose the initial concentrations in R and M are $C_2 = 1\cdot 5M$ and $C_1 = 0\cdot 5M$. For the ratio $V/V_m = 0\cdot 5$ the line joining the two concentrations intersects the vertical representing this ratio, at a value corresponding to a concentration $C = 1\cdot 1M$.

These devices are very easy to construct from ordinary laboratory glassware. As the two-vessel designs are all similar, they are shown schematically in Fig. 7.9. The three-vessel device is illustrated in Fig. 7.10 by the design by Bendich *et al.* [27], along with the gradient obtained.

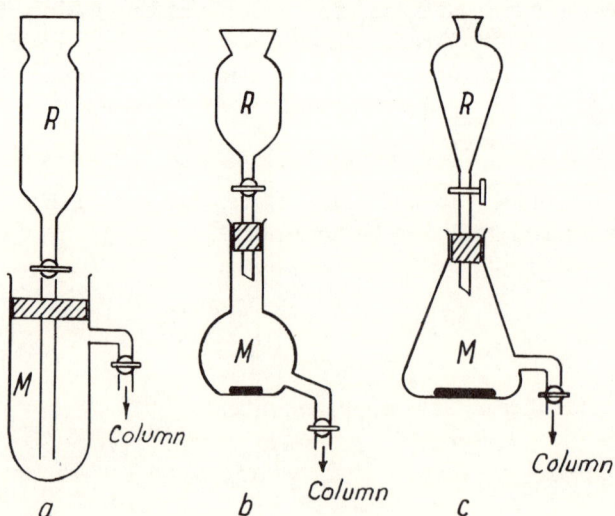

FIG. 7.9. Various arrangements of two vessels for producing exponential elution gradients.

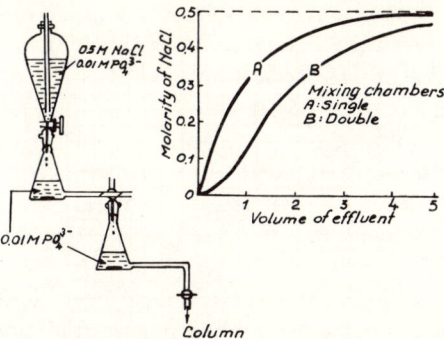

FIG. 7.10. Diagram of the apparatus used for elution with two mixing chambers. The inset shows the change of solution concentration as a function of the volume of eluent [27].

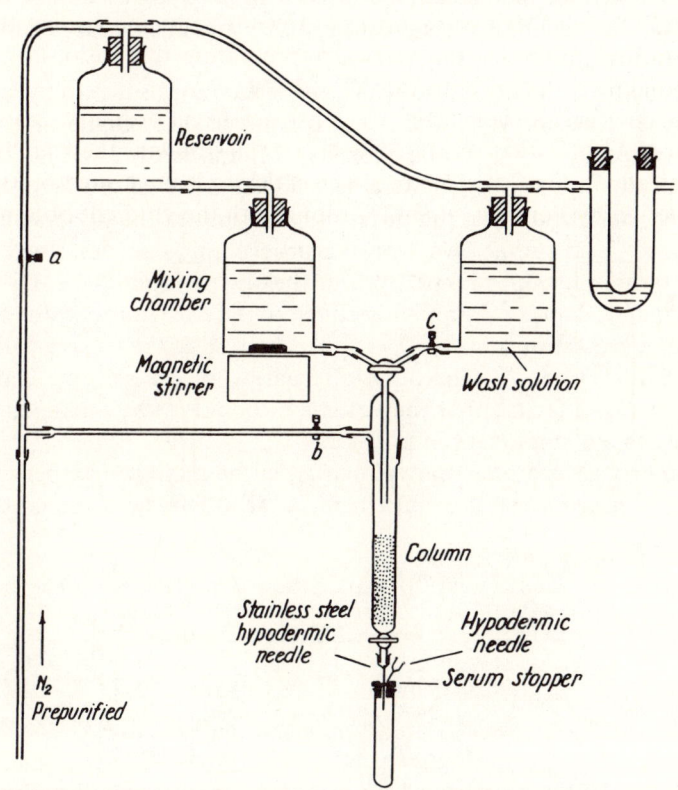

FIG. 7.11. Apparatus for anaerobic gradient elution; *a, b, c*—clamps [28].

Sakami [28] described an anaerobic device made of a reservoir and a mixing flask set vertically. The air is replaced by pure nitrogen. The apparatus (Fig. 7.11) allows the column and the space above the reservoir to be purged with nitrogen before the elution.

Sometimes in column chromatography the air dissolved in the liquid may cause certain difficulties, especially when the eluent is forced through the column under applied air-pressure. To avoid this drawback, Kesner *et al.* [29] suggested deaerating the eluent by heating the liquid in the last mixing flask.

The devices described in this section have been used extensively [29–70] in applications requiring mobile-phase gradients to give better separations.

To obtain better control of the flow-rate from the reservoir into the mixing flask, and hence of the eluent composition, Vestergaard [71] used a constant-speed motor-driven syringe instead of the reservoir. Six columns could be fed simultaneously with a gradient eluent by this apparatus. The technique has great flexibility and allows easy change of a gradient during an experiment.

Differing from the devices in which the eluent flow is due to hydrostatic pressure are those in which the eluent is pumped through the column, e.g. that designed by Wallach and Nordby [72] and shown in Fig. 7.12. By means of this apparatus an extremely wide variety of gradients can be produced, independent of the physical properties of the solvents and the packing of the column. Two working programmes are described below. In programme 1, both pumps work at the same output Q (i.e. $Q_1 = Q_2$). In programme 2, pump I supplying eluent to the chromatographic column works at a greater output than pump II which feeds the mixing flask from the reservoir, $Q_1 > Q_2$. The other alternative for programme 2, characterised by $Q_1 < Q_2$, so that the volume of liquid in the mixing chamber increases, is less important in practice.

In the first programme, the volume V_m in the mixing chamber remains constant. The concentration of eluent, C, tends towards the concentra-

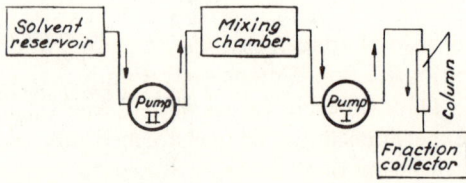

FIG. 7.12. Scheme of chromatographic gradient system with two metering pumps [72].

tion C_r of the solution in the reservoir, but cannot be identical to it. The concentration of the eluent as a function of time t is given by

$$C = C_{0_m} + \left[Q_1 C_r \int_{t_0}^{t} dt - \int_{t_0}^{t} C \, dt \right] / V_m \tag{7.9}$$

where C_{0_m} represents the original concentration in the mixing chamber. To obtain the explicit function of C, Eq. (7.9) must be differentiated with respect to time:

$$V_m(dC/dt) = Q_1(C_r - C) \tag{7.10}$$

or

$$\frac{dC}{C_r - C} = \frac{Q \, dt}{V_m} \tag{7.11}$$

By integrating between the limits C_{0_m} and C, and t_0 and t, we obtain

$$\ln\left[(C_r - C)/(C_r - C_{0_m})\right] = -Q_1(t - t_0)/V_m \tag{7.12}$$

or

$$C = C_r - (C_r - C_{0_m}) \exp\left[-Q_1(t - t_0)/V_m\right] \tag{7.13}$$

To make (7.13) indicate better the change in concentration as a function of time, we write it as

$$\Delta C = (C_r - C_{0_m})[1 - \exp(-Q_1 \Delta t/V_m)] \tag{7.14}$$

If $Q_1 \Delta t_{max}/V_m$ is denoted by μ (non-dimensional) and Δt_{max} is the maximum time used for the experiment, Eq. (7.14) may be transformed into a completely general form:

$$\Phi = (C_r - C_{0_m})[1 - \exp(-\mu\Omega)] \tag{7.15}$$

in which Φ is the ratio of concentrations $\Delta C/\Delta C_{max}$ and Ω is a time ratio $\Delta t/\Delta t_{max}$. Both ratios are non-dimensional. Figure 7.13 shows graphically the values of Φ vs. Ω for different values of μ, C_r and C_{0_m}. The coordinates for a particular experiment must be transformed into concentrations and times. The time, in terms of Ω on the abscissa, is transformed into real time by $\Delta t = \Omega \Delta t_{max}$, and the concentration given in terms of Φ on the ordinate, is transformed into a real concentration by $\Delta C = \Phi \Delta C_{max}$. For

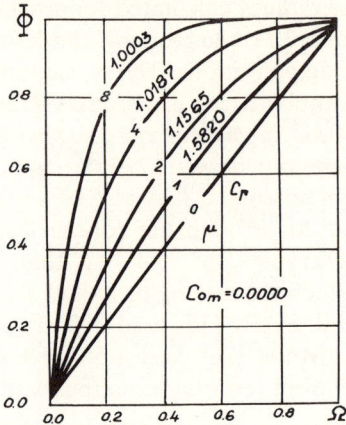

FIG. 7.13. Graph of the function (7.15) for stated values of the parameter μ (mode I gradient elution) [72].

instance, if $Q = 2$ ml/min, the total volume $V_t = 500$ ml, $\Delta t_{max} = V_t/Q = 250$ min, $V_m = 250$ ml, $C_r = 1\cdot 1565M$ and $C_{0_m} = 0\cdot 0M$, the value $\mu = Q \Delta t_{max}/V_m = 2$ is first calculated, then the maximum concentration ΔC_{max} is obtained from Eq. (7.14) written in the form

$$\Delta C_{max} = (C_r - C_{0_m})[1 - \exp(-\mu)] = 1\cdot 000M \qquad (7.16)$$

i.e. in 250 min the concentration at the outlet of the mixing chamber will increase from $C = 0\cdot 0M$ to $C = 1\cdot 000M$.

TABLE 7.1. Calculation of $\Phi, \Delta t, \Delta C$ and C (for $\mu = 2$, $C_r = 1\cdot 1565M$ and $C_{0_m} = 0\cdot 060M$, $\Delta t_{max} = 250$ min, $\Delta C_{max} = 1\cdot 000M$) [72]

Ω	Φ calculated, Eq. (7.15)	$\Delta t = \Omega \Delta t_{max}$ (min)	$\Delta C = \Phi \Delta C_{max}$ (M)	$C = \Delta C - C_{0_m}$ (M)
0·0	0·0000	0	0·0000	0·0000
0·1	0·2096	25	0·2096	0·2096
0·2	0·3813	50	0·3813	0·3813
0·3	0·5218	75	0·5218	0·5218
0·4	0·6369	100	0·6369	0·6369
0·5	0·7311	125	0·7311	0·7311
0·6	0·8082	150	0·8082	0·8082
0·7	0·8713	175	0·8713	0·8713
0·8	0·9230	200	0·9230	0·9230
0·9	0·9654	225	0·9654	0·9654
1·0	1·0000	250	1·0000	1·0000

Table 7.1 gives some values calculated for working programme 1. By means of Eqs. (7.10) and (7.12) the gradient can be very simply calculated as a function of working conditions. The procedure offers advantages in systematization and simplification of chromatographic experiments. We consider that this method of calculation can also be applied in the case of devices with a single pump, as well as in the case described by Anselmo [73] in which pump II is missing. It may also be applied to outfits in which the liquid is being moved through the column by gas-pressure applied to the liquid in the reservoir (Fig. 7.14), e.g. that described by Thomas and Thomas [74] in which the mixing flask had a capacity of 250 ml and the reservoir of 1000 ml. Skelly [75] used 250-ml mixing chamber and a 1- or 2-litre reservoir. Miller [76] used a 50-ml Erlenmeyer flask as the mixing vessel, and a 500-ml reservoir equipped with a manometer at the air inlet.

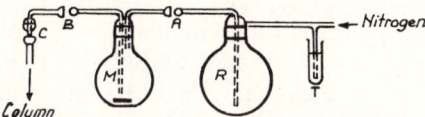

FIG. 7.14. Gradient elution apparatus with gas pressure for liquid displacement. A, B, C—ground-glass joints, M—mixing flask, R—reservoir, T—mercury back-pressure trap [74].

In the second programme [72] (Fig. 7.12), $Q_1/Q_2 = r > 1$. For $r > 2$, the form of the elution gradient produced, $f(\Phi, \Omega)$ is convex and similar to the one produced by programme 1. For $r = 2$ a linear gradient is produced. For $1 < r < 2$ a concave gradient is produced. The concentration of the eluent as a function of time is given by

$$C = \left[a_0 + Q_2 C_r \int_{t_0}^{t} dt - Q_1 \int_{t_0}^{t} C\, dt \right] \bigg/ \left[V_{0_m} + (Q_2 - Q_1) \int_{t_0}^{t} dt \right] \quad (7.17)$$

where a_0 is the initial *quantity* (not the concentration) of the component in the mixing chamber, in initial volume V_{0_m}. The numerator gives the change of the quantity of component in the mixing chamber in the time interval $(t - t_0)$, and the denominator represents the corresponding change in the volume of eluent in the mixing chamber. Their ratio evidently does not give $C = f(t)$.

By differentiating (7.17) with respect to time, we obtain

$$V_m(dC/dt) + Q_2 C + Q_2 t(dC/dt) - Q_1 C - Q_1 t - Q_1 t(dC/dt)$$
$$= Q_2 C_r - Q_1 C \quad (7.18)$$

or

$$dC/(C_r - C) = Q_2 \, dt/[V_m + (Q_2 - Q_1)t] \quad (7.19)$$

Integrating this equation between the limits C and C_0, and t and t_0, yields

$$\ln \frac{C_r - C}{C_r - C_0} = \frac{Q_2}{Q_2 - Q_1} \ln \frac{V_m + (Q_2 - Q_1)t}{V_m + (Q_2 - Q_1)t_0} \quad (7.20)$$

or

$$C = (C_r - C_0) \exp \left[\frac{Q_2}{Q_2 - Q_1} \ln \frac{V_m + (Q_2 - Q_1)t_0}{V_m + (Q_2 - Q_1)t} \right] \quad (7.21)$$

This equation can be expressed in a more general way in terms of the changes in concentration and time:

$$\Delta C = (C_r - C_0) \left\{ 1 - \exp \left[\frac{Q_2}{Q_2 - Q_1} \ln \frac{V_m + (Q_2 - Q_1) \Delta t}{V_m} \right] \right\} \quad (7.22)$$

If the following substitutions are made: $Q_2/(Q_2 - Q_1) = v$; $\Delta C/\Delta C_{max} = \Phi$; $\Delta t/\Delta t_{max} = \Omega$; $(Q_2 - Q_1) \Delta t_{max} = -V_m$ (this is the volume in the mixing chamber, which becomes zero at the end of the experiment), Eq. (7.22) may be written

$$\Phi = (C_r - C_0)\{1 - \exp[v \ln(1 - \Omega)]\} \quad (7.23]$$

Between parameters v and r, which are dimensionless, there exists the relationship $r = (v - 1)/v$ or $v = 1/(1 - r)$. The values of Φ calculated by means of Eq. (7.23) for different values of the variable Ω and of parameters C_r, C_0 and v are given in Fig. 7.15. The change from coordinates Φ and Ω is made by means of $\Delta C = \Phi \Delta C_{max}$ and $\Delta t = \Omega \Delta t_{max}$, as in the case of programme 1.

Wren [77] obtained a concentration gradient of methanol in chloroform by using an extremely simple device, shown in Fig. 7.16. The reservoir R is a 2-litre Erlenmeyer flask filled with methanol and the mixing chamber M is a 10-litre vessel containing 4 litres of chloroform. The contents of this vessel are mixed by means of a magnetic stirrer. The two vessels are connected by a siphon (S) and so are the mixing chamber and the column (D). Both vessels are closed by cotton-wool plugs. The reservoir is initially fitted in such a way as to ensure a constant flow of the methanol.

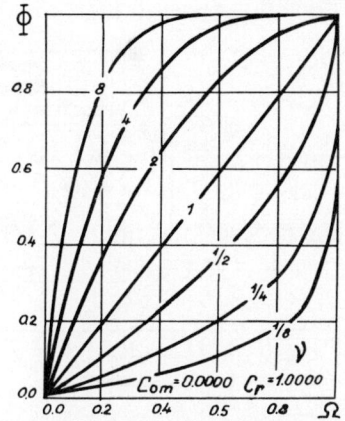

FIG. 7.15. Graph of function (7.23) for stated values of the parameter v (mode II gradient elution [72].

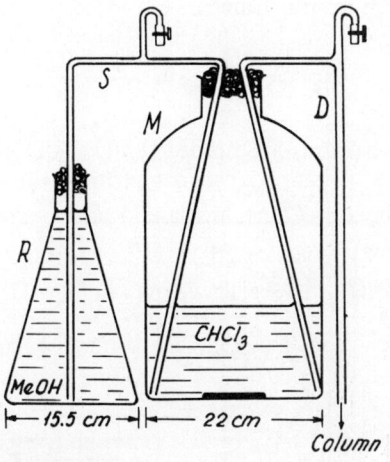

FIG. 7.16. Apparatus for producing a concave concentration gradient of methanol in chloroform [77].

Nelson [78, 79] replaced the siphons and the pressure heads in this system by pumps with controlled output, thus creating a device for concave gradient elution, the set-up of which is given in Fig. 7.17. The speed of the motor of the gradient pump is controlled by a circuit (Fig. 7.18) which increases it as a function of time, starting from very low speed. The consequence is a linear increase of the discharge from the pump. The

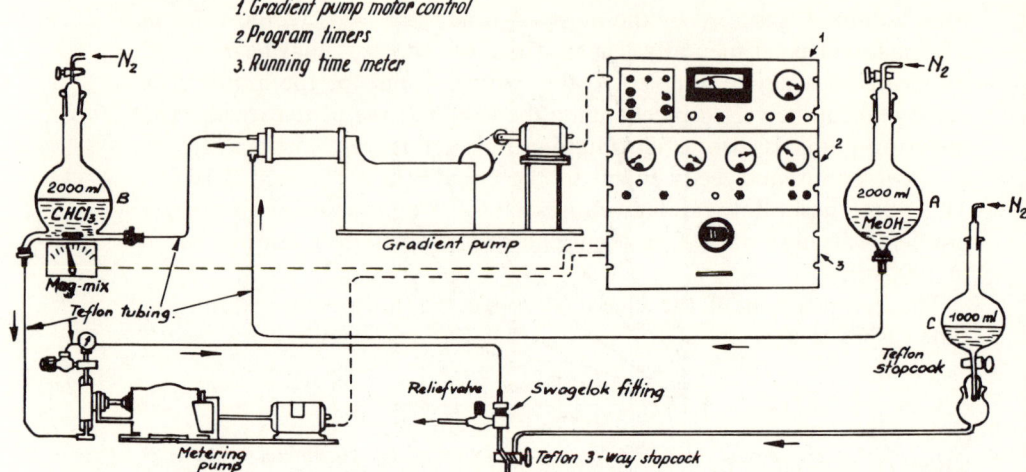

Fig. 7.17. Schematic diagram of the automated concave gradient elution column system; *A*—a solvent reservoir to facilitate charging of the gradient pump (not used during the run); *B*—the mixing chamber, *C*—an additional solvent reservoir for non-gradient elution [79].

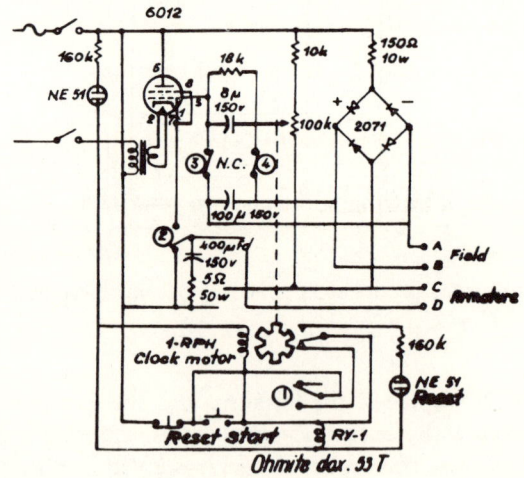

Fig. 7.18. Schematic circuit diagram for the gradient-pump motor control of the automated gradient elution column chromatographic system shown in Figure 7.17 [78].

DEVICES FOR ACHIEVING A MOBILE-PHASE GRADIENT   149

total volume V supplied by the pump in time t is $V = at^2/2$, the value of a being found experimentally. The maximal duration t_{max} may be modified by means of a device coupled to the control board for the motor. The value of a can be varied by changing the length of the piston stroke. The metering pump is also a 'chromatographic minipump' identical to the gradient pump, and feeds eluent to the column at a constant flow Q_1. This pump has a flow-rate and range of 1·0–1·8 ml/min when pumping a gradient varying from 100 per cent chloroform to 100 per cent methanol in 1000 min.

The composition of the eluent supplied by the mixing chamber at time t is given by

$$C = \left[\int C_r Q_2 \, dt - \int C Q_1 \, dt\right] \bigg/ \left[V_{0_m} + \int Q_2 \, dt - \int Q_1 \, dt\right] \quad (7.24)$$

where Q_2 is the flow from the gradient pump. This equation is analogous to Eq. (7.17) given by Wallach and Nordby [72] and both were published at about the same time. It is also similar to the equation given by Lakshmanan and Lieberman [80] with the difference that Q_2 is a variable of the form

$$Q_2 = at \quad (7.25)$$

where a is an arbitrary constant determined by the way the pump is fixed, and has units of ml/min. Differentiating (7.24) with respect to time gives

$$dC/(C_r - C) = (Q_2 \, dt)/[V_{0_m} + (Q_2/2 - Q_1)t] \quad (7.26)$$

which after substitution of Q_2 from (7.25) becomes

$$dC/(C_r - C) = (at \, dt) \bigg/ \left(V_{0_m} + \frac{at^2}{2} - Q_1 t\right) \quad (7.27)$$

By integration the following solutions are obtained.

(a) If $2aV_{0_m} = Q_1^2$:

$$C = C_r \left[1 - \frac{Q_1^2 \exp 2[1 + Q_1/(at - Q_1)]}{(at - Q_1)^2}\right] \quad (7.28)$$

(b) If $2aV_{0_m} < Q_1^2$:

$$C = C_r \left[1 - \frac{V_{0_m} \left(\frac{Q_1 + \beta}{Q_1 - \beta}\right)^\delta}{\left(V_{0_m} + \frac{at^2}{2} - Q_1 t\right)\left(\frac{\alpha - \beta}{\alpha + \beta}\right)^\delta}\right] \quad (7.29)$$

where $\alpha = at - Q_2$; $\beta = \sqrt{Q_1^2 - 2aV_{0_m}}$ and $\delta = Q_1/\beta$.

(c) If $2aV_{0_m} > Q_1^2$:

$$C = C_r \left[1 - \frac{V_{0_m} \exp\left[\gamma \tan^{-1}(-\gamma)\right]}{\left(V_{0_m} + \frac{a}{2}t^2 - Q_1 t\right) \exp\left[\gamma \tan^{-1}\left(\frac{at - Q_1}{\sqrt{2aV_{0_m} - Q_1}}\right)\right]} \right] \quad (7.30)$$

where $\gamma = 2Q_1/\sqrt{2aV_{0_m} - Q_1}$. The three solutions are necessary as the root $\pm [(Q_1^2 - 2aV_{0_m})]^{1/2}$ appears and is indeterminate when the expression within the square brackets is negative or zero. Figure 7.19 gives the curves obtained by means of the three solutions of Eq. (7.24). The curves C_1, C_2 and C_3 were obtained from Eqs. (7.28), (7.29) and (7.30) respectively for the parameters given in the figure.

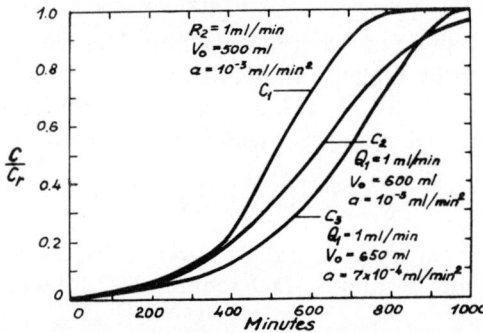

FIG. 7.19. Graphic presentation of gradient obtained under conditions indicated on the figure for the three solutions of Eq. (7.24). Curves C_1, C_2 and C_3 are solutions of Eqs. (7.28), (7.29) and (7.30) respectively [79].

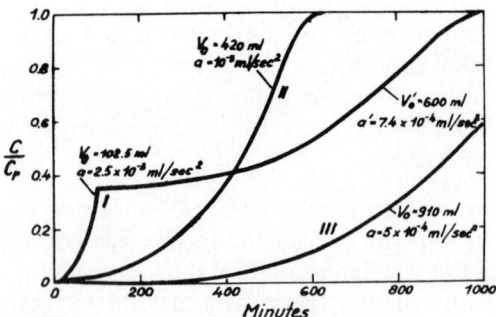

FIG. 7.20. Some additional types of gradients that can be produced by the system shown in Fig. 7.17, for three solutions of Eq. (7.29) [79].

Figure 7.20 gives three of the most commonly used gradients [Eq. (7.29)]. For curve I it is necessary to add non-polar solvent to the mixing chamber at the discontinuity (this can be achieved automatically). The values of the parameters are given in the figure. The expressions for the solutions of Eq. (7.24) are evidently fairly complicated, but the use of computers makes this no longer a problem.

The calculation of the pH gradient was first done by Piez [81]. In this case the problem is more difficult than for the concentration gradient, as the activity of the protons depends on the nature and concentration of the other ions present, and their buffering action. Calculation of the pH gradient requires knowledge of the relationship of the volumes in the reservoir and the mixing chamber. The flow-rates, Q_1 at the entry and Q_2 at the exit of the mixing chamber, may be different, but their ratio should be constant. If we write V_m = total volume of the mixing chamber, V_{0_m} = initial volume of solution in the mixing chamber, V'_m = volume of original solution left in the mixing chamber, and V = volume of liquid that has left the mixing chamber, we can write the following equations.

$$\frac{dV'_m}{V'_m} = -\frac{dV}{V_m} \tag{7.31}$$

and

$$V_m = V_{0_m} + \frac{V(Q_1 - Q_2)}{Q_2} \tag{7.32}$$

By substituting (7.32) in (7.31) and integrating V_m between the limits V_{0_m} and V'_m, and V between the limits 0 and V, and replacing $(Q_1 - Q_2)/Q_2$ by a, we obtain

$$V'_m = V_{0_m}(1 + aV/V_{0_m})^{-1/a} \tag{7.33}$$

This equation is equivalent to the one deduced by Lakshmanan and Lieberman [80] for concentration gradients. The pH can be changed in various ways: (a) by adding one buffer solution to another; (b) by adding the salt of a weak acid or a weak base to a buffer; (c) by adding a strong acid or base to a buffer.

(a) *The mixing chamber contains a buffer solution made of a weak monobasic acid and its salt with a strong base.* The reservoir contains the same buffer system but at a different pH, i.e. different proportions of the components. It is assumed that the buffers are not too dilute and that the changes in the acid and salt concentrations are proportional to the change in volume and are additive. The concentration of the acid in the buffer solution leaving the mixing chamber is given by

$$[HA] = \frac{V'_m}{V_m}[HA]_0 + \frac{V_m - V'_m}{V_m}[HA]_r \quad (7.34)$$

and $[A^-]$ by

$$[A^-] = \frac{V'_m}{V_m}[A^-]_0 + \frac{V_m - V'_m}{V_m}[A^-]_r \quad (7.35)$$

where $[HA]_0$ and $[A^-]_0$ are the initial concentrations in the mixing chamber and $[HA]_r$ and $[A^-]_r$ are the concentrations in the reservoir. The weak acid will dissociate: $HA \rightleftharpoons H^+ + A^-$, so that from the law of mass action we obtain:

$$K_a = \frac{[A^-]_0[H^+]_0}{[HA]_0} \quad (7.36)$$

for the start of the gradient and

$$K_a = \frac{[A^-][H^+]}{[HA]} \quad (7.37)$$

at any time thereafter. By equating the two we obtain:

$$\frac{[H^+]_0}{[H^+]} = \frac{[A^-][AH]_0}{[A^-]_0[AH]} \quad (7.38)$$

and after substituting for $[HA]$ and $[A^-]$ from Eqs. (7.34) and (7.35),

$$\frac{[H^+]_0}{[H^+]} = \frac{1 + \left(\frac{V_m}{V'_m} - 1\right)\frac{[A^-]_r}{[A]_0}}{1 + \left(\frac{V_m}{V'_m} - 1\right)\frac{[HA]_r}{[HA]_0}} \quad (7.39)$$

Taking the values for V_m and V'_m given by Eqs. (7.32) and (7.33), we obtain:

$$\frac{V_m}{V'_m} = \left(1 + a\frac{V}{V_{0m}}\right)^{(a+1)/a} \quad (7.40)$$

After substitution of this ratio in (7.39) and taking logarithms, we obtain the final expression for the change in pH,

$$pH - pH_0 = \log \frac{1 + \left[\left(1 + a\frac{V}{V_{0m}}\right)^{(a+1)/a} - 1\right]\frac{[A^-]_r}{[A^-]_0}}{1 + \left[\left(1 + a\frac{V}{V_0}\right)^{(a+1)/a} - 1\right]\frac{[HA]_r}{[HA]_0}} \quad (7.41)$$

The gradient obtained depends only on the flow-rates and on the concentrations of the buffers. To keep the pH within the buffer range of the system, the salt/acid ratios must lie between 0·1 and 10. For an ascending gradient $[A^-]_r/[HA]_r = 10$ and $[A^-]_0/[HA]_0 = 0.1$ if we wish to obtain the broadest pH range. Assuming that both buffers are of the same total molarity, Eq. (7.41) becomes

$$\text{pH} - \text{pH}_0 = \log \frac{1 + [(1 + aV/V_{0_m})^{(a+1)/a} - 1]10}{1 + [(1 + aV/V_{0_m})^{(a+1)/a} - 1]0.1} \quad (7.42)$$

A graph of Eq. (7.42) for three values of the flow-rate ratio Q_1/Q_2 gives the curves shown in Fig. 7.21, which shows that for $Q_1/Q_2 = 1$ the gradient is convex and tends asymptotically towards the pH of the solution in the reservoir. For $Q_1/Q_2 < 1$, the gradient is nearly linear at the beginning and then becomes concave.

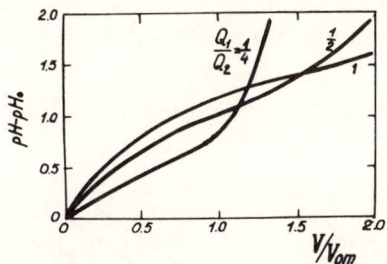

FIG. 7.21. Effect of different ratios of Q_1/Q_2 on the pH gradient obtained by addition of one buffer to another of equal molarity [Eq. (7.42)] [81].

For buffer solutions of different molarities, and the same salt/acid ratios as above, if $Q_1/Q_2 = 1/4$, Eq. (7.41) becomes

$$\text{pH} - \text{pH}_0 = \log \frac{1 + [(1 - 3V/4V_{0_m})^{-1/3} - 1]10R_m}{1 + [(1 - 3V/4V_{0_m})^{-1/3} - 1]0.1R_m} \quad (7.43)$$

where $R_m = ([A^-]_r + [HA]_r)/([A^-]_0 + [HA]_0)$. Equation (7.43) is shown graphically in Fig. 7.22 for different values of the total molarity ratio R_m. It shows that the gradient becomes more concave as R_m decreases. If the buffer solution in the reservoir has a lower pH than the buffer in the mixing chamber, a decreasing gradient is obtained.

(b) *The mixing chamber contains the same buffer as in (a) and the reservoir contains a solution of a weak-acid salt with alkaline hydrolysis.* This case differs from the preceding one because $[HA]_r \sim 0$ and thus

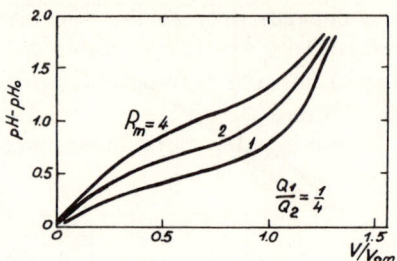

FIG. 7.22. Effect of different ratios of buffer strengths on the pH gradient obtained by addition of one buffer to another [Eq. (7.43)] [81].

Eq. (7.41) becomes

$$\text{pH} - \text{pH}_0 = \log\{1 - R_A[1 - \exp(V/V_{0_m})]\} \quad (7.45)$$

where $R_A = [A^-]_r/[A^-]_0$. Graphs of Eq. (7.45) for different values of R_A show that $R_A < 1$ gives a concave gradient, $R_A = 1$ gives a linear gradient, and $R_A > 1$ a convex gradient (Fig. 7.23). If $R_A = 1$, Eq. (7.44) becomes

$$\text{pH} - \text{pH}_0 = \frac{a+1}{a} \log(1 + aV/V_{0_m}) \quad (7.46)$$

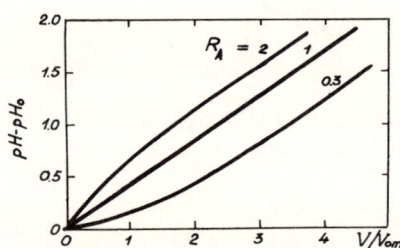

FIG. 7.23. Effect of different ratios of salt concentrations on the pH gradient obtained by addition of a salt of a weak acid to a buffer, for $Q_1 = Q_2$ [Eq. (7.45)] [81].

Plotting (7.46) for different values of Q_1/Q_2 yields for $Q_1/Q_2 < 1$ a concave gradient, for $Q_1/Q_2 = 1$ a linear gradient, and for $Q_1/Q_2 > 1$ a convex gradient (Fig. 7.24). If the reservoir contains a weak acid, $[A^-]_r \sim 0$, and Eq. (7.41) becomes

$$\text{pH} - \text{pH}_0 = -\log\{1 + [(1 + aV/V_{0_m})^{(a+1)/a} - 1]R_{HA}\} \quad (7.47)$$

where $R_{HA} = [HA]_r/[HA]_0$. We thus obtain negative ΔpH values, i.e. a descending gradient. In this case the decisive factor is the concentration of the acid.

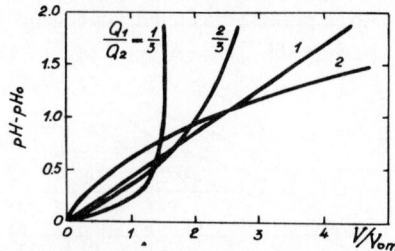

FIG. 7.24. Effect of different flow-rate ratios on the pH gradient obtained by addition of a salt of a weak acid to a buffer at equal salt concentrations [Eq. (7.46)] [81].

(c) *The mixing chamber contains a buffer consisting of a weak monobasic acid and its salt with a strong base, and the reservoir contains a solution of a strong base.* In this case we can write the equations

$$[A^-] = [A^-]_0 V'_m/V_m + [OH^-]_r (V_m - V'_m)/V_m \qquad (7.48)$$

and

$$[HA] = [HA]_0 V'_m/V_m - [OH^-]_r (V_m - V'_m)/V_m \qquad (7.49)$$

where $[OH^-]_r$ is the concentration of the strong base in the reservoir. With the same reasoning as in case (a) we obtain

$$pH - pH_0 = \log \frac{1 + [(1 + aV/V_{0_m})^{(a+1)/a} - 1][OH^-]_r/[A^-]_0}{1 + [(1 + aV/V_{0_m})^{(a+1)/a} - 1][OH^-]_r/[HA]_0} \qquad (7.50)$$

For $[A^-]_0/[HA]_0 = 0.1$ and $Q_1 = Q_2$, Eq. (7.50) becomes:

$$pH - pH_0 = \log \frac{1 - \dfrac{[OH^-]_r}{[A^-]_0}[1 - \exp(V/V_{0_m})]}{1 + \dfrac{[OH^-]_r}{10[A^-]_0}[1 - \exp(V/V_{0_m})]} \qquad (7.51)$$

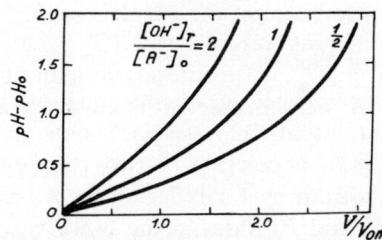

FIG. 7.25. Effect of different ratios of base to buffer salt on the pH gradient obtained by addition of strong base to a buffer, for $Q_1 = Q_2$ [Eq. (7.51)] [81].

Plotting equation (7.51) gives a concave pH gradient (Fig. 7.25). For $[A^-]_0/[HA]_0 = 0.1$ and $[OH^-]_r = [A^-]_0$ Eq. (7.51) is reduced to the form:

$$pH - pH_0 = \log \frac{\left(1 + a\dfrac{V}{V_{0_m}}\right)^{(a+1)/a}}{1.1 - \left(1 + a\dfrac{V}{V_{0_m}}\right)^{(a+1)/a}} \qquad (7.52)$$

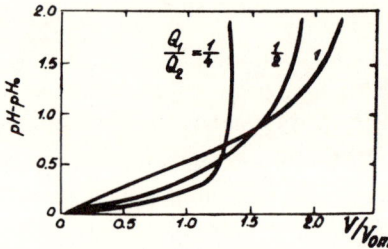

FIG. 7.26. Effect of different ratios of Q_1/Q_2 on the pH gradient obtained by addition of strong base to a buffer at equal concentrations of buffer salt and strong base [Eq. (7.52)] [81].

From a plot of this equation for different values of Q_1/Q_2 (Figure 7.26) we see that the concavity of the gradient increases when this ratio decreases.

If the mixing vessel contains a buffer solution as in the preceding case, and the reservoir a strong acid, descending gradients are obtained, the pH change is negative, and the concentration of the acid in the reservoir is the decisive factor. Buffer systems obtained from monobasic acids and their salts with strong bases may be used only over a range of 2 pH units. By use of a polybasic acid and one of its salts, or a mixture of buffer systems, the range of pH variation may be broadened.

The problem of the pH gradient was also tackled by Reiner and Reiner [82], who also took into account the corrections for ionic strength.

7.1.1.3 Linear Gradient Elution

As a rule the devices for linear gradients consist of two communicating vessels (cylinders) of equal cross-section. One of the simplest was described by Arcus [83]. It consists of two glass cylinders (2.5 cm diameter) joined at the bottom by a capillary tube of 1 mm bore (Fig. 7.27a). The rod of pistons P_1 and P_2 slides in the brass bearings B_1 and B_2, set vertically on the rod of a retort stand. The weights W_1 (1.36 kg) and W_2 (2.04 kg) are made of lead. B_3 is a brass arm fixed on P_2 so as to slide

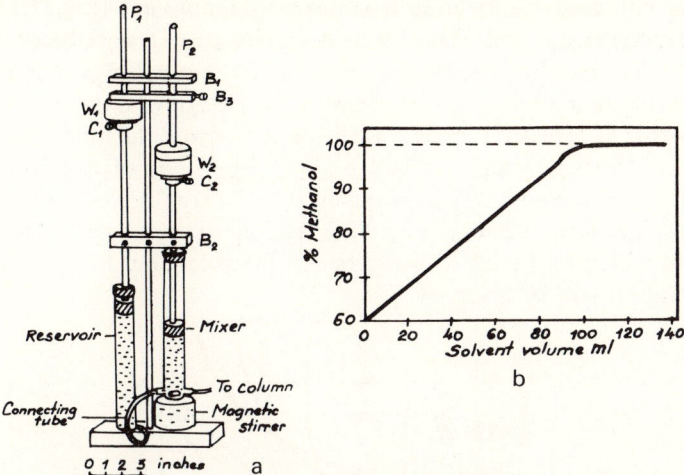

Fig. 7.27. Apparatus for producing a linear elution gradient. The inset shows the change of the composition of the solvent supplied by this apparatus [83].

easily on the rod P_1. The rings C_1 and C_2 are set in such a way that when they reach the bar B_2 the pistsons are 1·5 cm from the bottom of the vessel.

The reservoir contains 100 ml of deaerated methanol, and the mixing chamber 50 ml of 60 per cent methanol (v/v) in water. The flow of eluent is started by loosening the screws fixing the rods P_1 and P_2 in B_2, and will continue till the rings C_1 and C_2 reach bar B_2. The changing composition of the eluent is shown in Fig. 7.27b.

Another apparatus based on essentially the same principle was built by Choules [84] and consists of two cylindrical chambers (the reservoir and the mixing chamber) containing solutions of different concentrations. The liquid is transferred by means of two pistons, coupled to a common transmission moving at a constant rate. The solution in the reservoir passes into the mixing chamber where it is mixed with the second solution by means of a screw turbine which circulates the liquid through the mixing chamber and return channel. The liquid is transferred from the mixing chamber to the column through a side-arm.

Since the cross-sections of both cylinders are the same, Parr's equation [66] becomes

$$C = C_m + (C_r - C_m)V/(V_{0_m} + V_{0_r}) \tag{7.53}$$

(see p. 143 for the notation). This device also works with viscous liquids and was used to obtain linear gradients by mixing a 50 per cent sucrose solution with water at room temperature. As shown in Fig. 7.12 (p. 142), linear gradients may also be obtained with the apparatus described by Wallach and Nordby [72], Billimoria et al. [85] have shown that a linear gradient of methanol in chloroform is produced when the condition $\rho_m d_r^2 = \rho_r d_m^2$ is satisfied (where ρ is the density and d the inner diameter).

A linear gradient can also be obtained by means of the apparatus, consisting of two communicating vessels (Fig. 7.28) described by Bock and Ling [14]. The outer vessel is a cylinder, and the inner vessel is a cone, the radius of which (r) varies with the height (h) according to $r^2 = kh$. The concentration will be given by Eq. (7.53).

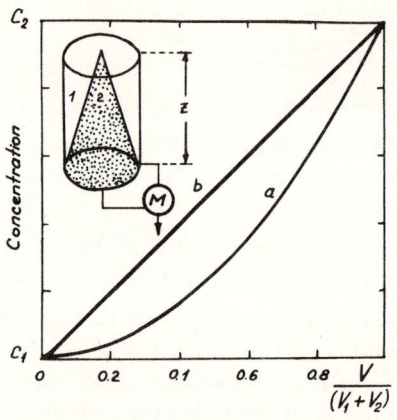

FIG. 7.28. Diagram of the device for producing an elution gradient, formed from one cylindrical and one conical vessel (M—mixing chamber). Curve $a: r = kz$ and $C = C_1 + (C_2 - C_1)[V/(V_1 + V_2)]^2$; curve $b: r^2 = kz$ and $C = C_1 + (C_2 - C_1) \times [V/(V_1 + V_2)]$ [14].

A linear gradient may also be obtained in a particular instance [see Eq. (7.68)] with the device designed by Peterson and Sober [86]. Chase [87] also showed the possibility of obtaining a linear gradient by combining some simple gradients.

Lakshmanan and Lieberman [80, 88] have shown that with the apparatus for which the equation

$$C = C_r[1 - (V/V_{0_m})^{Q_r/(Q_m - Q_r)}] \qquad (7.54)$$

is valid, the gradient becomes linear if $Q_m = 2Q_r$. A linear gradient was also used in other work [89–91].

7.1.1.4 *Concave and Convex Parabolic Gradients*

Devices for obtaining these gradients consist of two cylindrical vessels, the reservoir and the mixing vessel, initially containing volumes V_{0_r} and V_{0_m} respectively, the variable volumes being V_r and V_m. Let C_r and C_m be the initial concentrations of, for instance, the more polar component in the two vessels, and C its concentration in the volume V leaving the mixing vessel and entering the column. The reservoir and mixing vessel are connected by a tube (Fig. 7.29a) or a siphon (Fig. 7.29b).

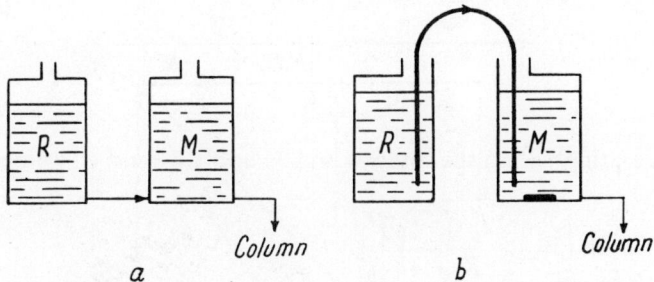

FIG. 7.29. Block diagram of apparatus for producing parabolic concentration gradients; *a*—with tubing between vessels, *b*—with siphon between vessels.

To find the function $C = f(V)$, Lebreton [1] suggested a calculation procedure in which the mass-balance for the mixing vessel is taken as the starting point:

$$CV_m = C_m V_{0_m} + C_r(V_{0_r} - V_r) - \int_0^V C \, dV \qquad (7.55)$$

The ratio of the volumes is constant:

$$K = V_r/V_m \qquad (7.56)$$

and the volume V leaving the system will be given by

$$V = (V_{0_r} - V_r) + (V_{0_m} - V_m) \qquad (7.57)$$

From Eqs. (7.56) and (7.57) we obtain

$$V = (K + 1)(V_{0_m} - V_m) \qquad (7.58)$$

or

$$V_m = V_{0_m} - \frac{V}{K + 1} \qquad (7.59)$$

Substituting for V_r and V_m from (7.56) and (7.59) in (7.55) gives

$$CV_{0_m} - C\frac{V}{K+1} = C_{0_m}V_{0_m} + \frac{KC_r}{K+1}V - \int_0^V C\,dV \quad (7.60)$$

After differentiating, we obtain:

$$V_{0_m}\frac{dC}{dV} - \frac{C}{K+1} - \frac{V}{K+1}\frac{dC}{dV} - \frac{KC_r}{K+1} + C = 0 \quad (7.61)$$

By separating the variables:

$$\frac{dV}{V_{0_m} - \frac{1}{K+1}V} = \frac{dC}{\frac{KC_r}{K+1} - \frac{K}{K+1}C} \quad (7.62)$$

and integrating within the limits 0 and V, and C_{0_m} and C, we obtain:

$$-(K+1)\left[\ln\left(V_{0_m} - \frac{V}{K+1}\right)\right]_0^V$$
$$= -\frac{K+1}{K}\ln\left(\frac{KC_r}{K+1} - \frac{K}{K+1}C\right)_{C_{0_m}}^C \quad (7.63)$$

or

$$C = C_r - (C_r - C_m)\left[1 - \frac{V}{(K+1)V_{0_m}}\right]^K \quad (7.64)$$

By the substitution $(K+1)V_{0_m} = V_{0_m} + V_{0_r}$, Eq. (7.64) may be written

$$C = C_r - (C_r - C_m)\left(1 - \frac{V}{V_{0_r} + V_{0_m}}\right)^K \quad (7.65)$$

(a) For $K = V_r/V_m = 1$, Eq. (7.65) becomes:

$$C = C_m + (C_r - C_m)\frac{V}{V_{0_r} + V_{0_m}} \quad (7.66)$$

and a linear gradient is obtained (Fig. 7.30).

(b) For $K > 1$, i.e. $V_r > V_m$, a convex parabolic gradient [Eq. (7.65)] is obtained (Fig. 7.30).

(c) For $K < 1$, a concave parabolic gradient [Eq. (7.65)] is obtained (Fig. 7.30).

The convexity and concavity are more marked when $V > (V_{0_r} + V_{0_m})/2$ [1, 14, 15].

Devices for Achieving a Mobile-phase Gradient

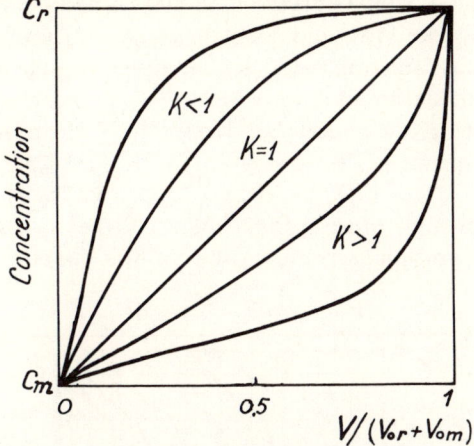

Fig. 7.30. Graph of Eq. (7.65).

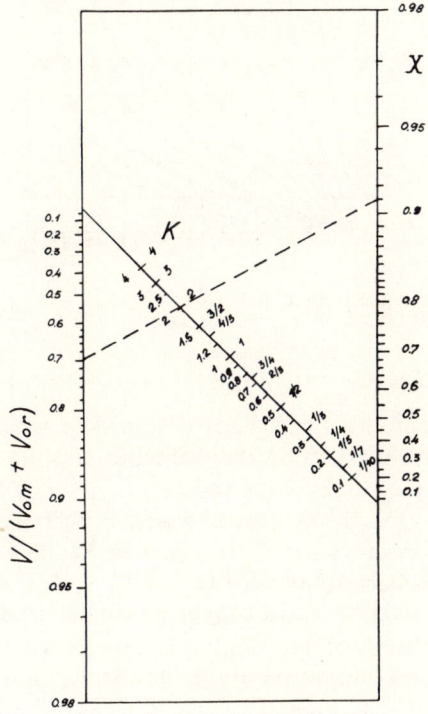

Fig. 7.31. Nomogram for computing the function $X = 1 - [1 - V/(V_{0_m} + V_{0_r})]^K$ [26].

Equations similar to (7.64) were also established by Parr [66], Bock and Ling [14], Drake [15], and Kenyon et al. [92] for liquids of equal density. In that case the mixing does not cause a change in volume.

For a rapid calculation of C vs. $V/(V_{0_m} + V_{0_r})$ for a given set of experimental conditions (K, C_m and C_r), Kocent [26] proposed a system of nomograms. By means of the nomogram in Fig. 7.31 we can calculate the function $X = 1 - [1 - V/(V_{0_m} + V_{0_r})]^K$, e.g. for $V/(V_{0_m} + V_{0_r}) = 0.7$ and $K = 2$, a line through these points intersects the X axis at $X = 0.91$. Alternatively, a graphic representation of the function can be used (Fig. 7.32).

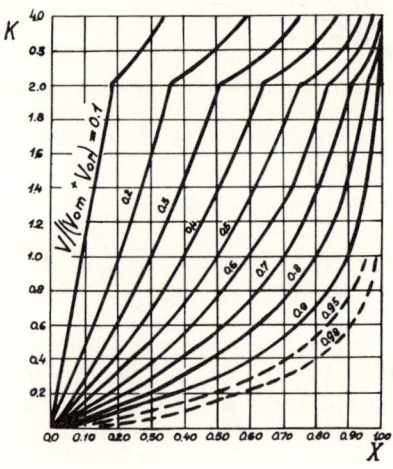

FIG. 7.32. Graph of the function $X = 1 - [1 - V/(V_{0_m} + V_{0_r})]^K$ for definite values of $V/(V_{0_m} + V_{0_r})$ [26].

An auxiliary nomogram has been designed in which the concentrations C_r and C_m are marked on the ordinates and the abscissa carries a scale representing the values of the ratio $V/(V_{0_m} + V_{0_r})$, for a certain value of K, e.g. $K = 0.4$ (Fig. 7.33). This nomogram permits very rapid determination of the concentration C for different values of C_r and C_m at a given K value and ratio of V and $(V_{0_r} + V_{0_m})$. In the example shown, $C_r = 1.0M$, $C_m = 0.5M$, $K = 0.4$, and at $V/(V_{0_r} + V_{0_m}) = 0.6$, $C = 0.83M$.

Peterson and Sober [86] have built a simple device by means of which linear, parabolic and composite gradients can be obtained (see Section 7.1.5). This apparatus is known as the 'Varigrad' and consists of nine identical chambers connected in series and arranged in compact form (Fig. 7.34). The main scheme is given in Fig. 7.35.

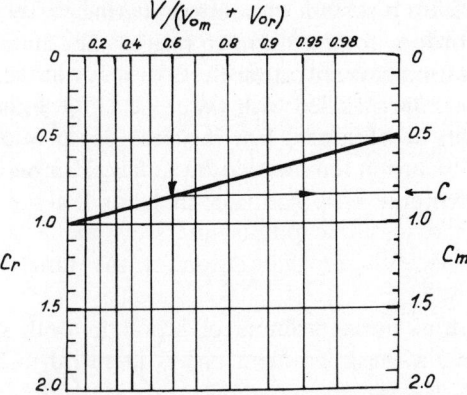

FIG. 7.33. Nomogram for a closed mixing chamber and for $K = 0.4$ [26].

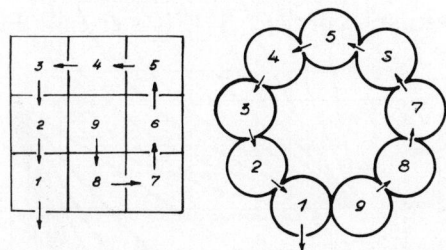

FIG. 7.34. Two arrangements of the nine mixing-chamber system for gradient elution, showing direction of flow [86].

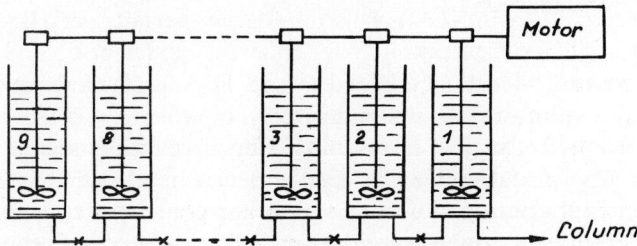

FIG. 7.35. Block diagram of nine-chambered system for gradient elution ('Varigrad').

The chambers form a system of communicating vessels in hydrostatic equilibrium. According to the number (N) of the chamber containing the solution of limiting concentration C_r, $N - 1$ systems can be obtained. The other chambers are all filled with water, i.e. $C_m = 0$. In this way up to eight gradients can be obtained. For instance, for $N = 2$ there is water in chamber 1 and solution (concentration C_r) in chamber 2. The concentration C of the volume V of solution that has left the system will be given by

$$C = C_r V / V_t \tag{7.67}$$

where V_t is the total initial volume of liquid in both chambers. This equation represents a linear gradient and is identical to Eq. (7.66) when $C_m = 0$ and $V_{0_r} + V_{0_m} = V_t$.

For $N = 3$, the solution of concentration C_r is in chamber 3 and water in chambers 1 and 2. For the general case, with the solution in chamber N and water in the others, Peterson and Sober established the equation

$$C = C_r (V/V_t)^{N-1} \tag{7.68}$$

A graphical representation of C/C_r vs. V/V_t is given in Fig. 7.36.

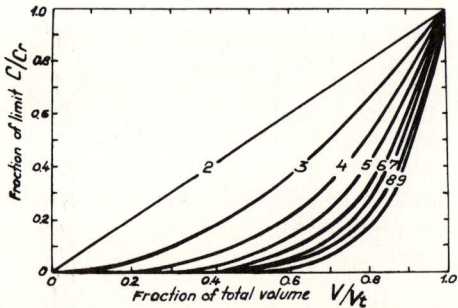

Fig. 7.36. Gradients produced by several 'Varigrad' systems with solution of concentration C_r in the chamber with number indicated on the curve (Fig. 7.35). Other chambers contain water [86].

The 'Varigrad' system is especially used in column chromatography with aqueous solutions of salts and buffers, or when it is difficult to use organic solvents because of errors caused by large differences in density.

Horton [93] has described a simple device made of two 1000-ml Erlenmeyer flasks placed mouth to mouth and connected by a three-way tap which also leads to the column (Fig. 7.37). The upper flask also has an outlet. By means of this device a descending pH gradient has been

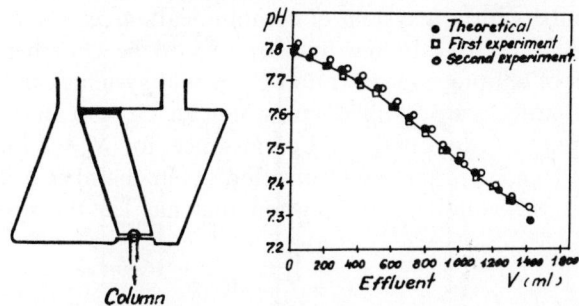

FIG. 7.37. A reciprocal cone apparatus for producing an elution gradient. The inset shows the graph of the pH gradient obtained by using 0·05M tris–HCl buffers [93].

obtained (Fig. 7.37) by use of buffers made with tris(hydroxymethyl)-aminomethane ('tris') and hydrochloric acid. A similar device was described earlier by van Tamelen and Taylor [94].

Wren [95] has shown that in the case of eluents with different densities, the gradients produced by devices of the type shown in Fig. 7.29 (p. 159) are better described by a modified empirical equation than by Eq. (7.65):

$$C = C_r - (C_r - C_m)[1 - V/(V_{0_r} + V_{0_m})]^{\rho_m d_r^2 / \rho_r d_m^2} \qquad (7.69)$$

where ρ_r is the density of the liquid in the reservoir, ρ_m that of the liquid in the mixing vessel, and d_r and d_m are the inner diameters of the cylinders forming the two vessels. The gradients are (a) linear for $\rho_m d_r^2 = \rho_r d_m^2$, (b) convex for $\rho_m d_r^2 > \rho_r d_m^2$ and (c) concave for $\rho_m d_r^2 < \rho_r d_m^2$. To find the dimensions of the vessels necessary to give a particular gradient, the ratio of the diameters must be determined. This is done by taking logarithms in Eq. (7.69), after replacement of C by C_H, the value of C when $V/(V_{0_r} + V_{0_m}) = 1/2$. Then

$$\frac{d_r}{d_m} = \left[\frac{\rho_r}{\rho_m} \cdot \frac{-1}{\log 2} \cdot \log \frac{(C_r - C_H)}{(C_r - C_m)} \right]^{1/2} \qquad (7.70)$$

and the ratio is determined by means of the nomogram given in Fig. 7.38. Bader and Morgan [96] have also shown that Eq. (7.70) predicts very well a concave gradient ($K = 0.222$) for methanol in chloroform.

The devices used [47, 92, 95, 97–104] to obtain parabolic gradients take one of the forms shown in Figs. 7.39–7.41. By varying the dimensions of the cylindrical parts it is easy to achieve the ratio of cross-sections necessary for a particular gradient.

Desreux [105] uses parts with conical profiles to obtain some parabolic gradients (Fig. 7.42). Consider the system formed by tubes 1 and 2,

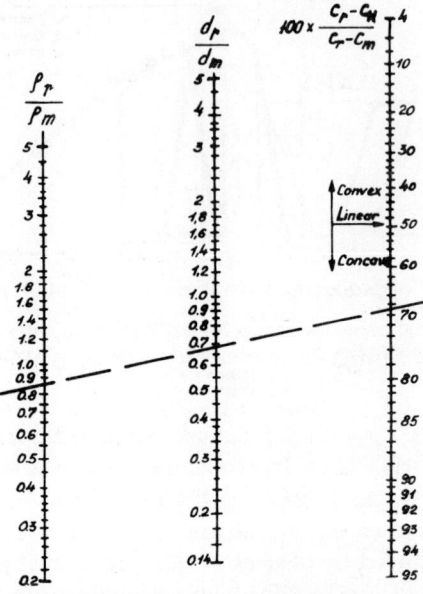

FIG. 7.38. Nomogram of Eq. (7.70). The dashed line illustrates the computation of the diameter ratio for a concave gradient of 0–30 per cent methanol in chloroform, $C_H = 10$ per cent [95].

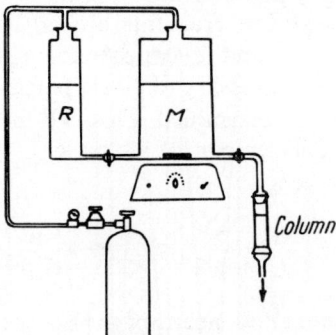

FIG. 7.39. Apparatus for producing a parabolic concentration gradient [92].

with equal base radius r and height Z. The radius r_1 of cone 1 as a function of height z is given by $r_1 = r(1 - z/Z) = r(1 - x)$, and the radius r_2 of cone 2 by $r_2 = rz/Z = rx$. The volume V_1 of the liquid delimited by the cylinder of height Δz and the frustrum of the cone of radii r_1 and r_2

Devices for Achieving a Mobile-phase Gradient

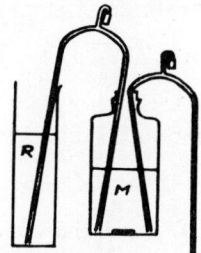

Fig. 7.40. Apparatus for producing a parabolic concentration gradient with eluents of unequal density [95].

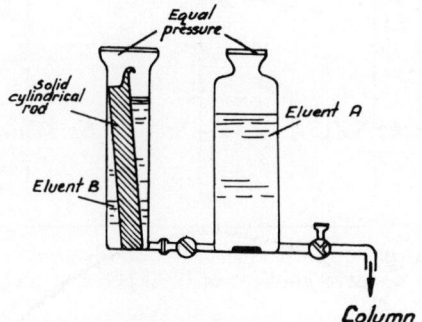

Fig. 7.41. Apparatus for producing a parabolic concentration gradient [104].

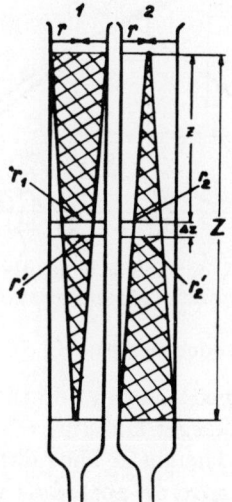

Fig. 7.42. Diagram of battery of two cylindrical vessels each containing a conical rod [105].

(at heights z and $z + \Delta z$) for cylinder 1 is given by the expression

$$V_1 = b - b\,\Delta z/3Z + (2a - b)x - ax^2 \qquad (7.71)$$

where $a = \pi r^2\, \Delta z$ and $b = a\, \Delta z/Z$.

Also, the volume corresponding to tube 2 will be

$$V_2 = a - b\,\Delta z/3Z - bx - ax^2 \qquad (7.72)$$

The curves 1 and 2 in Fig. 7.43 show the variation of volumes V_1 and V_2 as a function of z, curve 3 shows the variation of the total volume ($V_1 + V_2$) with height, and curve 4 shows the variation of the fractional concentration $F(0 \cdots 1)$ which can be established graphically or calculated by means of the equation

$$F = \frac{A' - bx - ax^2}{A'' + B'x - ax^2} \qquad (7.73)$$

where $A' = (a - b\,\Delta z/3Z)$, $A'' = (a + b - 2b\,\Delta z/3Z)$ and $B' = 2(a - b)$.

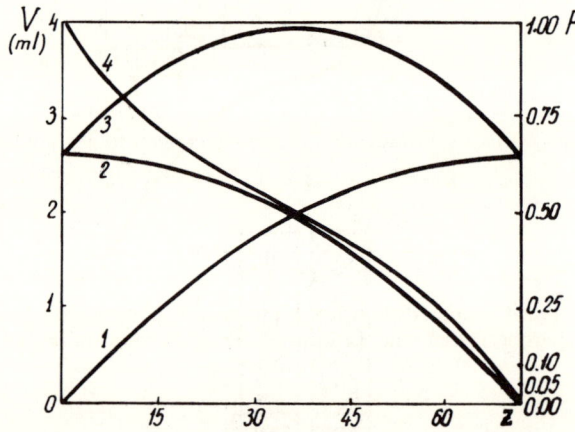

FIG. 7.43. The plot of volume V_1 of eluent leaving tube 1 (curve *1*), V_2 leaving tube 2 (curve *2*), $V_1 + V_2$ (curve *3*) and fractional concentration F, *vs.* height z (curve *4*) [105].

Such a device can be planned and built on the basis of these equations. If the experiment requires larger amounts of eluent, several vessels can be coupled in parallel.

Bock and Ling [14] designed an apparatus with a cylinder containing an empty cone into which the second liquid is put (see Fig. 7.28, p. 158). In this way the space required is reduced. These authors showed that in that case the gradient is parabolic, whereas if the inner vessel is shaped like a bell-jar with radius $r = \sqrt{kz}$, where z is the height and k a constant, then the gradient is linear. They also describe a series of communicating cylindrical vessels by means of which parabolic gradients could be obtained.

Gradients obtained by means of communicating vessels have been described in several papers [106–112].

7.1.1.5 *Composite Gradients*

Composite gradients can be obtained in the simplest way, as Bock and Ling [14] have shown, by using a rectangular vessel divided diagonally into two communicating compartments. The shape of the dividing surface determines the variation of the concentration with volume delivered (Fig. 7.44).

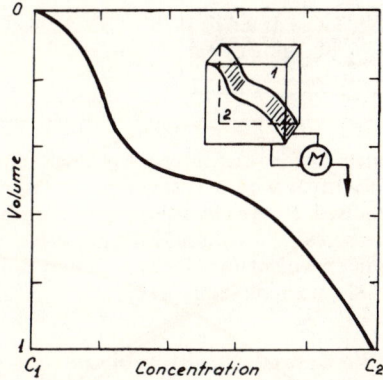

FIG. 7.44. Diagram of the device for producing a compound elution gradient (*M*—mixing chamber). Left, the concentration given as any desired function of volume [14].

Another apparatus, described by Watt [112], is able to provide a binary mixture of predetermined composition (Fig. 7.44). It consists of two cylindrical vessels fitted at the same height with side-arms which are joined in a T-piece. The vessels are filled with the two pure solvents (one in each) to the height of the side-arms. Polyethylene plungers are introduced into the two cylinders by screw-drive and the displaced solvents are mixed in the T-piece and pass into the chromatographic column. The apparatus is shown diagrammatically in Fig. 7.45.

The polyethylene blocks are square in section and their lengths are chosen according to the total volume of solution necessary for the elution.

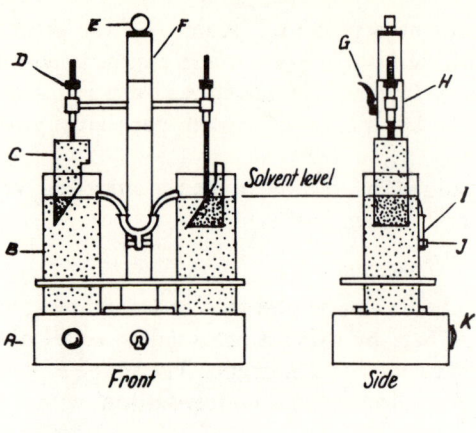

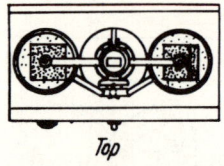

Fig. 7.45. A programmed solvent-dispenser for gradient elution. *A*—base containing drive-motor, *B*—solvent jar with overflow pipe, *C*—dispenser block, cut to suit concentration curve required, *D*—height-adjusting nut for displacer block, *E*—rotation indicator for drive screw, *F*—column-carrying drive screw, *G*—drive release trigger, *H*—yoke, carrying the displacer blocks, *I*—mixed solvent outlet (to column), *J*—Terry-clip support, *K*—warning lamp [112].

The curve showing the desired variation in concentration of components as a function of eluent volume is drawn on the face of one of the blocks, and to facilitate accurate cutting, a matching curve is drawn on the opposite face. The other block is treated similarly. The two blocks are cut to give a surface corresponding to the curve, and they are then set above the vessels in such a position that the total section of the two blocks is constant. The blocks are readily replaced with others, to give different gradients, which can be as complicated as desired.

Allen and Eggenberger [113] built an apparatus (Fig. 7.46) made of two cylinders, one (*A*) of constant section, and the other (*B*) cylindrically symmetrical but of varying section. Both are fitted at the top with a funnel for filling and a tap which is closed when they are full. Both are also connected at the top by capillary tubes to the distributor (*D*) which is connected to the chromatographic column. The liquid is displaced from the two cylinders by means of the mercury in vessel (*C*). The diameter and length of each segment of (*B*) are calculated from the desired gradient.

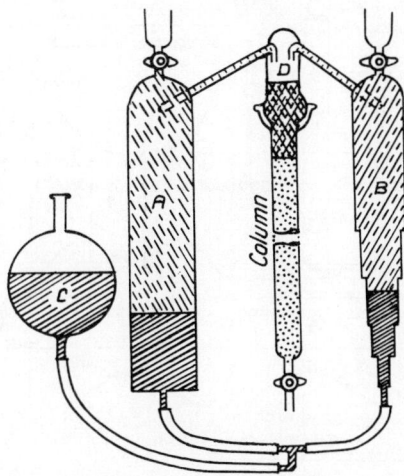

FIG. 7.46. Apparatus for automatically changing solvent polarity stepwise [113].

Sober and Peterson [58] have suggested the use of two or three communicating vessels of different forms, by means of which sinusoidal composite gradients can be obtained.

The 'Varigrad' (see Figs. 7.34 and 7.35, p. 163) can also provide compound gradients. Consider the contribution of one chamber (number n) in which there is a solution of concentration C_r (water in all the other eight chambers, $C_m = 0$) as a function of the number of chambers N. The concentration of the effluent, C, as a function of volume, V, can be determined empirically or calculated by means of the equation

$$C = C_r \frac{(N-1)!}{(N-n)!(n-1)} \left(1 - \frac{V}{V_t}\right)^{N-n} (V/V_t)^{n-1} \qquad (7.74)$$

This equation was demonstrated in a paper by Sorin and Vargue [17].

Plotting this equation for $N = 9$ and $n = 1, 2, \ldots, 9$ gives the curves in Fig. 7.47, where the full circles denote the chambers which carry the solution. It may be noted that the two curves are always symmetrical about a straight line perpendicular to the abscissa at $V/V_t = 0.5$. The maximum concentration appears when

$$V/V_t = (n-1)/(N-1) \qquad (7.75)$$

The formation of a compound gradient by use of the solution in four chambers of a nine-chamber system is illustrated in Fig. 7.48. In this case the concentrations of the solutions were $0.2C_r$ in chamber 3, C_r in

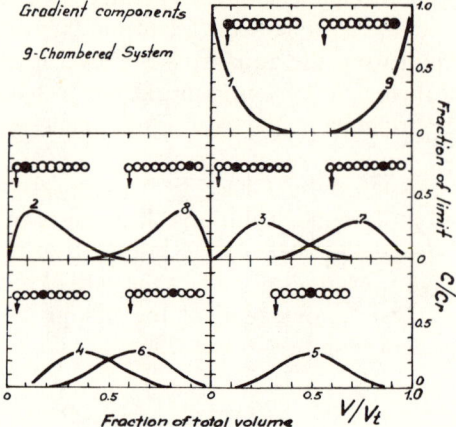

FIG. 7.47. Single-chamber contribution in nine-chambered system (Fig. 7.35). Filled circles represent the position of solution C_r; other chambers are filled with water [86].

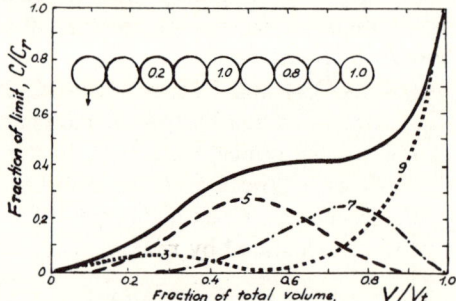

FIG. 7.48. Formation of a compound gradient by the summation of the contribution of four chambers in a nine-chambered system (Fig. 7.35) [86].

chamber 5, $0.8C_r$ in chamber 7 and C_r in chamber 9. The individual contributions from these chambers are first plotted for the system $N = 9$ (dotted curves) and then added to give the composite gradient (full curve).

Bombaugh et al. [114] described a new gradient elution chromatograph (Waters Associates ALC 100) based on a 'Varigrad'. Smith and Stahmann [115] grouped the nine chambers in three subunits coupled in series, thus obtaining gradients comparable with those described earlier [86].

Schmidtmann [116] described an apparatus built to yield salt and pH gradients. The apparatus consists of ten chambers linked at the bottom and each fitted with a mechanical stirrer operated by a low-speed electric motor. Each chamber has a capacity of 150 ml and they are placed in a line. During flow of liquid from the apparatus the level in the ten chambers decreases equally (principle of communicating vessels). The device is based on the same principle as the 'Varigrad', and gives the same types of gradient.

The apparatus described by Chase [87] yields gradients with both volatile and non-volatile buffer solutions. It consists of a reservoir and nine hermetically closed chambers connected in series in such a way that the density of the solutions does not affect the rate of transfer of the fluid between chambers. The arrangement of the mixing chambers and the method of linking them is shown in Fig. 7.49.

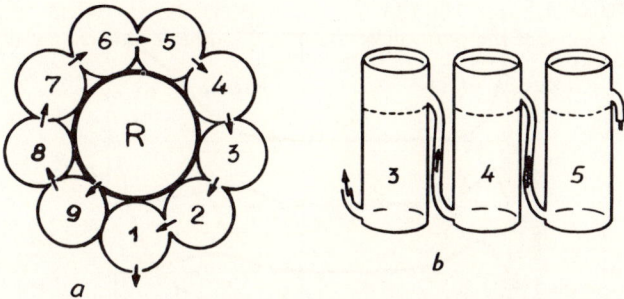

FIG. 7.49. Mixing-chamber system consisting of 10 airtight cylindrical chambers connected in series: *a*—schematic arrangement of chambers; *b*—arrows indicate direction of fluid transfer; dotted line indicates fluid levels [87].

The level of the liquid remains constant during the experiment, except in the reservoir, which must be big enough to supply the total elution volume needed. It can be refilled during the experiment without causing significant error. The nine chambers are fitted with magnetic stirrers operated by a single motor. A maximum volume of 250 ml can be introduced into each chamber at the start. The reservoir has a capacity of 4 litres and is fitted with a Mariotte bottle or peristaltic pump to ensure constant flow.

Consider the case where only one mixing chamber (volume V_m) contains solution of concentration C_m, and all the others and the reservoir contain pure solvent, e.g. water. According to which chamber (number n) contains the solution, nine gradients can be obtained. The concentration C as a function of the volume V that has left the system is given by [87]:

$$C = \exp(-V/V_m)\frac{C_m}{(n-1)!}(V/V_m)^{n-1} \qquad (7.76)$$

If all the mixing chambers contain water and the solution is in the reservoir, the equation is

$$C = C_r\{1 - [\exp(V/V_m)]\}\sum_{n=1}^{\infty}\frac{1}{(n-1)!}(V/V_m)^{n-1} \qquad (7.77)$$

From Eqs. (7.76) and (7.77) for $C_m = C_r = 1M$ and $n = 1, 2, \ldots, 9$, we obtain ten individual gradients representing the concentration of a single chamber or of the reservoir to the system (Fig. 7.50). The maximum concentration is obtained when $V/V_m = n - 1$. The experimental curves agree very well with theory for a 40 ml/hr flow-rate.

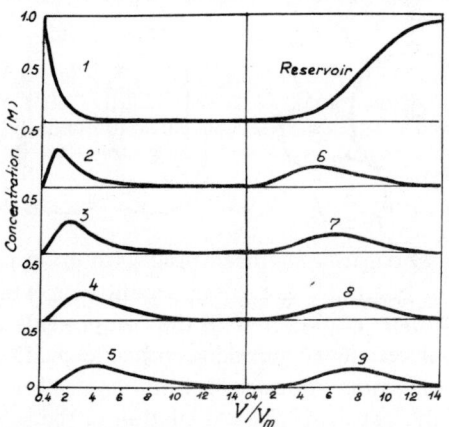

FIG. 7.50. Individual chamber gradients (apparatus from Fig. 7.49). Numbers above curves indicate the chamber containing solute at unit concentration; all other chambers filled with water. V/V_m is the ratio of effluent volume to mixing-chamber volume [87].

Chase [87] shows that by putting solution into more than one chamber, several simple gradients can be combined to give a composite gradient corresponding to a concentration profile defined by

$$C = [\exp(-V/V_m)]\sum_{n=1}^{\infty}\frac{C_n}{(n-1)!}(V/V_m)^{n-1}$$

$$+ C_r\left\{1 - [\exp(-V/V_m)]\sum_{n=1}^{\infty}\frac{1}{(n-1)!}(V/V_m)^{n-1}\right\} \qquad (7.78)$$

which is analogous to Eq. (7.77) and where C_n is the concentration of solution in chamber n. Figure 7.51 shows a series of simple gradients and the corresponding compound gradient described by Eq. (7.78).

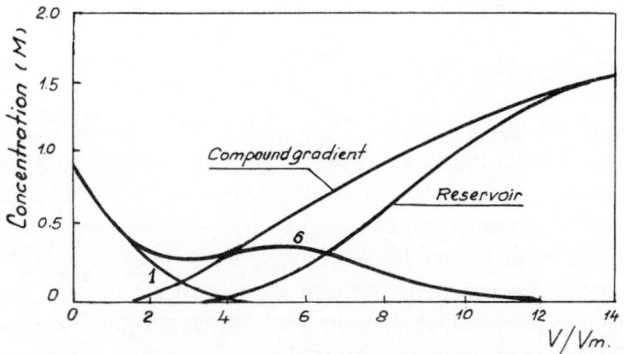

FIG. 7.51. Compound linear gradient (apparatus from Fig. 7.49). Linear gradient is the summation of individual gradients contributed by 0·60M NaCl in chamber 1, 2·2M in chamber 6 and 1·75M in the reservoir; all other chambers are filled with water (from data in [87]).

The 'Varigrad' apparatus has been used with good results for a great many separations, but there are some separations which require quite complicated composite gradients if all the components of the system are to be resolved. The use of the simpler composite gradients often permits separation of only a limited number of components of the system, within a certain region of the gradient. The calculation of the solution concentration needed in the different chambers of the 'Varigrad' to obtain certain complex gradients is extremely difficult by means of the usual algebraic methods. To deal with this difficulty Burns et al. [117] suggested a new method using a small analogue computer and an X–Y plotter. They started from the Varigrad pattern shown in Fig. 7.52, where $C_1, C_2, \ldots, C_9$ are the concentrations of a certain substance in the nine chambers of the 'Varigrad' after time t, the volumes being equal (V in each chamber). The chambers are connected by capillaries so that diffusion is avoided. The first chamber is connected to a pump operating at a constant rate Q (ml/sec) small enough for the level of the liquid to be the same in all chambers. The concentration C_n becomes $C_n + dC_n$ at time $t + dt$. The change in quantity of the substance in chamber 1 in time dt will be given by the difference between the amount entering from chamber 2 and the

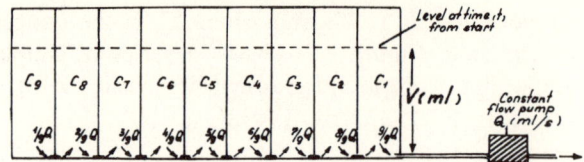

Fig. 7.52. Diagram of 'Varigrad' [117].

amount leaving chamber 1:

$$(V + dV)(C_1 + dC_1) - VC_1 = \frac{8}{9}Q\,dt\left[\frac{C_2 + (C_2 - dC_2)}{2}\right]$$
$$- \frac{9}{9}Q\,dt\left[\frac{C_1 + (C_1 - dC_1)}{2}\right] \quad (7.79)$$

where dV is the change in the volume in chamber 1 in time dt, and the expressions in square brackets represent the average concentrations in chambers 1 and 2 for that time interval.

If the terms containing second-order differentials are neglected, Eq. (7.79), on division by dt, becomes

$$V\,dC_1/dt + C_1\,dV/dt = 8QC_2/9 - 9QC_1/9 \quad (7.80)$$

As the net volume of liquid flowing from each chamber during a second is $Q/9$ ml, it follows that

$$dV/dt = -Q/9 \quad (7.81)$$

which on substitution in (7.80) gives

$$dC_1/dt = 8Q(C_2 - C_1)/9V \quad (7.82)$$

By repeating this reasoning for each of the other chambers we arrive at the following system of ten differential equations:

$$\frac{dC_1}{dt} = \frac{8}{9}\frac{Q}{V}(C_2 - C_1)$$

$$\frac{dC_2}{dt} = \frac{7}{9}\frac{Q}{V}(C_3 - C_2)$$

$$\vdots \quad (7.83)$$

$$\frac{dC_8}{dt} = \frac{1}{9}\frac{Q}{V}(C_9 - C_8)$$

$$\frac{dC_9}{dt} = 0$$

and

$$\frac{dV}{dt} = -\frac{1}{9}Q$$

Given the initial concentrations $C_1^\circ \cdots C_9^\circ$, this set of equations can be integrated with respect to time by an analogue computer and the concentration C_1 leaving the apparatus can be recorded as a function of time, this curve being the gradient resulting from the initial conditions.

To obtain a precise solution by analogue computer it is advantageous to introduce a new variable x for the integration, defined by $dx = (1/V)\,dt$ and $x = 0$ when $t = 0$. Then:

$$\frac{dC_1}{dx} = \frac{8}{9}Q(C_2 - C_1)$$

$$\frac{dC_2}{dx} = \frac{7}{9}Q(C_3 - C_2)$$

$$\frac{dC_8}{dx} = \frac{1}{9}Q(C_9 - C_8) \qquad (7.84)$$

$$\frac{dC_9}{dx} = 0$$

$$\frac{dt}{dx} = V$$

The system is now easy to integrate on a small analogue computer with fair precision. The values are chosen in the following way: the highest concentration in the initial chamber is set equal to one machine unit (MU, say 100 V) and the initial volume V° is also set to 1 MU. The time is scaled so that $t = 1.00$ when $V = 0$ (all the chambers are empty). For computation of the composition of the solutions in the chambers to give a particular gradient the absolute value of the flow-rate is irrelevant, but for ease of computation a half-life of 10 seconds for the emptying of the chambers is convenient. This requires that $Q/9 = (\ln 2)/10 = 0.06931$. The authors chose arbitrarily a value of 0.0696 for $Q/9$, which gives a very similar rate. The differential equations then become

$$\frac{dC_1}{dx} = 0.5568(C_2 - C_1) \qquad C_1^o$$

$$\frac{dC_2}{dx} = 0.4872(C_3 - C_2) \qquad C_2^o$$

$$\vdots \qquad (7.85)$$

$$\frac{dC_8}{dx} = 0.0696(C_9 - C_8) \qquad C_8^o$$

$$\frac{dC_9}{dx} = 0 \qquad C_9^o$$

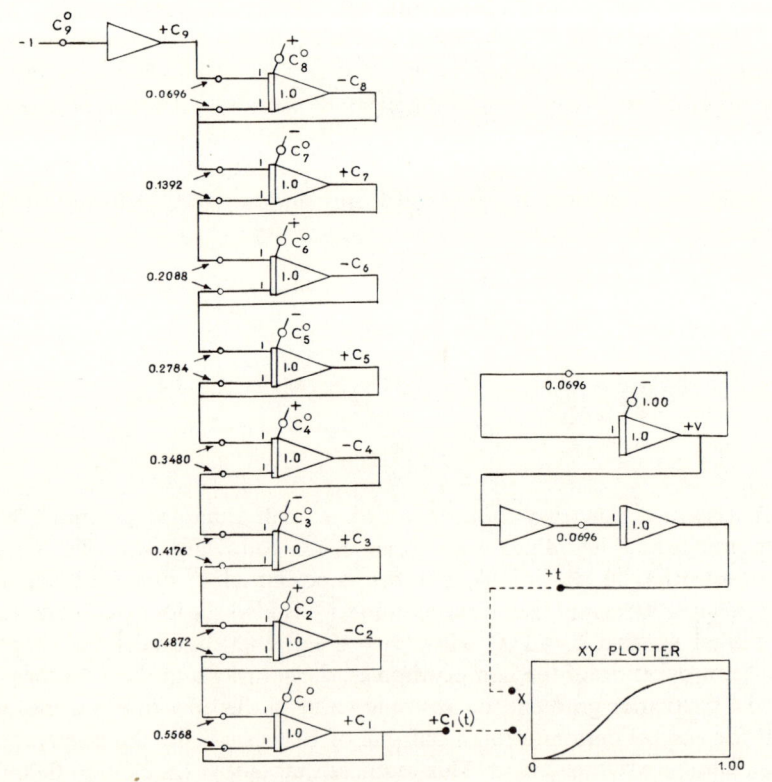

FIG. 7.53. Analogue circuit simulating the 'Varigrad' [117].

and the values for the other initial conditions are multiplied by the same factor so that none shall exceed 1 MU:

$$\frac{dV}{dx} = -0.0696V \qquad\qquad V° = 1.00$$

$$\frac{dt}{dx} = 0.0696V \qquad\qquad t = 0$$

The circuit of the analogue computer for solving this system of equations is given in Fig. 7.53.

To check the computation, a simple model was chosen with $C_5° = 1.000$ and all the other $C°$ values were set to zero. The results were in good agreement with the values given in Peterson and Sober's tables [86]. Only about a minute is needed to obtain the programme for a gradient. Table 7.2 gives the initial concentrations in the nine chambers of a 'Varigrad' for four gradients, as proposed by the Technicon Co. and by Burns et al. [117], and the gradients themselves are shown in Fig. 7.54.

Later, Hagerman [118] worked out a FORTRAN IV programme to compute some linear and composite gradients for a nine-chamber system; the programme is easily modified and adapted for a greater or smaller number of chambers by changing the calculation values.

Another group of automatic devices for obtaining composite gradients of widely varying forms is based essentially on one and the same principle,

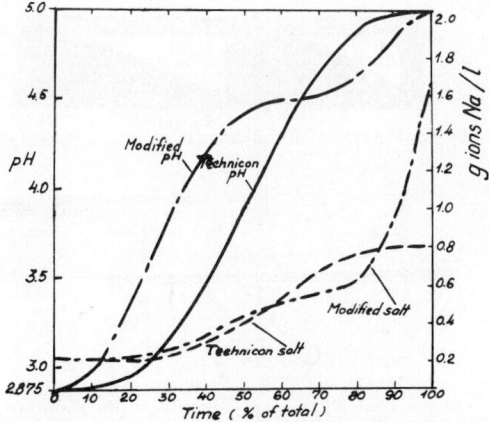

FIG. 7.54. The calculated pH and salt gradients obtained with computerised 'Varigrad' (analogue circuit in Fig. 7.53) for the initial chamber compositions shown in Table 7.2 [117].

180 Mobile-phase Gradients

Table 7.2. Initial conditions in the Varigrad chambers for four gradients [117]†

Chamber No.	Technicon, pH	Modified, pH	Technicon, [Na$^+$], M	Modified, [Na$^+$], M
1	2·875	2·875	0·20	0·20
2	2·875	2·875	0·20	0·20
3	2·875	2·875	0·20	0·20
4	2·875	5·0	0·20	0·20
5	3·890	5·0	0·24	0·23
6	4·70	4·16	0·68	1·16
7	5·0	4·16	0·50	0·218
8	5·0	5·0	0·80	0·33
9	5·0	5·0	0·80	1·67

† All solutions are 0·2M in sodium ions from the trisodium citrate and the sodium hydroxide used. Higher concentrations are obtained by adding sodium chloride.

the translation into electrical signals, by means of photocells, of a predetermined mixing programme for the solvents which produce the gradient. The resulting electric currents operate some electrochemical (electrolysis cells) or electromechanical (electrically-operated pumps) devices by means of which the programme is carried out. Curtain [119] suggested a device utilising the gas resulting from electrolysis: it is made of four parts, the programmer, a differential amplifier, the electrolysis cells, and a selector. The profile of the desired gradient is put on 35-mm film (Fig. 7.55) by the equipment shown schematically in Fig. 7.56, consisting of an electric bulb (1) giving an image which is focused on a slit (2) by means of a lens (3).

Fig. 7.55. Typical gradient functions represented photographically [119].

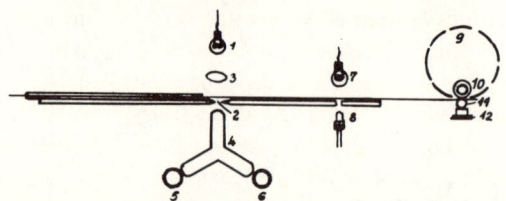

Fig. 7.56. Mechanical and optical arrangement of gradient generator using a photographed function from Fig. 7.55. 1—6 V, 4 W lamp, 2—20 × 2 mm slit, 3—lens, 4—Perspex conductor, 5, 6—photocells, 7—2·5 V, 2 W lamp, 8—photo transistor, 9—synchronous clock motor, 10—film drive roller, 11—idle roller, 12—polyurethane foam loading pad [119].

After the light-beam has passed through the slit it is divided into two beams by a 'Perspex' light-guide in the form of an inverted Y (cut from a cylindrical 3 cm Perspex bar). The film is moved across the slit by a synchronous motor (9) and a system of driving rollers (10–12). Two photocells (5, 6) are linked to the differential amplifier so that the optical signal reaching them is transduced into an electric signal. Two transistors are used for amplification. The first photocell is connected with the first transistor (T_1) so as to generate a voltage reducing the current supplied by T_1 to the first electrolysis cell, when the flux of light on the photocell increases. The second photocell and transistor T_2 are connected so that increase of the light flux on the photocell increases the electrolysis current supplied by T_2 to the second electrolytic cell. Reduction of the flux on the photocells reverses these effects. Thus the production of gas by two electrolytic cells is a function of the light and dark areas of the slit, i.e. of the position of the film in front of the slit. The gas generated in the two cells is used to displace different quantities of liquid from two reservoirs, thus giving the gradient corresponding to the profile registered on the film. The apparatus is provided with ten pairs of electrolysis cells, and by means of a selector formed by an electric bulb (7), a slit and a phototransistor (8), can switch the electrolysis currents by means of a relay from one pair of electrolysis cells to another, each pair being connected to a couple of reservoirs containing solvents of appropriate composition. The command to switch cells is recorded on the sound-track of the film or can be given independently.

The system proposed by Keck [120] is of this category, an electrical signal commanding the opening or closing of magnetic valves according to a preset programme. The concentration gradient is programmed by tracing the profile on a sheet of paper, blackening the area above the curve, and following the profile by means of a photocell.

Johnson et al. [121] built an automatic elution gradient programmer in which the programme is recorded on punched tape. The functional operation of the electronics system is shown in Fig. 7.57. The electronics, shown in block diagram form, consist of two parts. The first punches the preselected solvent ratio values on the tape, by means of a decimal-to-binary coded thumbwheel switch ('Data Input'), a transfer selector, and a storage circuit. Once punched, the tape is inserted into the tape-reader and then the 'Timing Synchroniser' is switched to the 'Read' position. The data on the tape are transferred via a 'Transfer Selector' to two memory units for controlling the storage reservoirs for petroleum ether (PE) and dichloromethane (DCM) whence these solvents are delivered upon command from the 'Timing Synchroniser' to the BCD decoders. The output of the BCD circuits determines the input voltage of the oscillator

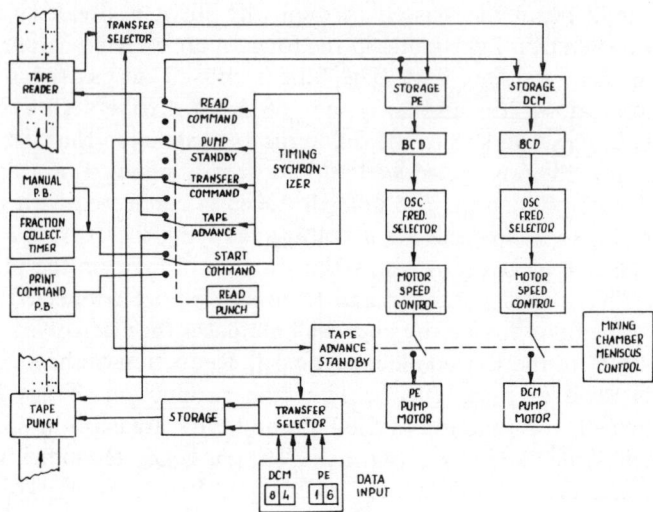

Fig. 7.57. Block diagram of electronic system for tape-controlled gradient elution chromatography [121].

frequency selectors which in turn provide the driving pulse for two separate stepping motors, and consequently determine the pumping rate of the solvents. By means of this apparatus linear or other gradients can be obtained (Fig. 7.58).

Blattner and Abelson [122] have built a gradient-generating system, the programme for which is recorded on a three-track tape. The programmes for the liquid feed from two reservoirs to the mixing chamber are punched

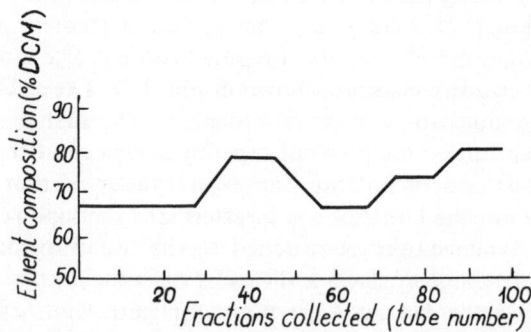

Fig. 7.58. Complex elution gradient obtained by a steroid analyser with a simple punched tape and reader system: per cent DCM–dichloromethane in petroleum ether.

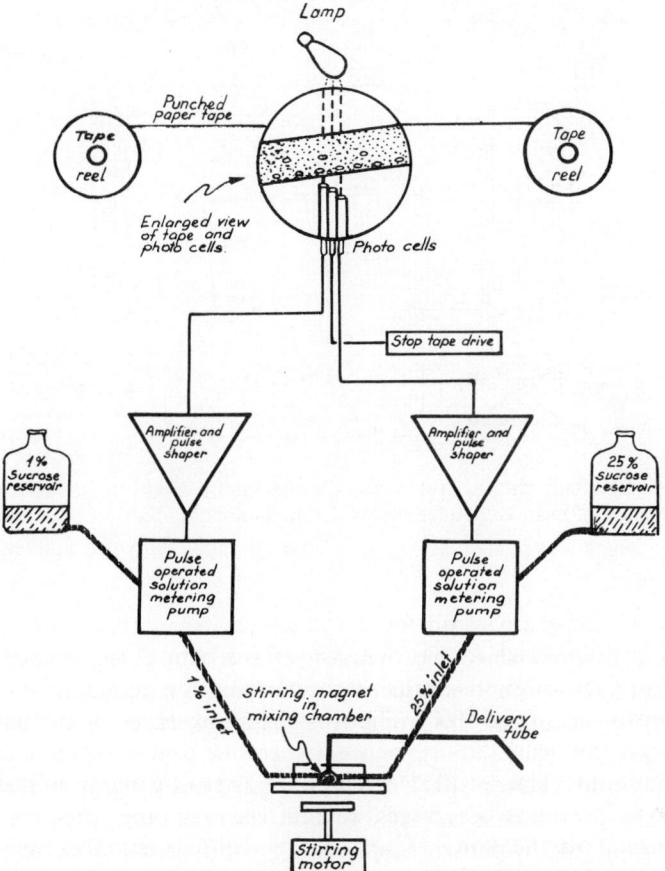

FIG. 7.59. Block diagram of gradient generating system [122].

on the two outside tracks, and the central track is used for stop signals. Three photocells are used as programme translators (Fig. 7.59) being illuminated according to the programme on the tape, and their outputs are amplified and used to operate the metering pumps that feed liquid to the mixing chamber. The tape can be punched by a digital computer using a specially established programme corresponding to the desired gradient.

Another group of devices does the translating by means of electromechanical systems via cams. Such an apparatus was built by Anderson et al. [123]. The control cam for the gradient has a triangular profile. The eluent composition is displayed on the ordinate and time on the abscissa. The arm of a linear potentiometer follows the changes in the

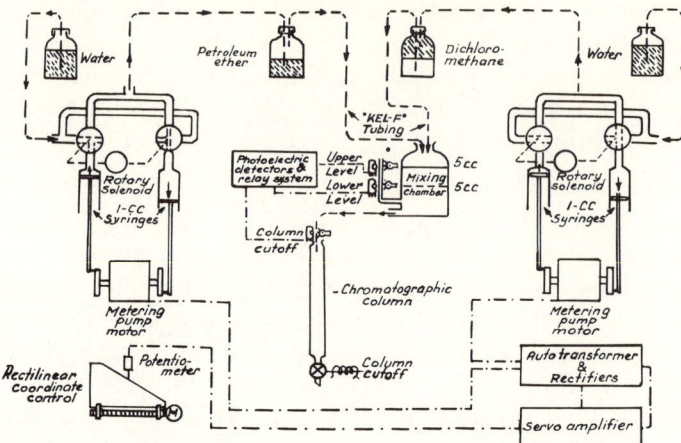

FIG. 7.60. Gradient elution system for the automatic quantitative separation and determination of individual adrenocortical hormones in mixtures (steroid analyser) [123].

position of the ordinate on the curve and converts them into a voltage directly proportional to the position of the arm. This voltage is then applied to a servo-amplifier, the motor of which is coupled to two variable autotransformers (Fig. 7.60). The a.c. output voltages of the autotransformers are connected in opposition (when one is at maximum, the other is at minimum). The rectified voltages are applied directly to the motors which operate the two syringes, so that the pumping rates are directly proportional to the voltages applied to the motors. The two eluents, petroleum ether and dichloromethane, are displaced into the mixing chamber by means of water pumped by the two motors. The apparatus is also equipped with a system of photoelectric relays to maintain the level of the liquid in the mixing chamber at about the same level as the liquid above the column. When the liquid in the mixing chamber has risen above a certain level, the photoelectric relay of the upper cell will be tripped and interrupt the pumps, which will be started only when the level in the chromatographic column falls below the level of the lower photocell. When the liquid in the column falls below a certain level, another photocell attached to the column will trip a relay causing a solenoid to close the bottom of the column.

Bailey [124] has described an apparatus for programming the eluent with a Sealelectroswitch, model No. 92-2066-203. A very low speed motor (1 revolution per day) is used to operate the camshaft. There are 60 positions for a cam under each contact, and at this rate each cam closes a contact for 24 min. The relays activated by the programmer are used for controlling the discharge valves for the liquid, the magnetic stirrers, the pump motor, the programmer itself, and the ancillary equipment. The gradient is obtained by programmed connection of reservoirs containing appropriate solvents.

7.1.2 Open Column Elution

The technique of mobile-phase gradient elution on open columns (paper and thin-layer chromatography) was developed at the same time as the closed column techniques. In fact, all the equipment for mobile-phase gradient elution described in Section 7.1.1 can be adapted and used with open columns. However, some novel chromatographic chambers for mobile-phase gradient elution have also been constructed.

We consider it best to classify these systems according to the mode of chromatography: paper or thin-layer; ascending, descending and horizontal development. The gradient may be stepwise or continuously variable (concave, convex or composite) for any of these techniques.

7.1.2.1 *Paper Chromatography*

The first research on stepwise mobile-phase gradient elution was done by Lederer [125], in the separation of a mixture of eight inorganic ions. The ascending technique was used. The starting eluent was butanol saturated with $1M$ hydrochloric acid, introduced into the chromatographic chamber (a glass jar 30 cm high) and stirred magnetically. The second solvent, concentrated hydrochloric acid, was added by means of a burette set in an aperture in the lid. After each 1 cm of migration of the solvent front, 2 ml of acid were added, up to a total migration of 11 cm, and 4 ml per cm were added thereafter up to 15 cm.

Muic and Meniga [126] described a chromatographic chamber for ascending development with a discontinuous gradient. The chamber was a rectangular glass tank (30 × 16 × 12 cm) with the bottom replaced by a thick glass plate provided with 10–20 rectangular grooves (50 × 3 × 2 mm) to act as reservoirs for eluent. The chromatographic paper was on a plastic support, and at the start was placed so that it dipped into the first eluent (saline solution) in the first groove. The other grooves contained saline solutions of different concentrations [the dilution factor in the case of ten concentrations being $f = (C_{10}/C_1)^{1/9}$], and by lateral shifts of the base plate the paper was moved from groove to groove at appropriate intervals, the eluent being changed to that of next highest concentration at each 1 cm travel of solvent front.

The first apparatus for continuous gradient horizontal elution on paper was introduced by Franks [127, 128] for the separation of fatty acids. The

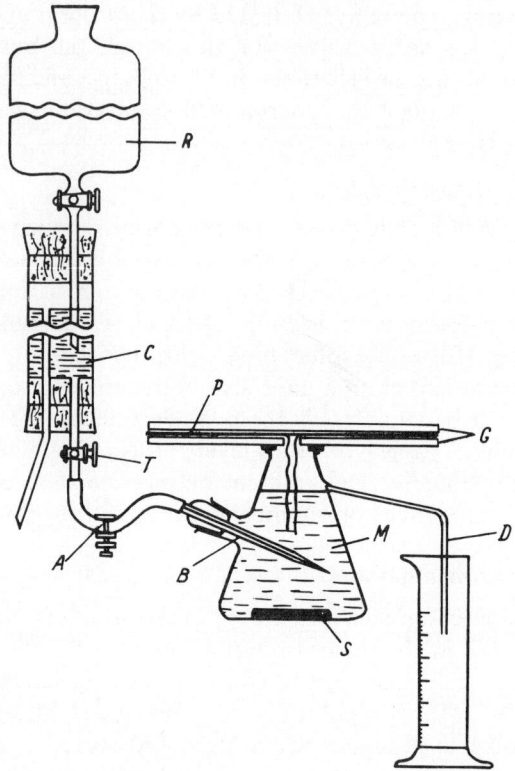

Fig. 7.61. Apparatus with continuous change in solvent composition for separation of fatty acids [127].

apparatus (Fig. 7.61) consists of an Erlenmeyer flask as mixing chamber (M), and a 500-ml separatory funnel as reservoir (R), connected by a constant-level reservoir (C) from which the solvent passes by rubber tube [equipped with a screw-clamp (A)] and a 0·5 mm capillary to a polished plug (B) fitted to the mixing chamber. The mixing chamber is also fitted with a tube (D) to keep the level constant (at 100 ml), and a magnetic stirrer (S). The chromatographic paper (P) is held between two glass plates (G) which are approximately $50 \times 20 \times 1$ cm (or $25 \times 25 \times 1$ cm for circular chromatography). The lower plate has a 12-mm diameter hole in it, and the joint between the lower plate and the mixing chamber is made watertight with 'Plastilin'. The shape and approximate dimensions of the paper strips are shown in Fig. 7.62. Two samples are applied on each strip.

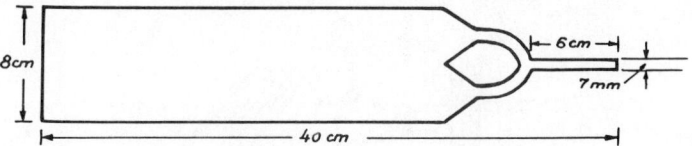

FIG. 7.62. Shape and approximate dimensions of filter-paper strips used in the separation of fatty acids with the apparatus shown in Fig. 7.61 [127].

The equation deduced by Franks [127] for the gradient produced is

$$C = C_1 \exp(-vt/V_m) \qquad (7.86)$$

and is equivalent to Eq. (7.3) for $C_2 = 0$, i.e. when the reservoir contains pure solvent, and V is the liquid volume flowing from the mixing chamber in unit time, i.e. $vt = V$.

Curtain [129] used descending chromatography with a gradient for the separation of proteins on DEAE-cellulose paper. A photoelectric scanner (1–6 in the block diagram shown in Fig. 7.63; for details see Figs. 7.55 and 7.56) is used as gradient generator, transforming the light and dark areas on the gradient profile strip into two variable voltages which are amplified (7, 8) and applied to two electrolysis vessels (9, 10). The gases from the electrolysis are led to two vessels (11, 12) where they displace buffer solutions into a micro mixing chamber (13) fitted with a magnetic stirrer (14). From here the eluent is led into a poly(vinyl chloride) canula (15) in which the upper end of the chromatographic paper is held (16).

De Wachter [130] showed that in the above-described devices an excess of solvent is used to obtain the gradient, and evidently only a small fraction is actually used in the chromatographic process, a large amount being wasted. The change in the composition of the solvent must be in

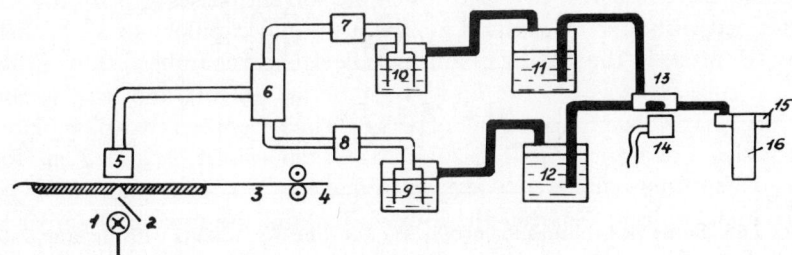

FIG. 7.63. Block diagram of photoelectrically-programmed electrolytic gradient generator [129].

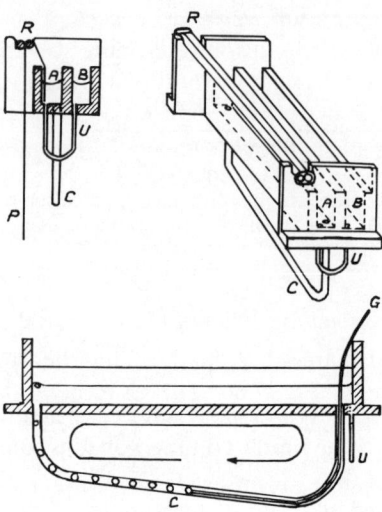

FIG. 7.64. Perspective view, transverse section and longitudinal section of the trough [130].

keeping with the flow-rate through the paper if reproducible results are to be obtained. These devices also need magnetic stirrers or pumps and thus limit the number of separations that can be performed simultaneously.

These disadvantages are avoided in the device described by de Wachter, which consists of two open grooves A and B (Fig. 7.64) linked at the bottom through a U-tube. Mixing is done in A by means of nitrogen led through tubes G and C. The chromatographic paper (P) is set between two glass rods (R) so that its upper end is dipping into the liquid in A. The

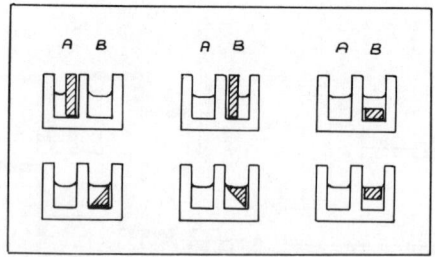

FIG. 7.65. Some possibilities for producing non-linear gradients with the apparatus from Fig. 7.64. By placing adapters of different section in mixing chamber A or reservoir B, convex (left), concave (middle), or stepwise (right) gradients are obtained [130].

two grooves form a communicating-vessel system and the gradient profile is controlled only by the migration rate of the solvent through the paper. The form of the gradient can be modified by putting into the grooves inserts with various profiles (Fig. 7.65). All considerations applying to the devices based on the principle of communicating vessels and described in Sections 7.1.1.3–7.1.1.5 apply to this apparatus.

Koman and Palo [131] have also made an apparatus for paper chromatography, allowing a continuous change of solvent concentration. This technique was used for the separation of fats from some long-chain acids.

7.1.2.2 *Thin-layer Chromatography*

Wieland and Determann [132] made a chamber for gradient ascending development, shown in Fig. 7.66. The chamber is cylindrical, 30 cm in height and 6 cm in diameter. The lower part is also the mixing chamber and is divided into two compartments by a removable filter plate (1) placed halfway between the bottom of the chamber and the eluent surface. There is a magnetic stirrer (2) in the lower compartment. The liquid in the reservoir penetrates into the lower compartment through the capillary (3). At a distance of 1 cm above the filter plate there is a side-arm (4) for maintaining constant level. The thin-layer plate is protected at the bottom by a band of filter paper (5) at a height of 1·5 cm, fixed at the top by a rubber band (6). The sample spots (7) are applied 2·5 cm from the lower edge of the layer (8). The concentration of the eluent in the mixing chamber as a function of the volume V of the eluent (equal to the sum of the volumes of eluent that has left the chamber and of eluent which has penetrated into the thin layer) is given by Eq. (7.3) (p. 138).

Rybica [133] achieved a gradient by still simpler means, using a conventional chromatographic chamber, the tank serving as mixing chamber, and a burette as reservoir from which the second component was added (diethyl ether was added drop by drop to a 10 per cent solution of diethyl

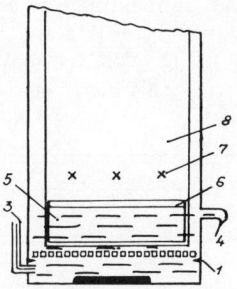

FIG. 7.66. Thin-layer chromatographic chamber for ascendent gradient elution [132].

ether in petroleum ether). The concentration in the mixing chamber is given [134] by the equation

$$C = \frac{C_1 V_{0m} + C_2 V}{V_{0m} + V} \tag{7.87}$$

where C_1 and C_2 are the initial concentrations in the mixing chamber (the chromatographic tank) and the reservoir, respectively, V_{0m} is the initial volume in the mixing chamber and V the volume passed from the reservoir into the mixing chamber. The contraction or expansion on mixing was ignored, and so was the volume of eluent imbibed by the thin layer.

Luzzatto and Okoye [135] described an apparatus for thin-layer separations with an elution gradient, using descending chromatography. The mixing chamber had a constant-level tube at the top and there was a drip-feed of the second component from a funnel. The eluent was led from the mixing chamber to the thin layer via a strip of chromatographic paper. An exponential gradient was obtained.

Stickland [136] built an apparatus for horizontal thin-layer gradient chromatography (Fig. 7.67). A polyethylene trough (1) is divided by a partition, and the two halves communicate by means of two apertures (2), carrying polyethylene tubes 3·5 mm in internal diameter. These tubes are plugged with rubber bungs, and their lengths determine the mixing rate (one tube is 4 mm long, the other 12 mm). The contents of the two halves of the trough are stirred by magnetic stirrers (3, 4). The thin-layer plate (5) is placed on two polyethylene supports (6) arranged so that the upper part of the plate is the same level as the top of the trough. The eluent in the right-hand half of the trough is led to the thin layer by means of a filter paper strip (7) held in place by two narrow glass plates (8). Excess of solvent is absorbed by a pad of filter paper (9) held down by a glass plate (10). The whole apparatus is placed in a polystyrene photographic dish (11) covered by a glass plate (12) with a hole (13) closed by a rubber bung and which allows access to the trough. The first solvent is put in the left-hand trough and the other in the right-hand one and the stirrers are started.

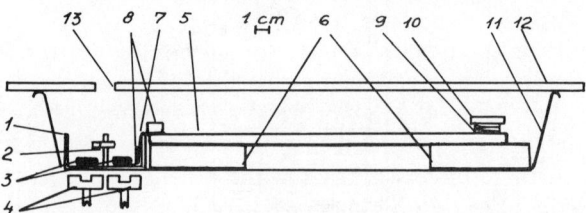

FIG. 7.67. Apparatus for gradient elution of thin-layer chromatograms [136].

The 12-mm tube is unplugged, and after 8 min of chromatography the plug in the 4-mm tube is transferred to the 12-mm tube. In this way the mixing rate is increased.

Two-dimensional chromatography can be done with this apparatus. It is advisable to put a smaller amount of solution in the left-hand chamber in order to increase the linearity of the gradient. For example, for the first direction of chromatography, 75 ml of 1·0M ammonium bicarbonate were put in the left-hand chamber and 85 ml of 1·5M ammonium bicarbonate in the other. The solutions (pH 7–8) were freshly prepared. The change in concentration of the solution in the right-hand chamber was followed by conductivity measurements, and is shown in Fig. 7.68. The

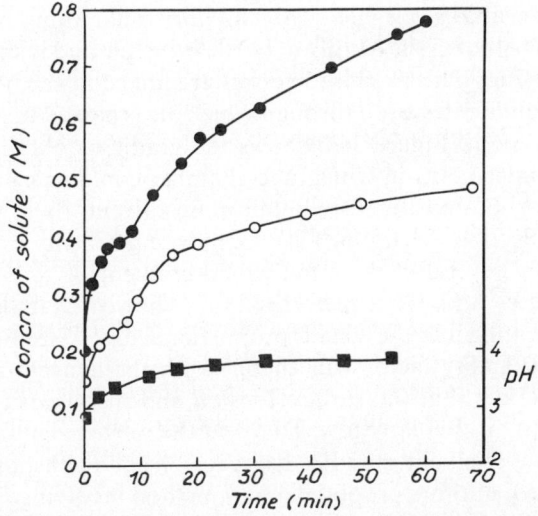

FIG. 7.68. Change of solute concentration and of pH with time in right-hand chamber of the apparatus shown in Fig. 7.67. The size of the hole connecting the two chambers was changed after 8 min. ○—change of NH_4HCO_3 concentration for first direction of chromatography, ●—change of $HCOONH_4$ concentration and ■—of pH for second direction [136].

solvent migrated 20–21 cm in 30 min, but chromatography was continued for 75 min, the surplus solvent being absorbed by the filter pad (9). For the second direction, 75 ml of a 2·0M formic acid–ammonium formate buffer (pH 4·2) were put in the left-hand chamber and 85 ml of 0·2M formic acid–ammonium formate buffer (pH 2·8) in the other. The vents were opened according to the same programme for both directions. The results are also shown in Fig. 7.68.

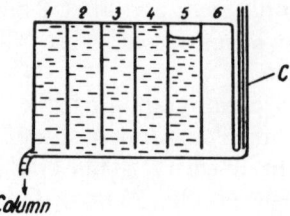

FIG. 7.69. Scheme of mixing battery with six closed chambers with constant flow-rate for obtaining a concentration gradient [137].

A device for obtaining various exponential gradients was described by Niederwieser and Honegger [137] in horizontal thin-layer chromatography. It is based on the Mariotte-bottle principle and is shown schematically in Fig. 7.69. The six closed vessels are linked at the bottom, and the last has a capillary tube (C) through which air is passed to force the liquid out of the system. Figure 7.70 shows the gradients obtained for the six possible arrangements in which one chamber contains solution of initial concentration C_r and the rest contain pure solvent. As in the case of the 'Varigrad' (Section 7.1.1.5, p. 171) composite gradients can be obtained by combining several of these individual gradients.

Niederwieser and Honegger [134, 137] showed that the quantity of eluent penetrating into the layer is proportional to $t^{1/2}$ [see Eq. (1.3), p. 138]. The proportionality factor will change with the concentration gradient owing to the fact that the surface tension and the viscosity change. The reproducibility is therefore low. To avoid the wastage entailed by use of a larger volume of eluent than necessary (e.g. from a constant-level mixing chamber) these authors proposed a new method involving introduction of a predetermined volume of eluent with the required concentration

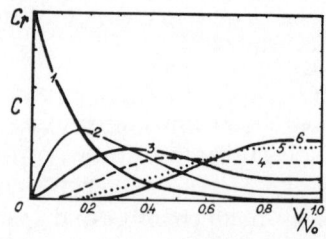

FIG. 7.70. Concentration gradients from a mixing battery with six closed chambers (Fig. 7.69). The number on the curve indicates the number of the chamber containing solution of concentration C_r, all other chambers are filled with pure solvent [137].

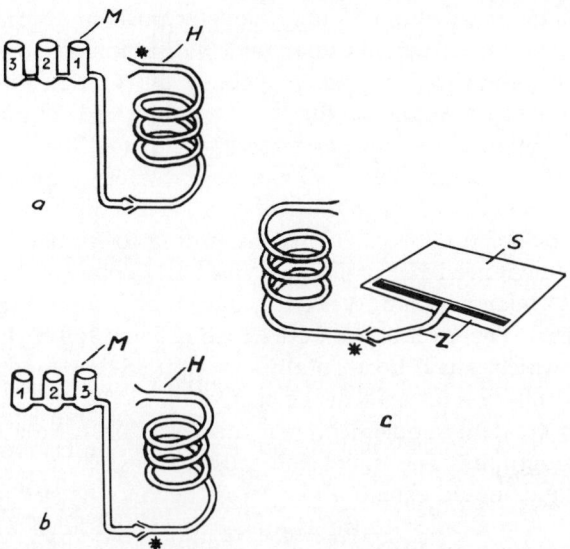

FIG. 7.71. Scheme of the arrangement for gradient elution by thin-layer chromatography. M—mixing battery, H—tube spiral reservo' (can be filled in two ways, a and b), Z—eluent distributor, S-thin layer, c-arrangement for chromatography [137].

gradient, into a capillary from which as much eluent as the layer can absorb is carried to it by means of a distributor (Fig. 7.71).

The use of this capillary as eluent vessel has the following advantages: the possibility of measuring the volume of eluent absorbed by the layer, adaptation to any type of vessel, reproducible gradients, and direct correlations between the volume absorbed by the layer and the composition of the eluent, and the time interval between production of the gradient and the actual chromatographic operation.

The mixing battery (Fig. 7.71) is composed of seven closed chambers (only three shown) fitted with ground-glass plugs. The inner dimensions of each are 7 × 40 mm. The chambers are set in a circle about 6 cm in diameter, and they are connected by means of 1-mm bore capillaries sealed in at the bottom and joined by short polyethylene tubes. The drain-tube (first chamber) is about 15 cm long and has a bore of 0·8 mm. The inlet tube (last chamber) is curved upwards. In each chamber there is a 3 × 18 mm magnetic stirrer. The volume of each chamber is 1·3 ml, and of the connecting tube 0·05 ml. The last chamber can be connected to a larger reservoir, and in that case the gradient is described by Eq. (7.78).

The capillary which forms the reservoir for the eluent may take different shapes: spiral [137] or zigzag in a plane [134]. In the latter it is made of

Teflon tubing carried round 13-mm diameter posts in a series of hairpin bends. The posts are fixed on a small table with runners like a sledge. The plane carrying the capillary can be inclined by rotation about the axis of the last horizontal of the capillary at the outlet end. The slope is fixed reproducibly by means of an eccentrically graduated disc at the back of the sledge. The capillary has 1–1·5 mm bore, corresponding to optimum flow-rate of eluent for 0·35-mm thick silica gel layers on 18 × 18 cm plates. The length of the tube is calculated according to its diameter and the quantity of eluent needed. For a Merck silica gel G plate of the dimensions given, about 4 ml of benzene are needed, and a tube 260 cm long and 1·4 mm in bore is used. The level of the outlet tube is 5 mm lower than the thin-layer plate, which is laid horizontally. About a 25-cm length of the inlet end of the capillary is fastened at the back of the sledge to a rail which can be raised to create a supplementary hydrostatic pressure at the start of the chromatography. The outlet end must be flexible for joining to the distributor and should extend about 10 cm beyond the sledge top. When the capillary is connected to the mixing battery and filled with eluent, a continuous gradient is obtained. The filling may also be done in other ways: with a single eluent conventional elution is obtained; with different solutions added in succession without separation by air-bubbles, gradient elution is obtained; with 3-mm long air-bubbles between the solutions, the gradient is discontinuous [134].

The distributor for the eluent [137] (Fig. 7.72) has the role of feeding eluent uniformly into the layer across its whole breadth. It is made of a Teflon plate (T) (180 × 10 × 1 mm) with a groove 1 mm in breadth and 170 mm in length cut in it (1). This groove is continued at both ends by a hole (2) 0·8 mm in diameter, drilled through the Teflon. Two steel capillaries (3), 0·7 mm bore, are inserted in these holes and their other ends are connected to a polyethylene T-piece (4) connected to the eluent reservoir. The distributor is placed so that the thin layer on the plate makes contact with the eluent.

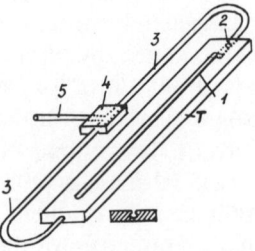

Fig. 7.72. Eluent distributor for thin-layer chromatography [137].

There are numerous commercial sources of equipment for gradient elution chromatography (see Appendix I).

7.2 Theory of Mobile-phase Gradient Chromatography

As shown in Chapter 4, the degree of separation of two components is given by the resolution [see Eqs. (4.65), (4.66), p. 78, and (4.76), p. 83] which depends on the partition coefficients of the two components. The greater the difference in the values, the better the resolution. The partition (or distribution) coefficient depends on a series of factors, among which the most important are the composition of the mobile and stationary phases, the temperature, and the pressure on the column. In this chapter we shall deal only with the influence of the mobile phase and of mobile-phase gradients, the other points being the subject of later chapters.

Generally speaking it is difficult to establish the form of the functions $\alpha = f$(composition of the mobile phase) and $\alpha = f$(concentration gradient, pH, etc.), even though the mobile-phase gradient is known (see Section 7.1). If these functions *are* known, however, the position of a peak in closed column chromatography can be predicted from the relation between the retention volume V_R (ml) and the distribution coefficient α' [138]:

$$V_R = \alpha' W + V° \tag{7.88}$$

where α' (in ml/g) is defined as the ratio of the concentrations in the stationary phase (g/g) and the mobile phase (g/ml), W is the total amount (g) of stationary phase and $V°$ (ml) is the volume of mobile phase held by the column. For open columns the position of the zone is given by the well-known relationship

$$R_f = \frac{1}{1 + \alpha V_S/V_M} \tag{7.89}$$

where α is the partition coefficient (dimensionless), defined as the ratio of the concentrations (g/ml) in the stationary and mobile phases, and V_S and V_M are the volumes of the stationary and mobile phases respectively (both in ml).

The form of the chromatographic peak is a function of the sorption isotherm. When this is linear, symmetrical peaks in the shape of the normal Gauss curve are obtained, and also a better resolution. The breadth of the peak is proportional to the equivalent height of the theoretical plate, H, which is a function of the separation conditions. Let us consider a certain component eluted with a weak eluent (i.e. one not too efficient), with an eluent of medium strength, and with a strong eluent. The corresponding elution curves are shown in Fig. 7.73, and as can be seen, the

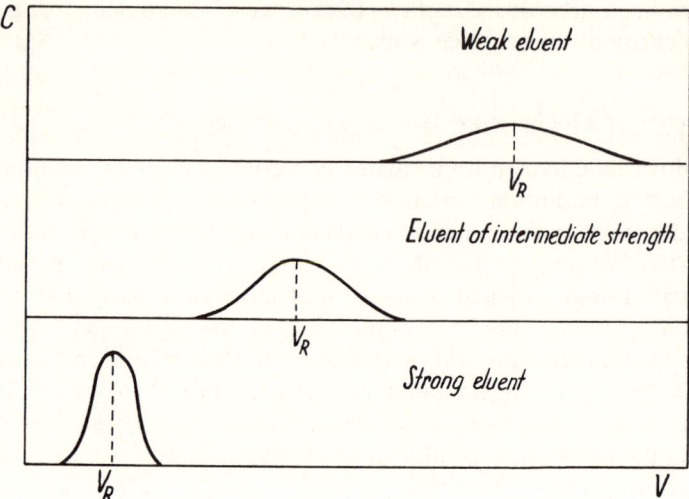

Fig. 7.73. Chromatographic elution of one component as a function of the strength of eluent.

chromatographic behaviour gradually improves from the weak eluent to the strong one.

These conclusions also apply to separations with a mobile-phase gradient (Fig. 7.74), which leads to the shortening of the analysis and hence to the optimisation of the separation.

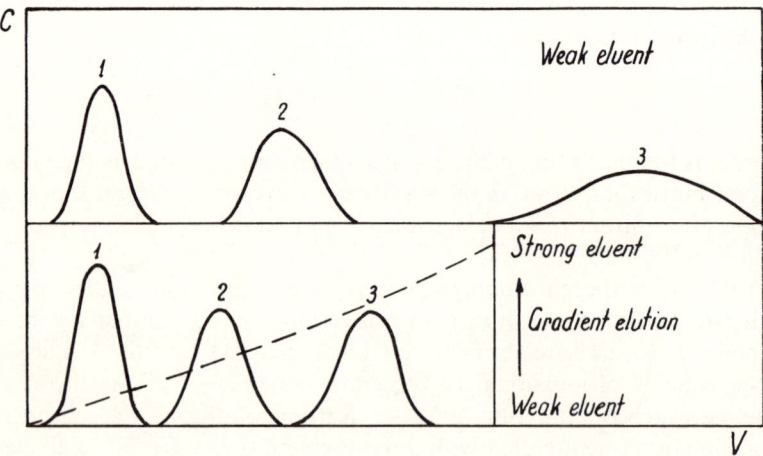

Fig. 7.74. Gradient elution of individual components in comparison with normal elution.

Sometimes a peak has a tail (Fig. 7.75), a phenomenon due to the non-linearity of the sorption isotherm. The use of a gradient with a steep enough profile will lead to asymmetry in the peaks. This situation was discussed as early as 1952 by Hagdahl, Williams and Tiselius [139]. They used the elution gradient to reduce the tails of the chromatographic zones. Nowadays enough is known about sorption isotherms for the separation conditions to be chosen so that the linear portion of the isotherm is used, and gradient elution has become a general separation technique.

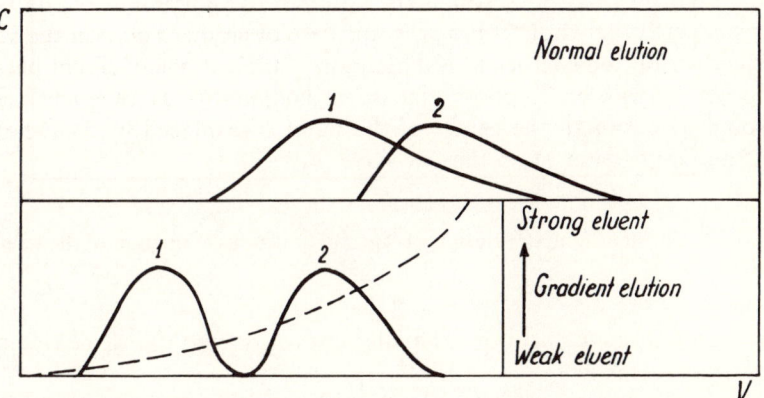

FIG. 7.75. Gradient elution of a pair of sample components in comparison with normal elution.

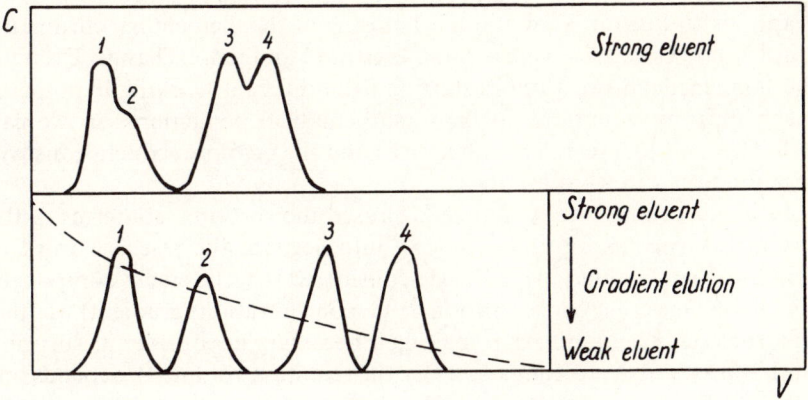

FIG. 7.76. Gradient elution of two pairs of sample components in comparison with normal elution.

Gradient chromatography offers a better resolution of a complex mixture by permitting the optimum conditions to be chosen for separating each neighbouring pair of components (Fig. 7.76).

7.2.1 Closed Column Chromatography

7.2.1.1. Calculation of the Position of the Peak

The displacement of the chromatographic zone, i.e. of the centre of maximum concentration, through a column is determined by the distribution coefficient α'. When a quantity of eluent equal to $V_R - V^\circ = \alpha' W$ [Eq. (7.88)] has passed through the column, the maximum of the band has traversed the whole column. In the case of gradient elution the value of the distribution coefficient is a function of the volume of eluent passing through the column. Suppose that when the quantity dV of eluent passes through the column the centre of the band is displaced by dx (where x has the limits 0 and 1), so that we have [4]:

$$\frac{dV}{\alpha' V W} = dx \tag{7.90}$$

or

$$\int_0^{V_R} \frac{dV}{\alpha' V W} = 1 \tag{7.91}$$

where $V'_R = V_R - V^\circ$ is the corrected retention volume. This formula is the basic equation for mobile-phase gradient chromatography, papers on which were published by Drake [15] and Freiling [140] simultaneously in 1955.

We believe that the best way of dealing with mobile-phase chromatography is to keep in view the mechanisms of the elementary chromatographic process, namely adsorption, partition and ion-exchange. Even for the same mechanism, authors starting from different working hypotheses reach different equations in their mathematical deductions, so we can treat them all together. The differences and the common conclusions will be pointed out in what follows.

In his theoretical study, Drake [15] used the sorption isotherms of the substances chromatographed, taking into account the research work of Williams and Tiselius [141, 142]. He considered that the eluent component which changes its concentration (the concentration gradient) is also adsorbed on the column, thus making it necessary to consider its sorption isotherm as well. According to Drake, the retention volume V_R depends on:

(a) the form of the mobile-phase gradient
(b) the migration of one eluent component through the column, with or without sorption (with or without developing a front)

(c) the sorption capacity for a sample as a function either of the gradient substance concentration (linear isotherm) or of the concentration of the sample component itself in the case of a non-linear isotherm.

If the substance producing the gradient (the gradient component) has a linear sorption isotherm, Drake proposes the general equation

$$L_s = \frac{L}{V°k} \int_{V_1}^{V_2} \frac{dV}{\alpha_s C_g(V) - \alpha_g} \qquad (7.92)$$

where L_s is the distance traversed on the column by the concentration maximum of the component being chromatographed, L is the length of the column, $V°$ is the volume of mobile phase in the column, k is the ratio between the stationary and the mobile phases measured in units such that $k\alpha$ is the mass distribution ratio of the component between the two phases (i.e. the ratio of the quantities in the two phases at a particular level in the column), α_s is the partition coefficient of the sample component, α_g the partition coefficient of the gradient component, and V the corrected volume corresponding to a constant value of C_g, the concentration of the gradient component in the mobile phase, which itself is a function $C_g(V)$ of the corrected volume.

The possibility of solving Eq. (7.92), as shown by Drake, depends on the complexity of the double function for the dependence of α_s and C_g on V. However, some simple arrangements give gradients that permit solution of the equation. For example, the linear gradient given by

$$C_g = C_r + (C_r - C_m)V/V_t \qquad (7.93)$$

where C_r and C_m are the initial concentrations in the reservoir and the mixing chamber respectively, and $V_t = V_r + V_m$, i.e. the sum of the volumes in the two vessels.

If the sample component is considered to be sorbed according to a Freundlich isotherm:

$$\alpha_s = p \exp(-qC_g) \qquad (7.94)$$

where p and q are functions of the concentration of sample component C_s, then Eq. (7.92) has the following solution:

$$L_s = \frac{L}{V°k}\left(\frac{V}{\alpha_g} + \frac{V_t}{q(C_r - C_m)\alpha_g} \ln \frac{\{p \exp(-q[C_r + (C_r - C_m)V/V_t])\} - \alpha_g}{[p \exp(-qC_m)] - \alpha_g}\right) \qquad (7.95)$$

In the special case in which the gradient component is not adsorbed, $\alpha_g = 0$, Eq. (7.95) becomes

$$L_s = \frac{LV_t \exp(qC_m)}{V°kpq(C_r - C_m)}(\{\exp[q(C_r - C_m)V/V_t]\} - 1) \qquad (7.96)$$

If the sample component is adsorbed according to a non-linear isotherm, the solution of the equation with partial differentials becomes more complicated. Drake suggested a graphical method which gives an approximate solution to the problem.

Snyder [143–149] has made an extremely important contribution to the theory of gradient elution in adsorption chromatography. In a series of papers [143, 148, 149] he discussed the role of interaction of the eluent and component in determining the retention volume for elution on alumina, silica gel and 'Florisil'. He also established a series of equations for prediction of the retention volume in adsorption chromatography with linear elution on these three adsorbents.

Snyder [143–145] established a relationship between the distribution coefficient and the eluent composition for linear isotherm systems in adsorption chromatography:

$$\alpha' = \alpha'_p 10^{-E\varepsilon A_s} \qquad (7.97)$$

where α' is the distribution coefficient (ml/g), α'_p is the distribution coefficient for elution with the 'standard' weak eluent pentane, E the surface energy of the adsorbent (as a function of the adsorbent activity), ε the eluting power parameter (eluent adsorption energy per unit area) and A_s the area necessary to adsorb the component on the surface. The values for E, ε and A_s for alumina, silica gel and Florisil were listed [143, 144].

In adsorption chromatography with a mobile-phase gradient, the eluent mixtures used have a composition which varies according to the gradient. To relate the gradient to the eluent composition and to the change in eluting power, Snyder [145] gave an equation expressing the power of a binary eluent, $\varepsilon_{1,2}$ as a function of the mole fractions X_1, X_2 and the eluting powers of the two solvents ($\varepsilon_1 < \varepsilon_2$):

$$\varepsilon_{1,2} = \varepsilon_1 + \frac{\log[X_2 10^{En_2(\varepsilon_2 - \varepsilon_1)} + 1 - X_2]}{En_2} \qquad (7.98)$$

where n_2 is the relative area covered by the eluent, which for strong eluents is usually A_s [143]. An equation for a ternary eluent was also deduced [146] (X_1, X_2, X_3; $\varepsilon_1 < \varepsilon_2 < \varepsilon_3$; $A_s = n_3$):

$$\varepsilon_{1-3} = \varepsilon_2 + \frac{\log[X_3 10^{En_3(\varepsilon_3 - \varepsilon_2)} + X_2]}{En_3} \qquad (7.99)$$

These equations were tested for a series of eluents on alumina [143, 145, 146] and silica gel [148]. As seen in Section 7.1, the mole fraction X or the concentration C of one component of the eluent is sometimes a very complicated function of the volume, and so $E\varepsilon_{1,2}$ may also take a complicated form as a function of eluent volume.

To find the retention of the component as a function of the mobile-phase gradient, we substitute the eluting power function (7.98) into the equation for the distribution coefficient (7.97) which in turn is substituted into the basic Eq. (7.91) which thus becomes:

$$\int_0^{V'_R} \frac{10^{\left\{A_s\varepsilon_1 E + \frac{\log[X_2(V)10^{En_2(\varepsilon_2-\varepsilon_1)} + 1 - X_2(V)]}{n_2}\right\}}}{V'_p} dV = 1 \qquad (7.100)$$

where $V'_p = W\alpha'_p$ is the corrected retention volume for pentane and $X_2(V)$ is the mole fraction as a function of volume. As can be seen, this equation is extremely complicated and does not permit an explicit algebraic solution.

Snyder [146] has shown that in the case of mobile-phase gradient adsorption chromatography, the so-called linear strength gradient has the form:

$$E\varepsilon = a + bV \qquad (7.101)$$

where a and b are constants for a particular gradient. He has also shown that even in cases in which $\varepsilon(V)$ corresponds to a curve, this can be approximated by a series of linear segments.

With these approximations Eq. (7.91) takes the form:

$$\int_0^{V'_R} \frac{10^{A_s(a+bV)}}{V'_p} dV = 1 \qquad (7.102)$$

the solution of which will be:

$$V'_R = \frac{\log(2.3 A_s b V'_p 10^{-EA_s} + 1)}{bA_s} \qquad (7.103)$$

This equation, expressing the corrected retention volume of the component in adsorption chromatography with a linear strength gradient was verified [104] for a series of components and different eluents on alumina and silica gel. Results are given in Table 7.3.

Snyder and Saunders [150] expressed the basic equation of mobile-phase gradient elution chromatography as:

$$\int_0^{V_R} \frac{dV}{V°\alpha^*} = 1 \qquad (7.104)$$

where α^* is the distribution coefficient (the ratio of the total quantities of component in the stationary and mobile phases). When α^* as a function

TABLE 7.3. Experimental test of Eq. (7.103) for binary systems [104]

Solute	V'_R (ml) in system							
	I		II		III		IV	
	Exptl.	Calc.	Exptl.	Calc.	Exptl.	Calc.	Exptl.	Calc.
Naphthalene	12	8						
Acenaphthalene	19	16						
Phenanthrene	24	27					8	7
Anthracene	27	27						
Fluoranthrene	35	40						
Triphenylene	51	53						
Chrysene	48	53						
Benzanthracene	51	53	3	4				
Perylene	63	65						
3,4-Benzpyrene	63	65	6	7				
Benzperylene	76	76						
Picene			15	13			16	22
1,2,4,5-Dibenzypyrene	96	84						
Coronene	86	84						
1,2-Benzcoronene			42	43	41	41		
1,2-Diphenylethane							8	10
Phenylethylsulphide							9	9
Nitrobenzene							30	37
Methylbenzoate							52	53
2-Methoxynaphthalene							22	25
1-Nitronaphthalene			7	5				
1-Cyanonaphthalene			8	7				
1-Acetonaphthalene			23	20	18	18		
2-Acetonaphthalene			20	22				
p-Diethoxybenzene			6	4			65	67
o-Nitroanisole					18	16		
p-Nitroanisole							71	71
m-Dinitrobenzene							82	84
o-Nitroaniline			75	69				
3,4-Benzacridine			8	8	7	8		
7,8-Benzquinoline					9	8		
Quinoline			30	29				
β-Naphthoquinoline			36	38	34	35		
Phenanthridine			38	37	37	35		
1-Azapyrene			40	40				
6-Nitroquinoline			53	58	51	53		
Isoquinoline			67	60	61	56		
Carbazole							50	55

TABLE 7.3. (*continued*)

System	Adsorbent	Eluent 1	Eluent 2	a	b	V^0 (ml)	W (g)
I	3.7% H_2O–Al_2O_3	Iso-octane	Ethyl ether	0.03	0.00190	133	10.2
II	4.0% H_2O–Al_2O_3	Carbon tetrachloride	Benzene	0.113	0.00070	109	2.0
III	4.0% H_2O–Al_2O_3	Carbon tetrachloride	Benzene	0.113	0.00078	95	2.0
IV	17% H_2O–SiO_2	Pentane	Methylene chloride	0.000	0.00125	124	1.7

of the volume V of the eluent creating the gradient is given by

$$\log \alpha^* = a - bV/V° \tag{7.105}$$

Eq. (7.104) can easily be integrated. By substituting Eq. (7.105) in (7.104) we obtain by integration the corrected retention volume:

$$V'_R = \frac{[\log(2\cdot3b10^a + 1)]V°}{b} \tag{7.106}$$

and the retention volume V_R will be

$$V_R = \frac{[\log(2\cdot3b10^a + 1)]V°}{b} + V° \tag{7.107}$$

The retention volume may also be expressed in column-volume units, $\bar{V}_R = V_R/V°$, and if $10^a = \alpha_1^*$, Eq. (7.107) becomes

$$\bar{V}_R = 1 + \frac{\log(2\cdot3b\alpha_1^* + 1)}{b} \tag{7.108}$$

$\bar{V}_R$ is shown in Fig. 7.77 as a function of $\log \alpha_1^*$ and b [150]. These remarks apply to all substances except those very feebly adsorbed.

The position of the chromatographic peak in mobile-phase gradient adsorption chromatography for certain special cases has been discussed in other papers [80, 151, 152].

The theory of mobile-phase gradient partition chromatography has not been dealt with by research workers, though some experimental material exists. Mader [31] showed that for the apparatus shown in Fig.

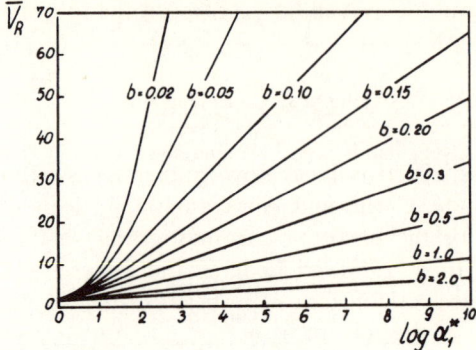

FIG. 7.77. Calculated retention volumes V_R for continuous solvent logarithmic programme as a function of $\log \alpha_1^*$ and b [150].

7.9 (p. 140) the eluent composition changes according to a relationship of the form $V_R = V_m \ln [X_r/(X_r - X_m)]$ where V_R is the retention volume (ml), X_r the mole fraction of butanol in the mixture with chloroform in the mixing chamber, and V_m the volume of the mixing chamber. By plotting $\log (X_r - X_m)$ vs. V_R for five acids, it was found that for each can be written an equation of the form

$$\log (X_r - X_m) = mV_R + d \qquad (7.109)$$

The values of m and d were obtained by the method of least squares and are given in Table 7.4.

TABLE 7.4. Values of the slope and intercept [31]

Acid	m	d
Aconitic	-0.00515	0.268
Oxalic	-0.00403	0.225
Malic	-0.00308	0.079
Citric	-0.00294	0.162
Tartaric	-0.00270	0.153

These equations can be used to identify the acids. It was also shown that the peaks are higher when the columns are shorter and the volumes of eluent smaller, but the best separation is obtained on a long column with a large volume of eluent.

Rony [153], dealing in a general way with the problem of gradient chromatography, started from the equation of mass conservation for an ideal chromatographic system, and obtained the following equation for the retention length:

$$L_R = \int_0^t v_{\text{eff}}(t)\, dt \qquad (7.110)$$

where $v_{\text{eff}}(t)$ is the effective linear flow-rate (cm/s) as a function of time t.

In a recent study, Carles and Abravanel [154] dealt with the mobile-phase gradient for the partition chromatography of organic acids. On the basis of Martin's [155] and Lederer and Lederer's [156] work, they showed that there is a constant ratio between the displacement rates of the zone (v_z) and of the solvent front (v_f):

$$v_z = v_f/(1 + \alpha^*) \qquad (7.111)$$

where α^* is the distribution coefficient.

The distribution of the zone, dL, after a time interval dt, will be

$$dL = v_z\, dt \qquad (7.112)$$

and the acid will be eluted when it has traversed the whole length of the column:

$$\int_0^L dL = \int_0^{t_R} v_z\, dt \qquad (7.113)$$

where t_R is the retention time for the acid.

This equation is analogous to (7.110) obtained by Rony. Taking into account that the rate of displacement of the zone is a function of the nature of the acid and the composition of the eluent, all other conditions being constant, each acid will obviously need a different elution time and have a different retention time. Carles and Abravanel used a chloroform–butanol mixture as eluent and a stationary phase composed of Celite 535 ($L = 12$ cm, cross-section = 0·5 cm^2) loaded with 1·5 ml of 0·5N sulphuric acid, and established a relationship (Fig. 7.78) between t_R and the eluent composition, for 19 organic acids. The relationship took the form

$$t_R = t_0 \frac{(a-b)}{(x-b)} \exp\left|\frac{a-x}{a-b}\right| \qquad (7.114)$$

where b is the limiting butanol concentration at which the acid is never eluted by partition chromatography ($x \to b$, $t_R \to \infty$), a is the butanol concentration at which the acid migrates with the solvent front ($a = x$, $t_R = t_0$), which is identical for all the acids ($t_0 = L/v_f$). Equation (7.114) describes the experimental curves well. The parameters a and b were determined for each acid from the curves. These relationships do not

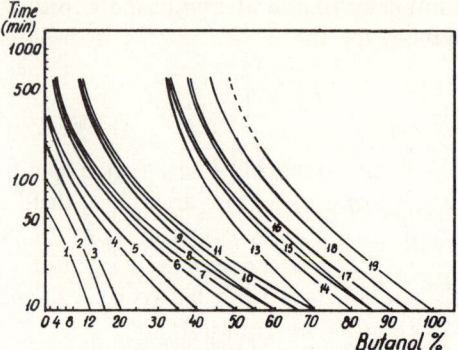

FIG. 7.78. Curves of the variation of the elution time of organic acids as a function of the solvent concentration in butanol [154].

completely solve the problem of gradient elution, but are useful as a guide for quick choice of the best gradient.

Marshall *et al.* [157] showed that the ratio S (ml/g) between the total quantity of the most polar eluent component and the mass of the stationary phase, needed for gradient elution of a component in a given system, is constant, being given by

$$S = \frac{V_R}{W} A \exp(-b) \tag{7.115}$$

where V_R is the retention volume, and A and b are constants in the equation for the change in density of the solvent ($\rho = Ae^{-yb}$, where y is the number of fractions collected) and W is the mass of the stationary phase. This equation has been used to calculate the position of the peaks for chromatography of some organic acids (fumaric, lactic, succinic, α-ketoglutaric, aconitic) in various eluent systems.

Another criterion for comparing the efficiency of two chromatographic techniques (e.g. normal and gradient elution) is the peak capacity, i.e. the maximum number of peaks resolved for a certain elution volume. The smaller the breadth of the peak-base and the distance between the peaks, the greater the number of peaks resolved (and the higher the peak capacity) by an elution volume $V_R^{(z)} - V_R^{(a)}$ (i.e. the difference between the retention volumes of the first and last peaks). Giddings [158] shows that for conventional elution chromatography the peak capacity is given approximately by

$$n = 1 + \frac{\sqrt{N}}{4} \ln [V_R^{(z)}/V_R^{(a)}] \tag{7.116}$$

where N is the number of plates. Horváth and Lipsky [159] deduced an analogous relationship for the peak capacity in mobile-phase partition chromatography:

$$n' = \frac{\sqrt{N}}{4}\left[\frac{V_R^{(z)}}{V_R^{(a)}} - 1\right] \qquad (7.117)$$

where N is the number of plates corresponding to the first peak, assuming it was eluted with an eluent of constant composition. Further it was shown that the ratio between the peak capacities in gradient and normal elution is given by

$$f = \frac{n'}{n} = \frac{[V_R^{(z)}/V_R^{(a)}] - 1}{\ln[V_R^{(z)}/V_R^{(a)}]} \qquad (7.118)$$

i.e. the number of plates necessary for a certain peak capacity decreases by a factor of f^2. For instance, for $V_R^{(z)}/V_R^{(a)} = 10$, 20 and 50, the number of plates is decreased by factors of 15, 40 and 150 respectively. The analysis time will be reduced in the same proportion, as it is directly proportional to the number of plates. It is worth mentioning that in the case of gel-chromatography, the ratio $V_R^{(z)}/V_R^{(a)}$ is limited to the value 2.3 ± 0.3. The advantage of using a mobile-phase gradient is shown in Fig. 7.79.

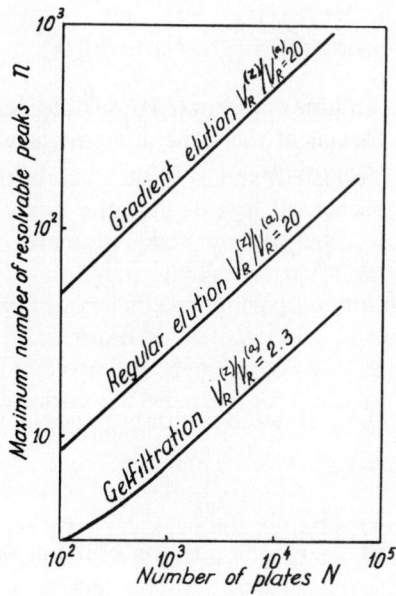

FIG. 7.79. Peak capacity as a function of number of plates in gel filtration as well as in elution chromatography with regular or gradient elution [159].

The first work dealing with the position and form of the peak in ion-exchange gradient elution chromatography was by Freiling [140, 160], who used the Mayer and Tompkins theory [161] of the stepwise-equilibrated bed with quasi-continuous flow.

In the case of the non-adsorbed component of the eluent which creates the gradient, Drake's equation for the peak position [15] becomes identical with Freiling's equation [140]. In his first paper, Freiling considered that the number of theoretical plates, N, remains constant during gradient elution. In that case the conclusions are obviously limited. This inconvenience is avoided in the second paper [160] where N is considered to be a function of the distribution coefficient, $N = N(K_d)$, where K_d is the (non-dimensional) distribution coefficient at a certain plate of the column.

Freiling [160] considered a column of length L containing a unit of adsorbed component at the upper end on an infinitesimally thin layer. Let $(\Delta V)_1$ be the increment in the volume of eluent introduced into the column, followed by a solution which does not affect the distribution of the solute. This volume increment is expressed in free column-volumes. Let N_1 be the number of plates and K_d the distribution coefficient with an eluent of given strength. If the adsorption isotherm is linear, the zone will be displaced by a distance given by

$$l_1 = \frac{L(\Delta V)_1}{K_{d_1}} \tag{7.119}$$

On introduction of a volume of eluent $(\Delta V)_2$ of different strength, for which the distribution coefficient of the same component is K_{d_2}, the total distance l_2 covered by the zone is

$$l_2 = L\left[\frac{(\Delta V)_1}{K_{d_1}} + \frac{(\Delta V)_2}{K_{d_2}}\right] \tag{7.120}$$

Generalizing, we obtain

$$l_n = L \sum_{i=1}^{n} \frac{(\Delta V)_i}{K_{d_i}} \tag{7.121}$$

If the volumes are infinitesimal, the sum becomes an integral and

$$l = L \int_0^V dV/K_d \tag{7.122}$$

If we put $L = 1$ and $V = \bar{V}_R$, the position of the peak will be given by

$$\int_0^{\bar{V}_R} dV/K_d = 1 \tag{7.123}$$

Freiling has verified this theory by graphically integrating Eq. (7.123), taking as model the elution of Na$^+$ and Cs$^+$ with a hydrochloric acid pH gradient on a Dowex-50 resin column. The position of the peak was correctly predicted within 3 per cent.

Maslova et al. [163] have studied the elution of Pm and Ce ions with a saline gradient, finding a good agreement between Eq. (7.123) and the experimental results.

Schwab et al. [46] have established a series of equations for prediction of the position of the peak in linear gradient chromatography, or exponential concentration gradient chromatography, at constant pH. The equations can also be applied to elution with variable pH or variable concentration of a complexing agent. To deduce the equations for ion-exchange and gradient elution chromatography, they started from the relationships

$$V_R = K_d V° + V° \tag{7.124}$$

$$K_d = WC_e K/V°C^z \tag{7.125}$$

where W is the weight of resin in the column, C_e the exchange capacity (in meq per g of resin), z is the charge on the ion concerned (z is always positive), and K is the equilibrium constant for the ion-exchange reaction BR + A $\rightleftharpoons$ AR + B (R = resin). From Eqs. (7.124) and (7.125) we obtain

$$V_R - V° = WC_e K/C^z = V_R' \tag{7.126}$$

In gradient elution, an infinitesimal volume of eluent, dV, will displace the position of maximum concentration of the ion by a certain distance corresponding to dW g of resin. We can therefore write

$$dV = C_e K \, dW/C^z \tag{7.127}$$

but $C = f(V)$, i.e. the concentration of one component of the eluent is a function of the volume V leaving the mixing chamber, and to integrate (7.127) we must replace C by the explicit expression for this function.

Consider the elution of a univalent ion B ($z = 1$) with a linear concentration gradient at constant pH, given by Eq. (7.66), which for the initial concentration in the mixing flask $C_1 = 0$, becomes

$$C = C_2 V/V_t \tag{7.128}$$

where $V_t = V_{0_r} + V_{0_m}$.

After substitution of (7.128) in (7.127) we have

$$dV = C_e K V_t \, dW/C_2 V \tag{7.129}$$

which on integration between the limits 0 and V_R and 0 and W gives

$$V'_R = \sqrt{2C_e K V_t W / C_2} \tag{7.130}$$

For a bivalent ion ($z = 2$) we obtain

$$V'_R = \sqrt[3]{3C_e K V_t^2 W / C_2^2} \tag{7.131}$$

and for a tervalent ion ($z = 3$)

$$V'_R = \sqrt[4]{4C_e K V_t^3 W / C_2^3} \tag{7.132}$$

In all three cases the ion A in the eluent is taken as univalent (e.g. sodium).

Now consider an exponential gradient [Eq. (7.3)] with $C_1 = 0$:

$$C = C_2[1 - \exp(-V/V_m)] \tag{7.133}$$

After substitution in (7.127) for the case of elution of a univalent ion, this gives, on integration,

$$V'_R + V_m[\exp(-V'_R/V_m) - 1] = W C_e K / C_2 \tag{7.134}$$

Though this equation is not explicit regarding V'_R, it can easily be solved graphically.

For bivalent and tervalent ions, the following equations are obtained:

$$V'_R - V_m[1 - \exp(-V'_R/V_m)]\{2 - [1 + \exp(-V'_R/V_m)]/2\}$$
$$= W C_e K / C_2^2 \tag{7.135}$$

and

$$V'_R - 3V_m[1 - \exp(-V'_R/V_m)] + \tfrac{3}{2}V_m[1 - \exp(-2V'_R/V_m)]$$
$$\frac{V_m}{3}[1 - \exp(-3V'_R/V_m) = W C_e K / C_2^3 \tag{7.136}$$

A pH gradient can be achieved, the concentration of the major constituent (e.g. sodium nitrate) remaining constant, by means of the device [46] shown in Fig. 7.9 (p. 140). In the mixing chamber C_1 is the molar concentration of each of the two buffer components and C is the molar concentration of the major component of the eluent. The reservoir contains the acid component of the buffer and the major component of the eluent, both at molar concentration C_2. As Piez [81] has shown, there are also other arrangements for varying the pH.

Obviously, according to Eq. (7.3) the concentrations of the weak base B and its conjugate acid A in the mixing chamber, after volume V of eluent has passed into the column, are given by

$$C^A = C_2^A - (C_2^A - C_1^A)\exp(-V/V_m) \qquad (7.137)$$

and

$$C^B = C_1^B \exp(-V/V_m) \qquad (7.138)$$

Accordingly, the pH gradient will be given by the concentration of hydrogen ions in the mixing chamber:

$$[H^+] = K_a C^A/C^B = K_a[C_2^A \exp(-V/V_m) - C_2^A + C_1^A]/C_1^B \qquad (7.139)$$

where K_a is the dissociation constant of the weak acid. Such a gradient may be used for the elution of a monobasic acid (dissociation constant K_1) from an ion-exchange resin. The distribution constant for the univalent ion is given by

$$K = \frac{WC_e K}{V^\circ C} \frac{K_1}{(K_1 + [H^+])} \qquad (7.140)$$

By separating the terms and considering the passage of an infinitesimal volume through the column we can write

$$\frac{(K_1 + [H^+])}{K_1} dV = \frac{C_e K}{C} dW \qquad (7.141)$$

After substituting for $[H^+]$ from (7.139) in (7.141) and integrating between the limits 0 and V'_R and 0 and W, we obtain

$$\left[1 - \frac{K_a}{K_1 C_1^B}(C_2^A - C_1^A)\right] V'_R + \frac{V_m K_a C_2^A}{K_1 C_1^B} \exp(V'_R/V_m)$$

$$= \frac{WC_e K}{C} + \frac{V_m K_a C_2^A}{K_1 C_1^B} \qquad (7.142)$$

Schwab et al. [46] checked this equation for the position of the peaks of chloride, bromide and oxalate, obtaining good agreement between the experimental and calculated values.

Inczédy [164] gave a method of calculating the elution volume in the case of ion-exchange chromatography with a linear concentration gradient. He examined the separation of bivalent ions on a cation- or anion-exchange column, using as eluent a solution with increasing concentration of univalent ions. He had earlier established [165] an approximate relationship between the distribution coefficient of the bivalent ion B and the concentration of the ion A in the eluent:

$$K_d^B = K_{BA} C_e'^2 (C^A)^{-2} \qquad (7.143)$$

where K_{BA} is the equilibrium constant for the exchange reaction (BR + A $\rightleftharpoons$ AR + B), C'_e is the exchange capacity of the exchanger (meq/ml), and C^A the concentration in the eluent of the ion that generates the gradient. For a linear gradient

$$C^A = bV \tag{7.144}$$

in which V is the volume (ml) of the eluent solution. After substitution of (7.144), (7.143) may be written

$$\frac{1}{K_d^B} = \frac{b^2}{K_{BA} C_e^{\prime 2}} V^2 \tag{7.145}$$

By putting this value in the basic equation for gradient chromatography, written in the form

$$\int_0^{V'_R} dV/K_d = \int_0^{V_c} dx \tag{7.146}$$

we obtain

$$\frac{b^2}{K_{BA} C_e^{\prime 2}} \int_0^{V'_R} V^2 \, dV = \int_0^{V_c} dx \tag{7.147}$$

or, after integration,

$$V'_R = \left[\frac{3 K_{BA} C_e^{\prime 2} V_c}{b^2} \right]^{1/3} \tag{7.148}$$

since $V_R = V'_R + V° = V'_R + F_I V_c$ (where F_I is the free fraction and V_c the volume of the column). The retention volume of ion B will decrease as the concentration gradient b increases. If K_{BA} and the data for the column are known, the position of the peak for B can be calculated. It was also shown that equations of this type can be inferred for univalent and tervalent ions etc.

The separation of ions of the same charge by varying only the eluent concentration is rarely achieved, as the values of the exchange constants do not differ sufficiently and the increase or decrease of the gradient does not very much influence the quality of separation. In this case the use of complexing agents is recommended for improving the separation.

Inczédy [164] also gave a method of calculating the position of the peak in the case of separation of bivalent metal ions on a cation-exchanger with an eluent containing a complexing agent at constant concentration but having a pH gradient. If a weak acid (dibasic) is used as the complexing agent, the ligand concentration and hence the complex formation can be influenced by variation in the pH. The distribution coefficient of the metal ion to be separated (M) is calculated from

$$K_d^M = K_{MA} C_e'^2 F_M (C^A)^{-2} \tag{7.149}$$

where K_{MA} is the ion-exchange equilibrium constant for $MR_2 + 2A \rightleftharpoons 2AR + M$, F_M is the mole fraction of free M in the solution, and C^A the concentration of A in the eluent. F_M can be calculated from the dissociation constants of the complexing agent. By plotting the logarithm of the distribution coefficient vs. pH, curves are obtained which approximate to straight lines:

$$\log K_d^M = b + n\text{pH} \tag{7.150}$$

The constants b and n are easily determined graphically.

If the initial pH is 2 and the gradient is linear, i.e. $\text{pH} = b + aV$, then taking into account that the slope n in (7.150) is negative, we obtain by substitution

$$1/K_d^M = 10^{anV} 10^{-(b-2a)} \tag{7.151}$$

Substituting this in (7.147) and integrating, we obtain

$$V_R' = \frac{\log(2.3 \times 10^b \times 10^{-2n} anV_c + 1)}{an} \tag{7.152}$$

This equation allows us to predict the position of the peak in ion-exchange chromatography with a pH gradient and may be used in conjunction with the resolution equation to find the optimum pH gradient.

Molnár et al. [68] showed that in the separation of rare earths on Amberlite IRA-400 (NO_3^- form), with a solution of ammonium nitrate in 65 per cent methanol in water as eluent, the distribution coefficient depends linearly on the nitrate concentration in the eluent:

$$K_d = a[NO_3^-] \tag{7.153}$$

where a is a constant characteristic for each ion and can be determined experimentally. These authors used a simple device to produce exponential gradients (Fig. 7.9, p. 140) and chose the experimental conditions so as to have the nitrate concentration vary according to the equation

$$C = C_1 \exp(-V/V_m) \tag{7.154}$$

In that case we may write

$$K_d = aC_1 \exp(-V/V_m) \tag{7.155}$$

and the variation of the distribution coefficient with the concentration gradient is relatively simple, which allows us to integrate the equation:

$$\int_0^{V'_R} dV/K_d = m \qquad (7.156)$$

where m is the mass of the resin in the column. On substituting for K_d in (7.156) from (7.155) and integrating, we obtain

$$V'_R = V_m \ln\left(\frac{aC_1 m}{V_m} + 1\right) \qquad (7.157)$$

In every case the calculated values are smaller than the experimental ones but they agree to within 5 per cent.

Another method was suggested by Koguchi et al. [67] for specifying the position of the elution peaks of some phosphorus oxyacids in elution gradient chromatography on an anion-exchanger. A solution of ammonium acetate was used as eluent, its concentration being given by an exponential equation of the form

$$C = C_r - (C_r - C_m)\exp -\left(\frac{V - V_a}{V_m}\right) \qquad (7.158)$$

where V_a is the volume of solution (ml) in the connecting tube between the mixing chamber and the column.

The basic equation in gradient chromatography may be written in the form

$$L = \int_0^{V_a} dV/K_{0_d} + \int_{V_a}^{V} dV/K_d \qquad (7.159)$$

where L is the distance traversed by the peak (in terms of ml of resin bed), K_{0_d} the distribution coefficient of the ion in the sample at the initial eluent concentration C_m in the mixing vessel, and K_d the distribution coefficient for this ion at an eluent concentration C. When neither complexation nor polymerization of the ions in the sample takes place, the following relation is valid for the whole range of variation of eluent concentration:

$$K_d = aC^{-n} = aC_r^{-n}\left\{1 - \frac{(C_r - C_m)}{C_r}\exp[-(V - V_a)/V_m]\right\}^{-n} \qquad (7.160)$$

where a and n are constants. After substitution of this value for K_d, Eq. (7.159) becomes:

$$L = \frac{V_a}{K_{0_d}} + \frac{C_r^n}{a}\int_{V_a}^{V}\left\{1 - c\exp[-(V - V_a)/V_m]\right\}^n dV \qquad (7.161)$$

where $c = (C_r - C_m)/C_r$.

When L corresponds to the value of the resin bed volume V_b, the concentration of the variable component of the eluent is at its maximum. If the corresponding effluent volume V_R is substituted for V in Eq. (7.161), we have

$$[V_b - (V_a/K_{0d})](a/C_r^n) = \int_{V_a}^{V_R} \left\{1 - c\exp[-(V - V_a)/V_m]\right\}^n dV \quad (7.162)$$

If n is a whole number, Eq. (7.162) can be developed thus:

$$[V_b - (V_a/K_{0d})]K_{dr} = V_R - V_a + V_m \sum_{j=1}^{n} (-1)^j n\mathbf{C}_j(1/j)c^j$$
$$\times \{1 - \exp[-j(V_R - V_a)/V_m]\} \quad (7.163)$$

where $K_{dr} = a/C_r^n$ is the distribution coefficient for the concentration C_r in the reservoir, and the symbol $\mathbf{C}$ means combination. Replacing K_{dr} by $K_d C^n/C_r^n$, we obtain:

$$[V_b - (V_a/K_{0d})]K_d(C/C_r)^n = V_R - V_a$$
$$+ V_m \sum_{j=1}^{n} (-1)^j n\mathbf{C}_j(1/j)c^j\{1 - \exp[-j(V_R - V_a)/V_m]\} \quad (7.164)$$

If $V_a/K_{0d} \ll V_b$, Eqs. (7.163) and (7.164) become:

$$V_b K_{dr} \quad \text{or} \quad V_b K_d(C/C_r)^n$$
$$= V_R + V_m \sum_{j=1}^{n} (-1)^j n\mathbf{C}_j(1/j)c^j[1 - \exp(-jV_R/V_m)] \quad (7.165)$$

If there is pure water in the mixing vessel $C_0 = 0$, $K_{0d} = \infty$, and Eqs. (7.163) and (7.164) can be written:

$$V_b K_{dr} \quad \text{or} \quad V_b K_d(C/C_r)^n = V_R - V_a$$
$$+ V_m \sum_{j=1}^{n} (-1)^j n\mathbf{C}_j(1/j)c^j\{1 - \exp[j(V_R - V_a)/V_m]\} \quad (7.166)$$

In Eqs. (7.162)–(7.165) for a given elution system, K_{dr}, n, V_b, V_m, c and V_a are considered constants for a certain ion. For each integral value of n Eq. (7.163) may be developed as follows:

$$n = 0: \quad K_{dr}[V_b - (V_a/K_{0d})] = V_R - V_a \quad (7.167)$$

$$n = 1: \quad K_{dr}[V_b - (V_a/K_{0d})]$$
$$= V_R - V_a - V_m c\{1 - \exp[-(V_R - V_a)/V_m]\} \quad (7.168)$$

$n = 2$: $K_{dr}[V_b - (V_a/K_{0d})]$
$$= V_R - V_a - 2V_m c\{1 - \exp[-(V_R - V_a)/V_m]\}$$
$$+ (1/2)V_m c^2\{1 - \exp[-2(V_R - V_a)/V_m]\} \quad (7.169)$$

$n = 3$: $K_{dr}[V_b - (V_a/K_{0d})]$
$$= V_R - V_a - 3V_m c\{1 - \exp[-(V_R - V_a)/V_m]\}$$
$$+ (3/2)V_m c^2\{1 - \exp[-2(V_R - V_a)/V_m]\}$$
$$- (1/3)V_m c^3\{1 - \exp[-3(V_R - V_a)/V_m]\} \quad (7.170)$$

Though these equations do not give the explicit value of V_R, the authors show that they can easily be solved graphically. The corresponding curves are shown in Fig. 7.80. To obtain these curves the values of K_{dr} and V_b are calculated by introducing the actual values of V_R into Eq. (7.165). All the other dimensions are constant, determined by the experimental conditions. For non-integral values of n an interpolation is made between two neighbouring curves.

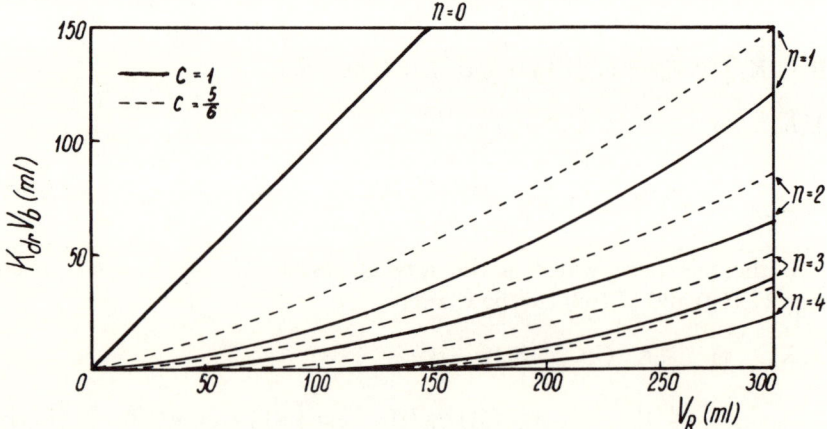

FIG. 7.80. Relation between V_R and $K_{d_r}V_b$ calculated from Eq. (7.165) [67].

Further on, these authors demonstrated that the average oxidation number i of phosphorus in its oxyacids is equal to the slope obtained by plotting $\log K_d$ against the logarithm of the ammonium acetate concentration. The results were in good agreement with the values obtained from the dissociation constants (Table 7.5).

TABLE 7.5. Calculated and observed values of i [67]

Oxyacid of phosphorus		Calc.	Obs.
P^I	$NaPH_2O_2 \cdot 2H_2O$	1.0	0.88
P^{III}	$Na_2PHO_3 \cdot 5H_2O$	1.7	2.0
P^V	Na_2HPO_4	1.5	1.7
$P^{III}-O-P^{III}$	$Na_2P_2H_2O_5$		2.1
$P^{IV}-P^{IV}$	$Na_2H_2P_2O_6 \cdot 6H_2O$	2.1	2.4
$P^{II}-P^{IV}$	$Na_3P_2HO_5 \cdot 12H_2O$		2.8
$(-P^{III}-)_6$	$[K(Na)PO_2]_6 \cdot xH_2O$		3.9
$P^{IV}-P^{III}-P^{IV}$	$Na_5P_3O_3 \cdot 14H_2O$		4.0

The position of the peaks may be calculated by means of Fig. 7.80, using the experimental values of K_d and i. The experimental and the calculated values for V_R are in good agreement with the values calculated in Table 7.6.

TABLE 7.6. Calculated and observed values of V_R [67]

	Run. No.	1		2		3	
Oxyacid of phosphorus	$C_m, (M)$	0		0.1		0.1	
	$C_r, (M)$	1		0.6		0.5	
	$V_b, (ml)$	5.54		5.51		10.1	
				V_R			
		Calc	Obs	Calc	Obs	Calc	Obs
P^I		88	76	60	54	93	108
P^V		138	113	136	105	206	198
P^{III}				158	135	238	249
$P^{IV}-P^{IV}$		202	178	260	265		
$P^{II}-P^{IV}$				313	311		
$P^{III}-O-P^{III}$		245	275				

Ohashi and Koguchi [69] continued their research, succeeding in establishing more general equations for the calculation of the position of the peak when an elution gradient is used for separation of ions bearing average charges which are not necessarily whole numbers. Aiming to deduce a basic equation, they assumed that the plate theory applies to all ion-exchange processes, and that the relationship between the distribution coefficient of the ions and the eluent concentration is given by Eq. (7.160). Starting from Eq. (7.162) deduced above, and neglecting the volume V_a as well as the interstitial volume and the dead volume of the resin bed in comparison with the elution volume of the peak, V_R, they write

$$K_{dr}V_b = \int_0^{V_R}[1 - c\exp(-V/V_m)]^n\,dV \qquad (7.171)$$

where K_{dr} is the distribution coefficient for an eluent of concentration C_r. Starting from this equation, and by a series of rather toilsome calculations, the authors succeeded in giving practical and convenient methods for calculating the approximate positions of the elution peaks for ions with an average charge of n_0, $n_0 + 1/4$, $n_0 + 1/2$, and $n_0 + 3/4$, where n_0 is a whole number (see Appendix II).

Massart and Bossaert [166] showed that by using the gradient technique, the separation time can be much shortened in comparison with conventional elution, when α-hydroxyisobutyric acid is used for the ion-exchange separation of some rare-earth elements. The dependence of the distribution coefficient on the ligand concentration C is given by

$$\log K_d = a \log C + b \qquad (7.172)$$

where K_d is defined according to Kraus and Moore [167, 168] as the ratio of the quantity of component on 1 g of dry stationary phase to the amount in 1 ml of mobile phase, i.e.

$$K_d = \rho\left(\frac{V_R}{AL} - F_I\right) \qquad (7.173)$$

where ρ is the specific volume of the column (volume per unit mass of resin), V_R is the retention volume, A the cross-sectional area of the column, L the column length, F_I the free fraction of column volume (the volume fraction occupied by the mobile phase), and a and b are characteristic constants for each eluent. The values of a and b were calculated by least-squares from the results published by Deelstra [169] and are given in Table 7.7.

The concentration of the ligand (C) is given by

$$C = C_m + (C_r - C_m)\exp(-V/V_m) \qquad (7.174)$$

and

$$C = C_m + (C_r - C_m)[V/(V_{0_r} + V_{0_m})]^{A_r/A_m} \qquad (7.175)$$

TABLE 7.7. Value of constants a and b

Element	a	b	Element	a	b
Lu	−5·73	−6·15	Gd	−5·54	−4·07
Yb	−5·67	−5·89	Eu	−5·45	−3·76
Tm	−5·66	−5·66	Sm	−5·43	−3·46
Er	−5·60	−5·37	Pm	−5·69	−3·38
Ho	−5·81	−5·36	Nd	−5·19	−2·65
Y	−5·56	−4·89	Pr	−4·97	−2·32
Dy	−5·46	−4·73	Ce	−5·12	−2·19
Tb	−5·27	−4·19	La	−4·88	−1·67

where A_r and A_m are the cross-sectional areas of the reservoir and mixing chamber respectively. Substitution for C from (7.174) and (7.175) in (7.172) gives

$$\log K_d = a \log [C_m + (C_r - C_m) \exp(-V/V_m)] + b \quad (7.176)$$

$$\log K_d = a \log \left[C_m + (C_r - C_m) \left(\frac{V}{V_{0_r} + V_{0_m}} \right)^{A_r/A_m} \right] + b \quad (7.177)$$

It can be seen from Eqs. (7.176) and (7.177) that $K_d = f(V)$ is a complex function, so it is easier to calculate an apparent distribution coefficient K'_d from Eq. (7.173). We may consider a small increment ΔV in the volume of eluent passing through the column. The migration of the maximum of the chromatographic zone caused by this, (ΔL_1), is given by

$$(\Delta L)_1 = \frac{\rho (\Delta V)_1}{A[(K_d)_1 + F_1 \rho]} \quad (7.178)$$

Similarly, for a second volume increase it is:

$$(\Delta L)_2 = \frac{\rho (\Delta V)_2}{A[(K_d)_2 + F_1 \rho]} \quad (7.179)$$

so that the total length after the two volume increases will be

$$(\Delta L)_1 + (\Delta L)_2 = \frac{\rho}{A} \left[\frac{(\Delta V)_1}{(K_d)_1 + F_1 \rho} + \frac{(\Delta V)_2}{(K_d)_2 + F_1 \rho} \right] \quad (7.180)$$

and after n volume increases, the total length is:

$$\sum_{i=1}^{n} (\Delta L)_i = \sum_{i=1}^{n} \frac{\rho}{A} \left[\frac{(\Delta V)_i}{(K_d)_i + F_1 \rho} \right] \quad (7.181)$$

If $\sum_{i=1}^{n} (\Delta L)_i = L$,

$$L = \frac{\rho}{A} \int_0^{V_R} \frac{dV}{K_d + F_I \rho} \tag{7.182}$$

From Eq. (7.173),

$$L = \frac{\rho}{A} \frac{V_R}{(K_d' + F_I \rho)} \tag{7.183}$$

TABLE 7.8. Values of K_d' for different gradients [166]

$C_m = 0$; $V_{om} = 250$ ml; $C_r = 0.375M$;
$V_{or} = 250$ ml; $AL = V_c = 5.40$ ml;
$\rho = 2.11$; $F_I = 0.44$

$C_m = 0$; $V_{om} = 250$ ml;
$C_r = 0.300M$; $V_{or} = 250$ ml;
$V_c = 4.53$ ml; $\rho = 2.11$:
$F_I = 0.44$

	K_d'			K_d'	
		Experimental			
	Calculated	Exp. 1	Exp. 2	Calculated	Experimental
Ho	44.6	43.7	46.0	54.0	50.1
Tb	54.8	—	54.1	66.25	59.5
Gd	63.2	60.3	62.2	76.5	69.4
Eu	68.4	64.5	67.2	82.8	74.8
Sm	75.4	72.2	75.2	91.6	83.1
Ce	110.1	109.1	114.6	133	117.5

$C_m = 0.0334M$; $V_{om} = 550$ ml; $C_r = 0.292M$; $V_{or} = 550$ ml; $AL = 4.53$ ml;
$\rho = 2.11$; $F_I = 0.44$

	K_d'				K_d'	
	Calculated	Experimental		Calculated	Experimental
Lu	42	43	Gd	150	142
Yb	49	48	Eu	168	—
Tm	59	58	Sm	194	184
Er	70	68	Pm	228	214
Ho	84	81	Nd	258	241
Y	94	90	Pr	275	266
Dy	96	—	Ce	312	—
Tb	117	113	La	363	—

By equating the last two equations, we obtain:

$$\frac{V_R}{K'_d + F_l\rho} = \int_0^{V_R} \frac{dV}{K_d + F_l\rho} = \frac{AL}{\rho} \qquad (7.184)$$

K'_d is the root of the function $\phi(z)$:

$$\phi(z) \equiv \int_0^{(AL/\rho)(z + F_l\rho)} \frac{dV}{K_d + F_l\rho} - AL/\rho \qquad (7.185)$$

and can be calculated by various methods. As can be seen, these authors [166] used a procedure analogous to that of Freiling [140, 160] to obtain Eq. (7.184). The equations deduced by Drake [15] and by Schwab et al. [46] may be brought to the same form.

The calculation of the function $\phi(z)$ by analytical procedures would be possible only in the case of some gradients for which $K_d = f(V)$ is simple in form and the integration does not create any problems. These authors [166] did not try analytical solutions for estimation of the integrals, but wrote a FORTRAN programme for the numerical calculation on an IBM-1620 computer. This method, shown in Appendix III, makes it possible to calculate $\phi(z)$ even for some of the more complex gradients.

Various separations of rare earths were made by use of a linear gradient. The calculated and experimental values of K'_d are given in Table 7.8 and show good agreement. The retention volume V_R can be calculated by putting the data from Table 7.8 into Eq. (7.173). For the gradient separations (Table 7.8) the apparent number of plates is between 1000 and 5000, whereas only 200 plates are obtained for the same column by elution with constant ligand concentrations.

A pH gradient was used by Galeffi et al. [170] for the counter-current separation of alkaloids, and the corresponding theory was worked out.

7.2.1.2 Resolution

The problem of resolution in mobile-phase gradient chromatography has been discussed by Snyder and Saunders [150]. As shown (see Section 4.3, p. 70), the separability of two components is given by the resolution R_s [Eq. (4.59)] or by the separation function F [Eq. (4.73)]:

$$R_s = \Delta L/4\sigma \qquad (7.186)$$

$$F = R_s^2 \qquad (7.187)$$

where ΔL is used instead of Δz in Eq. (4.59). This relation shows that the resolution is influenced by the thermodynamic properties of the chromatographic system, by the term ΔL, and by the separation efficiency of the

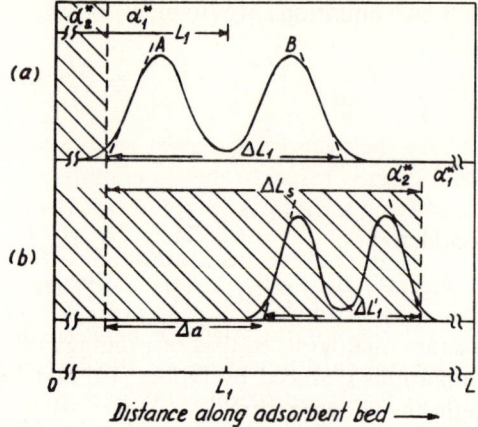

FIG. 7.81. Illustration of band compression in solvent programming. (a) adjacent bands immediately before being overtaken by solvent 2 (cross-hatched region), (b) adjacent bands immediately after being overtaken by solvent 2 [150].

column through the term σ ($\sigma = \sigma_A \sim \sigma_B$). If the composition of the mobile phase is changed, the value of the resolution is also modified.

Snyder and Saunders considered the gradient elution of two components. The effect of the first eluent on the position of the two components in the column is given in Fig. 7.81a. After displacement of the bands an average distance L_1 along the column, their breadth is ΔL_1. The elution is continued with the second eluent (denoted by hatching), as seen in Fig. 7.81b. The action of the second eluent decreases the breadth of the bands to $\Delta L'_1$, and the tail of component A has covered a distance Δa while the front of component B has reached a distance ΔL_s. It is considered that all of strip B migrated in the first eluent. By supposing that the two bands are narrow and adjacent, we can approximate $\alpha^*_A \sim \alpha^*_B = \alpha^*_1$ in the first solvent, and $\alpha^*_A \sim \alpha^*_B = \alpha^*_2$ for the second solvent (α^* is the distribution coefficient, in ml/g). We can also make the approximations $\alpha'_A W/V^\circ \sim \alpha'_B W/V^\circ = \alpha'_1 W/V^\circ = \alpha^*_1$ and similarly $\alpha'_2 W/V^\circ = \alpha^*_2$ and $\alpha'_2 W/V^\circ = \alpha^*_2$ where W is the quantity of adsorbent in the column, V° is the free volume of the column, and $\alpha_A \sim \sigma_B = \sigma_1$. As the tail of A has migrated entirely in the second solvent, we may write

$$\bar{R}_{f_A} = \frac{\Delta a}{\Delta L'_1 + \Delta a} = \frac{1}{1 + \alpha'_2 W/V^\circ} = \frac{1}{1 + \alpha^*_2} = \bar{R}_{f_2} \qquad (7.188)$$

and as the tail of B had migrated entirely in the first solvent:

$$\bar{R}_{f_B} = \frac{\Delta L'_1 + \Delta a - \Delta L_1}{\Delta L'_1 + \Delta a} = \frac{1}{1 + \alpha_1^*} = \bar{R}_{f_1} \qquad (7.189)$$

where $\bar{R}_{f_1}$ and $\bar{R}_{f_2}$ are the average R_f values for the two components in the first and second eluents respectively. By eliminating Δa between the two equations and solving for $\Delta L'_1/\Delta L_1$, we obtain the expression for the band compression factor G:

$$\Delta L'_1/\Delta L_1 = (1 - \bar{R}_{f_2})/(1 - \bar{R}_{f_1}) \equiv G_1 = \alpha_2^*(1 + \alpha_1^*)/\alpha_1^*(1 + \alpha_2^*) \qquad (7.190)$$

If we write $\Delta z_1 = z_B - z_A$ for the distance between the centres of the zones and σ_1 for the standard deviation for the first eluent, and correspondingly $\Delta z'_1$ and σ'_1 for the situation with the second eluent after both zones have moved completely into the α_2^* region (Fig. 7.81b), we can write

$$\Delta z'_1/\Delta z_1 = \sigma'_1/\sigma_1 = \Delta L'_1/\Delta L_1 = G_1 \qquad (7.191)$$

The zones will migrate a distance L_1 with the first eluent and become separated by Δz_1, then the zones become compressed and are separated by $\Delta z'_1$, and then there will be a further separation Δz_2 caused by the second eluent during travel over the distance L_2. When the separation is regarded as complete, the total bed-length used, L, is the sum of L_1 and L_2, and the final distance between the zone centres, Δz_g, is given by

$$\Delta z_g = \Delta z'_1 + \Delta z_2 = G_1 \Delta z_1 + \Delta z_2 \qquad (7.192)$$

An analogous expression is obtained for the variance σ_g^2 at the end of the separation:

$$(7.193)$$

As the number of plates corresponding to a standard deviation σ_i and a column length L_i is $N_i = (L_i/\sigma_i)^2$, the HETP will be $H_i = L_i/N_i$, or $H_i = \sigma_i^2/L_i$ (see Section 4.1).

By means of these equations we can deal with the separation function $F = (R_s)^2$ [see Eq. (4.73), p. 82] in the case of gradient elution with only two solvents:

$$F_g = (R_s)_g^2 = \frac{(\Delta z_g)^2}{16\sigma_g^2} = \frac{(G_1 \Delta z_1 + \Delta z_2)^2}{16(G_1^2 H_1 L_1 + H_2 L_2)} \qquad (7.194)$$

The treatment can then be extended. When the volume V_i of the eluent of composition for which the component has the distribution coefficient α'_i (given by $\alpha'_i W = \alpha_i^* V°$) is passing through the column, i.e. through the ith position of the zone, the zone centre will be displaced along a length

L_i of column, such that $V_i/\alpha_i^* V° = L_i/L$, or

$$L_i = LV_i/V°\alpha_i^* \tag{7.195}$$

In that case, the solvent front, which initially coincides with the zones of A and B which have migrated a distance L_i, will migrate on the column a distance

$$(L_s)_i = L_i/R_{f_i} = L_i(1 + \alpha_i^*) \tag{7.196}$$

The distance between the zone centres (initially not separated) after migration by a distance L_i in the solvent of ith composition, may be expressed as

$$\Delta z_i = (z_B - z_A)_i = (L_s)_i(R_{f_B} - R_{f_A})_i \tag{7.197}$$

or, after substitution:

$$\Delta z_i = L_i[(\alpha_A^*/\alpha_B^*) - 1][\alpha_i^*/(1 + \alpha_i^*)] \tag{7.198}$$

Subsequently, Snyder showed that for a certain separation, if the solvent viscosity is kept between 0·2 and 0·4 cP, we can make the approximation that the HETP is the same for all components [171, 172]. He also showed that the selectivity in liquid–solid chromatography may also be considered constant, as the distribution coefficients do not vary much with change in the solvent composition during gradient elution. Thus, combinings Eqs. (7.19) and (4.73), we obtain for two eluents ($n = 2$):

$$(F_g)_2 = (R_s)_g^2 = \frac{[(\alpha_A^*/\alpha_B^*) - 1]^2}{16}$$
$$\times \frac{\{G_1 L_1[\alpha_1^*/(1 + \alpha_1^*)] + L_2[\alpha_2^*/(1 + \alpha_2^*)]\}^2}{(G_1^2 H L_1 + H L_2)} \tag{7.199}$$

If both numerator and denominator are divided by L^2 and we write $L_1/L = \bar{L}_1$, $L_2/L = \bar{L}_2$ and $H/L = 1/N$, we obtain

$$(F_g)_2 = (R_s)_g^2 = \frac{N[(\alpha_A^*/\alpha_B^*) - 1]^2}{16}$$
$$\times \frac{\{G_1 \bar{L}_1[\alpha_1^*/(1 + \alpha_1^*)] + \bar{L}_2[\alpha_2^*/(1 + \alpha_2^*)]\}^2}{(G_1^2 \bar{L}_1 + \bar{L}_1)}$$
$$= \frac{N[(\alpha_A^*/\alpha_B^*) - 1]^2 E_2^2}{16} \tag{7.200}$$

If three eluents are considered, the equations necessary to express the resolution can easily be obtained by extending those established for two eluents. Thus:

$$(\Delta z_g)_3 = G_2(G_1\Delta z_1 + \Delta z_2) + \Delta z_3 = G_1 G_2 \Delta z_1 + G_2 \Delta z_2 + \Delta z_3, \quad (7.201)$$

and after substituting the values of Δz_i (for $i = 1, 2, 3$) from Eq. (7.198) in Eq. (7.201), we obtain:

$$(\Delta z_g)_3 = [(\alpha_A^*/\alpha_B^*) - 1]\{G_1 G_2 L_1[\alpha_1^*/(1 + \alpha_1^*)] \\
+ G_2 L_2[\alpha_2^*/(1 + \alpha_2^*)] + L_3[\alpha_3^*/(1 + \alpha_3^*)]\} \quad (7.202)$$

The same procedure is adopted to find the expression for the dispersion σ_g^2:

$$(\sigma_g^2)_3 = G_1^2 G_2^2 H_1 L_1 + G_2^2 H_2 L_2 + H_3 L_3 \quad (7.203)$$

After substituting in the equation giving the separation function, we obtain

$$(F_g)_3 = \frac{(\Delta z_g)_3^2}{16(\sigma_g^2)_3} = \frac{N\left(\frac{\alpha_A^*}{\alpha_B^*} - 1\right)^2}{16}$$

$$\times \frac{\left[G_1 G_2 \bar{L}_1\left(\frac{\alpha_1^*}{1+\alpha_1^*}\right) + G_2 \bar{L}_2\left(\frac{\alpha_2^*}{1+\alpha_2^*}\right) + \bar{L}_3\left(\frac{\alpha_3^*}{1+\alpha_3^*}\right)\right]^2}{(G_1^2 G_2^2 \bar{L}_1 + G_2^2 + \bar{L}_2 + \bar{L}_3)}$$

$$= \frac{N\left(\frac{\alpha_A^*}{\alpha_B^*} - 1\right)^2 E_n^2}{16} \quad (7.204)$$

Generalised for n eluents, the equation of the separation function will have the following form:

$$(F_g)_n = \frac{N[(\alpha_A^*/\alpha_B^*) - 1]^2 E_n^2}{16} \quad (7.205)$$

where E_n^2 is given [150] by the expression:

$$E_n^2 = \frac{\left[\sum_{}^{n}\left(\prod_{}^{i \leq j \leq n} G_j\right)\bar{L}_i \alpha_i^*/(1 + \alpha_i^*)\right]^2}{\sum_{}^{n}\left(\prod_{}^{i \leq j \leq n} G_j^2\right)\bar{L}_i} \quad (7.206)$$

and as no compression takes place in the last eluent, $G_n = 1$. When a continuous gradient is used, i.e. $V_i \to 0$, the summations in Eq. (7.205) are transformed into integrals.

Taking the ratio between the separation functions in the case of n eluents and of only one, we obtain:

$$\frac{(F_g)_n}{F} = \frac{E_n^2}{E_1^2} \quad \text{or} \quad (F_g)_n = F\frac{E_n^2}{E_1^2} \tag{7.207}$$

where $E_1^2 = [\alpha_1^*/(1 + \alpha_1^*)]^2$.

It results from the last equation that the separation of two components with a mobile-phase gradient is so much the better than that by conventional chromatography, as the ratio E_n^2/E_1^2 increases.

Snyder [173], in a comparative study between the different techniques in liquid chromatography, reached the conclusion that the resolution per unit time decreases in the following order: mobile-phase gradient (the best) > coupled-column > programmed temperature > programmed flow > normal elution.

In the selection of an optimal mobile-phase gradient programme, Snyder and Saunders [150] considered two aspects of the programme, viz. the mathematical form of the mobile-phase gradient and its effect on the distribution coefficient, $\alpha_i^* = f(V)$, and the rate of change of α_i^* with V. For the general case of separation of a mixture of unknown components, the programme must be chosen in such a way as to have the same resolution for the components, i.e. for E_n^2 to be constant. This is achieved if $\alpha_{i+1}^*/\alpha_i^*$ and V_i are constant for all values of i. If V_i is very small (continuous mobile-phase gradient), this condition may be expressed by

$$\log \alpha^* = a - bV/V^\circ \tag{7.208}$$

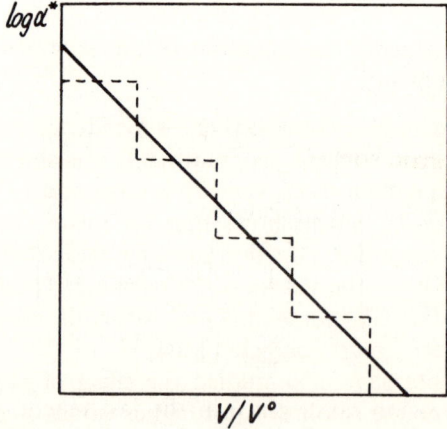

FIG. 7.82. Discontinuous (---) and continuous (———) logarithmic solvent programme with the same values of b from Eq. (7.208) [150].

The two types of programme (continuous and stepwise) are given in Fig. 7.82; b appears as the standard value of the slope. To illustrate all this, Snyder and Saunders considered two pairs of components, A–B and C–D, for which they took α_1^* as 1000 for the first pair and 8000 for the second, and set $\alpha_{i+1}^*/\alpha_i^*$ equal to 2. The results obtained for E_n^2 by means of Eq. (7.206) are given in Table 7.9. This table shows the equality of the resolution for the two pairs, 0·726 for A–B and 0·728 for C–D. These conclusions can be generalised for any pair of compounds in a solvent gradient of logarithmic form (except for high values of V_i and low values of α_1^*).

TABLE 7.9. Value of E_n^2 for two pairs of components in a logarithmic solvent programme
$(V_1 = V_2 = V_i = 2V°)$ [150]

Solvent	Pairs of components A–B					Pair of components C–D				
	α_i^*	$\bar{L}_i$	G_i	$\dfrac{\alpha_i^*}{(1+\alpha_i^*)}$	E_n^2	α_i^*	$\bar{L}_i$	G_i	$\dfrac{\alpha_i^*}{(1+\alpha_i^*)}$	E_n^2
1	1000	0·002	0·999	0·999		8000	0·000	1·000	1·000	
2	500	0·004	0·998	0·998		4000	0·000	1·000	1·000	
3	250	0·008	0·996	0·996		2000	0·001	1·000	1·000	
4	125	0·016	0·992	0·992	0·726	1000	0·002	0·999	0·999	
5	62·5	0·032	0·984	0·984		500	0·004	0·998	0·998	
6	31·3	0·064	0·970	0·969		250	0·008	0·996	0·996	0·728
7	15·6	0·128	0·943	0·939		125	0·016	0·992	0·992	
8	7·8	0·256	0·997	0·886		62·5	0·032	0·984	0·984	
9	3·9	0·490†	1·000	0·795		31·3	0·064	0·970	0·696	
10						15·6	0·128	0·943	0·939	
11						7·8	0·256	0·897	0·886	
12						3·9	0·489†	1·000	0·795	

† The pair was eluted by this solvent $\left(\sum\limits_{i}^{n}\bar{L}_i=1\right)$.

Snyder and Saunders also computed the dependence of E_n^2 on α_1^*, b, and $V_i/V°$ for a logarithmic solvent programme. It was shown above that the value of E_n^2 is independent of α_1^* when α_1^* is large and $V_i/V°$ small. Results are given in Fig. 7.83, which shows that for lower values of $V_i/V°$ the curve of E_n^2 tends towards a constant limiting value of 0·46 for $\alpha^* > 100$. At high values of $V_i/V°$ the values of E_n^2 oscillate. These oscillations are explained by the fact that for high $V_i/V°$ ratios the dimension $\bar{L}_i$ varies with α^* and hence E_n^2 must implicitly vary.

The same authors have also studied the effect of parameter b in Eq. (7.208) on E_n^2 and on the resolution, for higher values of α_* and low values of $V_i/V°$ (in this case, E_n^2, as already shown, is independent of these two parameters). The results are given in Fig. 7.84.

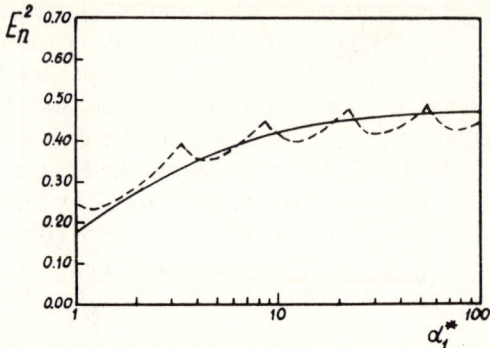

Fig. 7.83. Variation of E_n^2 with α_1^* (logarithmic solvent programme); $b = 0.38$, ——— $(V/V^\circ) = 0.05$, - - - - $(V/V^\circ) = 1.00$ [150].

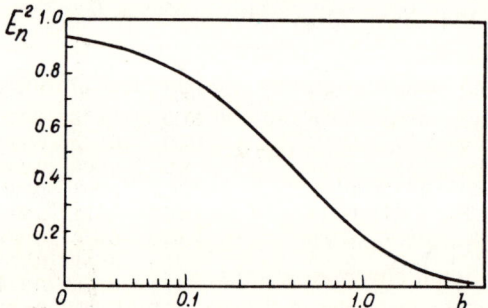

Fig. 7.84. Variation of E_n^2 with b; continuous logarithmic solvent programme and α_1^* large [150].

It was also shown that for maximum resolution (NE_n^2), $b = 0.38$ (obtained by plotting the calculated values of $E_n^2 V^\circ / V_R$ vs. b). Figure 7.84 shows that the optimum value of E_n^2 in gradient separations is 0.46 when $b = 0.38$. The deviation of E_n^2 from the standard value $\bar{E}_n^2$ (i.e. $\Delta \bar{E}_n^2 / \bar{E}_n^2$) has also been studied as a function of V_i/V° and b (Fig. 7.85). The results allow us to estimate the minimum number of components to be included in the solvent eluent programme to eliminate large fluctuations of the value of E_n^2.

7.2.2 Open Column Chromatography

The theory of mobile-phase gradient elution in open column (paper and thin-layer) chromatography is much less well developed than it is for closed columns. Indeed, we are almost at the beginning of its development. Niederwieser and Honegger [134] have systematised a vast amount

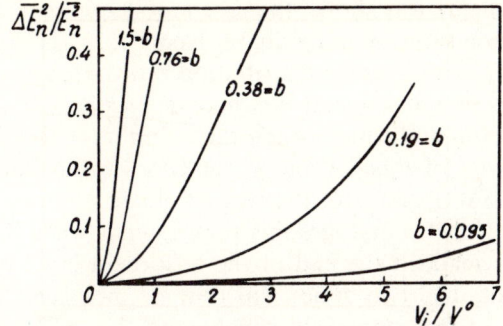

Fig. 7.85. Variability of $\Delta \bar{E}_n^2/\bar{E}_n^2$ with $V_i/V°$ and b [from Eq. (7.208)] [150].

of experimental material and outlined some theoretical problems. Snyder and Saunders [174] have also discussed fairly recently some of the theoretical problems.

In conventional chromatography, the simplest mobile-phase gradient is achieved by dehomogenising the eluent during the development (Sections 1.3 and 1.4). The reasons for this dehomogenisation lie in the preferential adsorption of some components of the eluent by the paper or thin layer, as well as the vaporisation of the more volatile components. We shall not dwell on these 'natural' gradients, as they are discussed in every book on conventional chromatography.

The behaviour of substances in thin-layer multizonal chromatography, in which a natural gradient also appears, has been amply described [134]. In the same paper the authors examined a parallel between multizonal and mobile-phase gradient chromatography, showing the superiority of the latter. The technique of multiple development was also discussed in Chapter 2.

In Snyder and Saunders' theoretical considerations [174] on the deduction of the resolution equation for mobile-phase gradient chromatography, the secondary factors which appear during the elution were not included, namely the eluent profile, dehomogenisation, heat of adsorption, and change in adsorbent activity during the separation. It was considered that in the case of a closed column the ratio V_S/V_M was constant at any point behind the front. The final conclusions do not appear to be seriously affected by these approximations.

Snyder and Saunders [174] considered elution with a mobile-phase gradient on a thin-layer plate to be similar to elution on a closed column (see Section 7.2.1.2), i.e. at the lower part of the plate there enters by capillary ascent (Chapter 1) a quantity of solvent of which the composition changes with time. It was shown that the mobile-phase gradient may be approximated by a series of individual solvents $1, 2, \ldots i \ldots n$, of volumes $V_1, V_2, \ldots V_i \ldots V_n$, with average distribution ratios $\alpha_1^*, \alpha_2^*, \ldots \alpha_i^* \ldots \alpha_n^*$ for two components in zones A and B (where $\alpha^* = \alpha' W/V°$). On passage of volume V_1 of the first eluent through the column (or thin layer), and so through the centres of zones A and B, the zones will be moved a certain distance (Fig. 7.86). Passage of the second eluent (volume V_2) shifts the centres further. The chromatographic process ends when the front of the first solvent has reached the end of the adsorbent layer or some other preselected point. The two zones will now be in solvent j. The rate of migration of the zones obviously depends on the values of α_i^*. If α_1^* is big enough, the rate of movement of the zone will be very small, and if the polarity of the eluent increases with the total volume passed, the value of α_i^* will decrease and the zones move faster. Taking into account the expression for resolution (Chapter 4) it is necessary that $\alpha_1^* > 0$ and $\alpha_n^*/(1 + \alpha_n^*) < 1$.

Snyder and Saunders adopted the same procedure to calculate the resolution in open column as in closed column chromatography (Section 7.2.1.2). When the initial eluent volume V_1 passes through the centres of the two bands, these will be displaced along the open column by a distance L_1 given by

$$L_1 = LV_1/V°\alpha_1^* \qquad (7.209)$$

where L is the length of the column (the distance between the starting line and the final front of the eluent) and $V°$ is the volume of eluent needed to soak that length of adsorbent layer. Passage of the second volume of eluent, V_2, displaces the zones by $L_2 = LV_2/V°\alpha_2^*$, and so on, so V_i displaces them a distance L_i or a fractional distance $\bar{L}_i = L_i/L$, i.e.

$$\bar{L}_i = V_i/V°\alpha_i^* \qquad (7.210)$$

The total distance L_f migrated by the eluent front is given by the average distance traversed by the centres of the zones and the distance traversed by the eluent which has passed through the zone centres (Fig. 7.86):

$$L_f = L\left[\sum^i (V_i/V°) + \sum^i \bar{L}_i\right] \qquad (7.211)$$

The separation is finished when $L_f = L$, a fact that can be expressed thus, according to Fig. 7.86:

$$\sum^{j-1} [(V_i/V°) + \bar{L}_i] < 1 < \sum^j [(V_i/V°) + \bar{L}_i] \qquad (7.212)$$

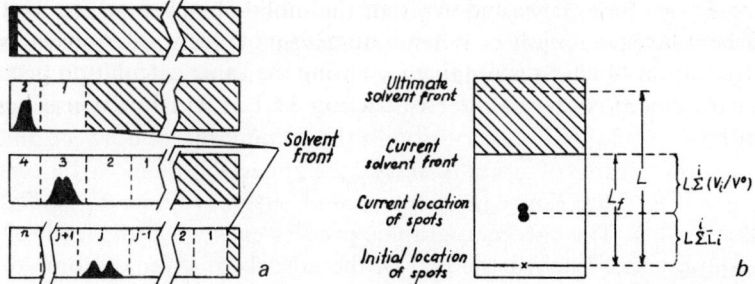

FIG. 7.86. (a) Separation of a pair of sample components in gradient elution thin-layer chromatography (TLC). (b) Gradient elution TLC after passage of the ith solvent through the two bands of interest [174].

When the jth eluent passes through the zone centres, these will be displaced a distance L_j and the front of the first eluent will be a distance ΔL from the end of the bed, where ΔL is given by

$$\Delta L = L - L \sum_{}^{j-1} [(V_i/V^\circ) + \bar{L}_i] \qquad (7.213)$$

The mean R_f value of the two zones in solvent j will be

$$(\bar{R}_f)_j = L_j/\Delta L = 1/(1 + \alpha_j^*) \qquad (7.214)$$

and after substituting $\bar{L}_j = L_j/L$ in (7.214) we obtain

$$\bar{L}_j = \left(\frac{\Delta L}{L}\right)\frac{1}{1 + \alpha_j^*} = \frac{1 - \sum^{j-1}[(V_j/V^\circ) + \bar{L}_i]}{1 + \alpha_j^*} \qquad (7.215)$$

After the centres of the two zones have been traversed by the fractions of eluent $(1, 2, \ldots, j)$ and the front of the first eluent has reached the total distance $L = \bar{L}_1 + \bar{L}_2 + \cdots + \bar{L}_{j-1} + \bar{L}_j$, the R_f value of the two zones will be:

$$\bar{R}_f = \left(\sum^{j-1} \bar{L}_i\right) + \bar{L}_j \qquad (7.216)$$

The resolution R_s, or the separation function $F = (R_s)^2$, in the case of one eluent for which the two components have distribution coefficients α_A^* and $\alpha_B^* (\alpha_A^* \sim \alpha_B^* = \alpha_i^*)$, may be written [150, 174]:

$$F = (R_s)^2 = \frac{N[(\alpha_A^*/\alpha_B^*) - 1]^2 E_1^2}{16} \qquad (7.217)$$

where $E_1^2 = \alpha_1^*/(1 + \alpha_1^*)$ and N is the number of plates in the bed of adsorbent layer of length L. When a number n of eluent fractions is used, the separation function is obtained by using the same calculation process as for the closed column [150] (see Section 7.2.1.2). The function is hence given by

$$(F_g)_n = (R_s)_n^2 = \frac{N[(\alpha_A^*/\alpha_B^*) - 1]^2 E_n^2}{16} \tag{7.218}$$

where:

$$E_n^2 = \frac{\left[\sum_{}^{j} \left(\prod_{}^{i \leq m \leq j} G_m\right) \bar{L}_i \alpha_i^*/(1 + \alpha_i^*)\right]^2}{\sum_{}^{j} \left(\prod_{}^{i \leq m \leq j} G_m^2\right) \bar{L}_i} \tag{7.219}$$

and the compression factor of the band G_m is given by

$$G_m = \alpha_{m+1}^*(1 + \alpha_m^*)/\alpha_m^*(1 + \alpha_{m+1}^*) \tag{7.220}$$

The ratio between (7.218) and (7.217) gives us $(F_g)_j = FE_n^2/E_1^2$, a relationship which enables us to appreciate the efficiency of using the mobile-phase gradient, by comparison of the values of the ratio E_n^2/E_1^2.

Another way of estimating the efficiency of the mobile-phase gradient in thin-layer chromatography, as Snyder and Saunders have shown [174], is by means of the number of real plates NE_n^2. By assuming a logarithmic function between the distribution coefficient α^* and the mobile-phase gradient, the same resolution is obtained for all the components except those weakly adsorbed, i.e.

$$\log \alpha_i^* = \log \alpha_1^* - bV/V^\circ \tag{7.221}$$

where V is the total volume of solvent used before α_i^* is attained, α_1^* is the value of α^* for the first solvent in the programme, and b is a constant showing how quickly α^* varies with V, i.e the slope of $\log \alpha^* = f(V)$ and a measure of the gradient.

Snyder and Saunders calculated by means of a computer the values of E_n^2 given by Eq. (7.219) as a function of b, for different logarithmic solvent programmes in thin-layer chromatography with a mobile-phase gradient. The results obtained are given in Figs. 7.87 and 7.88. Taking $N = 1000$, which is a common value in thin-layer chromatography [175], they calculated the numbers of effective theoretical plates NE_n^2, which are given in the same figures.

On examining these figures it is noticed that the number of effective plates, i.e. the separation function $(F_g)_n$ decreases with increase in the slope

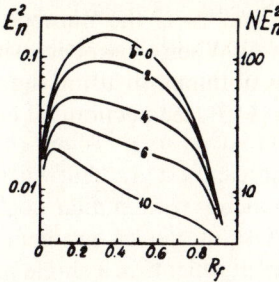

FIG. 7.87. Resolution in gradient elution TLC (E_n^2 vs. R_f) as a function of gradient steepness b from Eq. (7.221) (logarithmic programme with V_i small) [174].

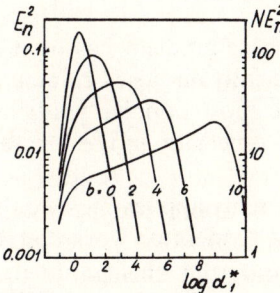

FIG. 7.88. Resolution in gradient elution (E_n^2 vs. $\log \alpha_1^*$) as a function of gradient steepness b from Eq. (7.221) (logarithmic programme with V_i small) [174].

b in Eq. (7.221). However, in this case the range of values for α_1^* is bigger and therefore the range of separability of sample components also increases, while NE_n^2 is nevertheless maintained at an acceptable value (Fig. 7.88).

It is also seen that a logarithmic solvent programme does not give equal values of NE_n^2 for all components of the mixture, especially not for the R_f values nearest to zero and to unity. For $b > 4$ we notice a systematic decrease of the NE_n^2 values with increase in the R_f values (Fig. 7.87). This can be corrected and maintained at an approximately constant value by modifying the gradient in such a way as to let $\log \alpha^*$ vary more rapidly with V in the last part of the programme. In mobile-phase gradient thin-layer chromatography, for $b > 4$ there are generally less than 30 effective plates. This is much lower than the number obtained in closed column chromatography (>400) for components with comparable retention times [150, 173].

7.2.3 Choice of the Form of Gradient

The success of a separation with a mobile-phase gradient depends to a large extent on the form of the gradient. This must be chosen according to the components which have to be separated and in correlation with the chromatographic column to be used. When the components have very different distribution coefficients, a gradient with a large slope is preferred, unlike the case of components with similar coefficients, when a gradient with low slope is used. Use of a steep gradient in the latter case would lead to migration of the components in a single band, with practically zero resolution. These problems will be illustrated in the next section.

7.3 Uses of Mobile-phase Gradients

Till now the mobile-phase gradient has been much more widely used than other gradients, especially on closed column ion-exchange chromatography on cellulose. The most popular gradients seem to be based on change in ionic strength, especially in the separation and purification of proteins.

Systematization of the many applications would far exceed the scope of this book, and we propose to mention work that illustrates certain specific features of gradient elution, or in which comparisons are made of results obtained with and without gradients, or with different gradients.

7.3.1 Organic Acids

The increasing polarity gradient has been applied to the separation of some organic acids. Donaldson *et al.* [33], using a 30-cm column 1 cm in diameter, filled with silica gel, and an exponential polarity gradient obtained with chloroform in the mixing chamber and chloroform and n-butanol mixture in the reservoir of the apparatus shown in Fig. 7.9 (p. 140), obtained good resolution with lower elution volume (Fig. 7.89), separating acetic acid cleanly from fumaric acid. The HETP was also much diminished for aconitic and oxalic acids. A mixture of chloroform and n-pentanol was used to separate lactic acid from succinic acid.

Lawson and Purdie [102] combined elution-gradient column ion-exchange microchromatography with paper chromatography. The column (10 cm long, 1·5 cm diameter) was filled with Dowex 1-X10 (100–200 mesh) anion-exchanger, and a 0–90 per cent formic acid gradient was used for elution. The fractions were collected as 1-cm diameter spots at 2-cm intervals on Whatman No. 1 paper and eluted with ethyl acetate–acetic acid–water (2:1:1 v/v) containing sodium acetate (0·05 per cent) and an indicator (Bromophenol Blue, 0·015 per cent) according to the method of

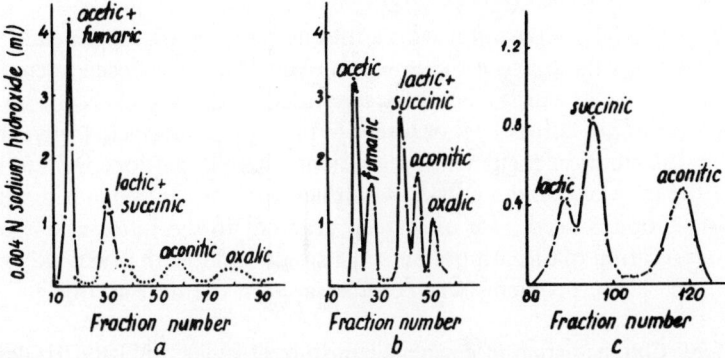

FIG. 7.89. Chromatograms of a mixture of six organic acids of physiological interest. a—Without gradient: eluent of fixed composition (10 per cent n-butyl alcohol in chloroform, v/v) is employed; b—with gradient; at start, reservoir vessel (R) contains 50 per cent chloroform–n-butyl alcohol and mixing vessel (M) pure chloroform (Fig. 7.9a); c—with gradient: at start in reservoir vessel, 50 per cent chloroform–n-amyl alcohol and in mixing vessel pure chloroform [33].

Hartley and Lawson [176]. The results are shown in Table 7.10. The method is quick and simple and allows the separation of 200-μg quantities of each acid.

Hauton [49] used a polarity gradient in reversed-phase partition chromatography for the separation of mono-, di- and trihydroxycholanic acids. The stationary phase was silaned kieselguhr impregnated with a chloroform–octanol mixture (1:1 v/v). The polarity gradient was achieved

TABLE 7.10. R_f values of some organic acids. Whatman No. 1 paper, eluent: ethyl acetate + acetic acid + water (2:1:1 v/v) [176]

Acid	R_f	Fractions
Glycollic	0·66–0·72	16–18
Succinic	0·80–0·87	17–24
DL-Malic	0·61–0·68	18–26
meso-Tartaric	0·39–0·46; 0·48–0·52	22–26
DL-Tartaric	(0·43)–0·47–0·52	27–29
Citric	0·54–0·59	26–32
Propan-1,2,3-tricarboxylic	0·74–0·80	25–29
Diglycollic	0·67–0·74	26–30
Malonic	0·72–0·78	29–37
Ethylmalonic	0·93–	31–38
Tartronic	0·40–0·47; 0·52–0·57	42–47
α-Ketoglutaric	0·65–0·70	44–52

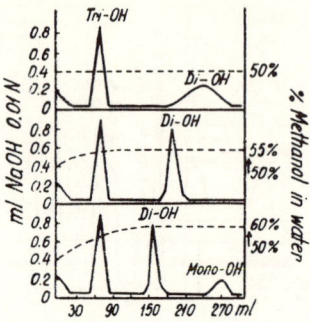

FIG. 7.90. Chromatogram of a synthetic mixture of cholic acid (tri–OH) desoxycholic acid (di–OH) and lithocholic acid (mono–OH). a—Without gradient, mobile phase methanol 50 per cent in water; b and c—with methanol gradient [49].

by using the apparatus shown in Fig. 7.9, with the high-polarity eluent in the mixing chamber and the low-polarity one in the reservoir. The results obtained with and without a gradient are shown in Fig. 7.90. The acid in the eluate was titrated with $0.01N$ methanolic sodium hydroxide (Bromothymol Blue as indicator). The elution order was characterised by paper chromatography [177] with the solvent system described by Sjövall [178].

Figure 7.90 clearly shows the advantages of the elution gradient over conventional chromatography. With a total eluent volume less than 300 ml all three components have been eluted, whereas without a gradient only two are eluted. The plate height is also much smaller than in conventional chromatography.

7.3.2 Carbohydrates

Alm [110] used an elution gradient on a column filled with charcoal that had been treated with stearic acid, for the separation of oligosaccharides and clearly showed that adjacent components can be separated. Hoban and White [97], using a linear elution gradient (0–6 per cent ethanol solution in water) on a column of charcoal treated with stearic acid, separated some disaccharides (Fig. 7.91) which in paper chromatography have practically the same R_f value (e.g. turanose 0·27, sucrose 0·26; melibiose 0·14, lactose 0·15).

Miller [76] worked out a chromatographic method for the analysis of oligosaccharides on a microcolumn filled with Celite 545 treated with stearic acid. Elution was with a convex exponential gradient (0–45 per cent aqueous ethanol). Jermyn separated saccharides on a charcoal column by gradient elution and discussed the effect of the shape of the gradient on

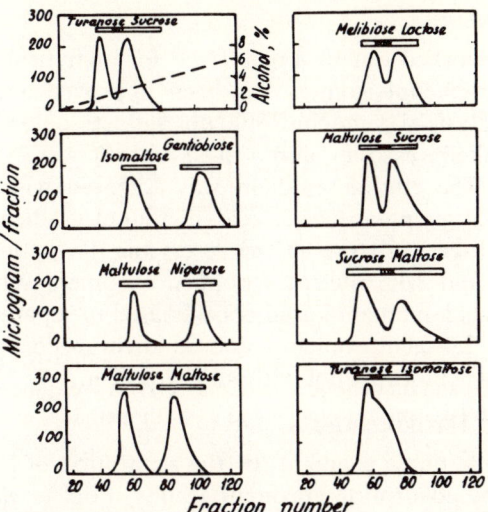

Fig. 7.91. Separation of disaccharides by gradient elution in microgram quantities. (----) Ethyl alcohol gradient of effluents [97].

the peak height and the band breadth [151], and later used DEAE-cellulose to separate acid polysaccharides, eluting with a sodium chloride gradient [179]. Barker *et al.* [180] separated carbohydrates on charcoal columns in the presence of borate, using a linear gradient of aqueous ethanol (0–8 per cent). The borate favoured the separation of the pairs melibiose–maltose and isomaltotriose–maltose.

To obtain reproducible results for separation of saccharides, Kimura [181] recommends that the charcoal used for packing the columns should have the fines removed by water-flotation. The column may be used repeatedly if it is regenerated after each separation by washing with water to remove the gradient eluent completely. The charcoal is further conditioned, after removal of fines, by washing with $1M$ hydrochloric acid and then with water to remove the acid, and is dried at 100°C. The dried material is mixed with Celite 560 in 4:3 ratio, and a suspension of this mixture in water is used to pack the column [182]. Using a butanol–water gradient (7 per cent aqueous butanol solution in the reservoir, water in the mixing chamber) Kimura [181] separated a mixture of sucrose, maltose and α- and β-cyclodextrine. The method may be extended to the preparative scale.

7.3.3 CHLOROPHENOLS

The method worked out by Logie [183] for analytical separations of di- and trichlorophenol isomers has been improved and extended by Skelly to the analysis of pentachlorophenol and tetrachlorophenol isomer mixtures [75]. Skelly used a column filled with Dowex 2-X8 (Cl^- form), 200–400 mesh. The elution gradient was achieved with an apparatus (Fig. 7.14, p. 145) giving an acetic acid–methanol gradient (methanol in the mixing chamber, mixture in the reservoir). With this gradient the 2,3,4,5-, 2,3,4,6- and 2,3,5,6-tetrachlorophenol isomers were separated. A different acetic acid–methanol gradient was used to separate a mixture of 4-chlorophenol, 2,4-dichlorophenol and 2,6-dichlorophenol, followed by elution of 2,4,6-trichlorophenol without a gradient.

7.3.4 AROMATIC HYDROCARBONS

Using a mobile-phase gradient for the separation of higher aromatic hydrocarbons by adsorption chromatography, Popl et al. [184] showed that by means of gradients it is possible to avoid the difficulties experienced in conventional chromatography of such mixtures, such as long duration of the analysis and broadening of the peaks when high-activity adsorbents are used. Use of low-activity adsorbents often gives incomplete separation.

With a pentane–ether mobile-phase gradient the analysis is shorter and the zones much narrower. Woelm Eschwege neutral alumina was used as adsorbent, being first activated at 400°C for 8 hr and then deactivated by addition of 2 per cent of water. An alumina having a specific surface of

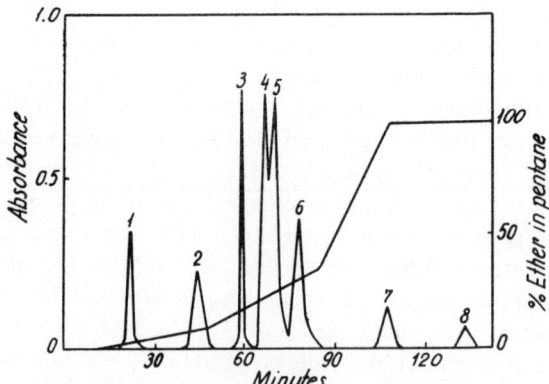

FIG. 7.92. Gradient elution adsorption chromatography of a mixture of standards and the course of the gradient. Adsorbent: alumina–2 per cent H_2O. Eluent: pentane–ether. *1*—Indane, *2*—naphthalene, *3*—acenaphthene, *4*—fluorene, *5*—phenanthrene, *6*—pyrene, *7*—chrysene, *8*—carbazole [184].

100 m^2/g was thus obtained. A 1-m column (4-mm bore) filled with 16 g of adsorbent was used for the separation, at a flow-rate of 45 ml/hr. The results obtained [184] for a standard mixture of indane, naphthalene, acetophenone, fluorene, pyrene, chrysene and carbazole are given in Fig. 7.92. Good results were also obtained for analysis of 1 μl of creosote oil with the same conditions.

The method was developed and the theory established by Snyder [146] for adsorption chromatography with a linear mobile-phase gradient.

7.3.5 POLYPROLYLENE GLYCOL AND POLYETHYLENE GLYCOL

Samples of polypropylene glycol and Triton X_{45} (an ethylene oxide derivative of octylphenol) have been fractionated by linear gradient elution with solvent programming by means of the 'Varigrad'. The column was packed with Porasil 60 (30–100 μm) and a mixture of polar and non-polar solvents (isopropyl alcohol and n-hexane) was used as eluent. The results obtained by Bombaugh et al. [114] with and without a gradient are shown in Fig. 93. Calzolari et al. [185], using column chromatography with silica gel deactivated with water (1 per cent) or with trimethylchlorosilane (0·5, 1·3 and 6 per cent), and acetone and methylene chloride as the

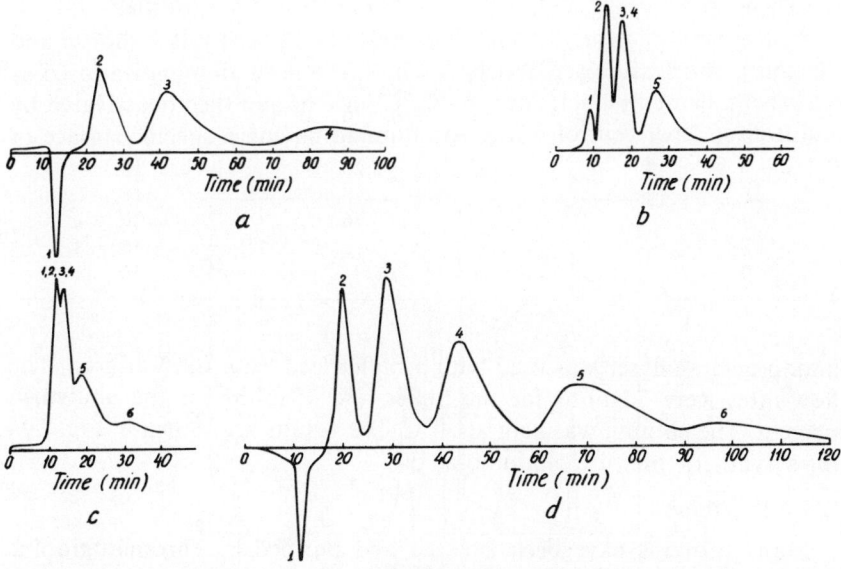

FIG. 7.93. Shape of elution curves of the polypropylene glycol (UCON$_{50}$HB$_{55}$). a—Eluent 8 per cent isopropyl alcohol in n-hexane; b—30 per cent isopropyl alcohol in n-hexane; c—70 per cent isopropyl alcohol in n-hexane; d—solvent programme; e—30 per cent isopropyl alcohol in n-hexane after 120 min [114].

solvents for obtaining the gradient, have also succeeded in separating polyethylene glycol derivatives.

7.3.6 Amino-acids

The chromatographic techniques for the separation of amino-acids are continually being improved. Substantial progress has been made by using ion-exchange resins, which has allowed the automation of liquid column chromatography. The Amberlite IR-4B, IRC-50 and 100, and the Dowex 1-X2, 1-X8, 50-X4, 50-X8 and 50-X12 resins have given the best results. The use of stepwise or continuous pH gradients and the effect of temperature has made it possible to separate some very complex mixtures.

Starcher *et al.* [186] described a rapid method for resolving on a single column all the amino-acids from the proteins of the conjunctive tissue, without use of a temperature programme. A 69 × 9 cm column was packed with a 50-cm length of UR-30 resin. The pH gradient was obtained by means of a 'Varigrad', as shown in Table 7.11. A Beckman model 116

TABLE 7.11. Composition in the Varigrad chambers [186]

Chamber	p = 2·91 (ml)	pH = 3·25 (ml)	0·4N Na citrate (ml)
1	36	—	—
2	36	—	—
3	—	36	—
4	—	30	6
5	—	26	10
6	—	18	18
7	—	6	30
8	—	—	36
9	—	—	36

amino-acid analyser was used with a buffer feed from the Varigrad. The flow-rates were 70 ml/hr for the buffer and 30 ml/hr for the ninhydrin reagent. The column was kept at 50°C. The results are shown in Fig. 7.94 for a synthetic mixture of amino-acids.

7.3.7 Proteins

Many proteins have been isolated and purified by chromatographic methods, and some new types of protein discovered. The use of chromatographic techniques for the separation of protein mixtures has grown steadily, notwithstanding the lack of precise information about the factors on which the separations depend. Success in chromatographic separation

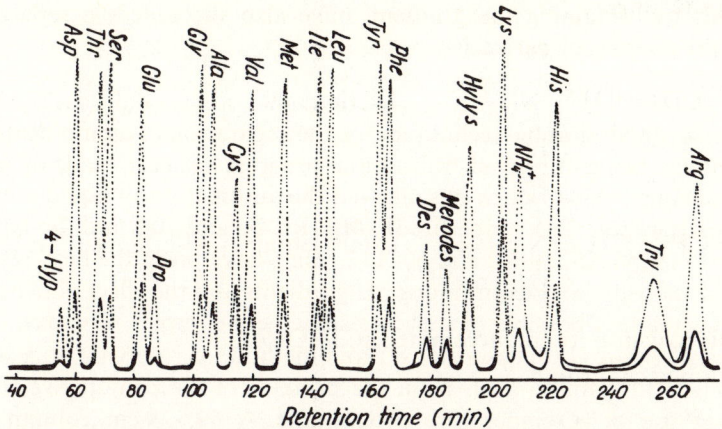

FIG. 7.94. Chromatogram of a synthetic mixture of amino-acids, obtained by using a Beckman Model 116 amino-acid analyser with the starting-buffer line (pH = 2·91) connected to a nine-chambered 'Varigrad' gradient device. A change to the pH 5·25 buffer (Na citrate) was made after 130 min [186].

of proteins depends on the choice of procedure, the best being with ion-exchangers, especially those specific for large molecules (ion-exchange cellulose). The reason why normal resins are inadequate for the separation of proteins (except in a few cases) is because numerous strong bonds are formed with the protein, so that simultaneous dissociation of all of them occurs only under conditions drastic enough to disrupt the molecule. These bonds are probably electrostatic in nature, but there are also undoubtedly van der Waals and hydrogen bonds.

As Sober and Peterson showed [187], two types of adsorbents have been developed for the chromatography of proteins: calcium phosphate gels [188] and ion-exchange cellulose. Both types present a large surface, well hydrated, and ionic interaction seems to be the main process. These adsorbents also give good selectivity.

The calcium phosphate gels are very much in use for the purification of enzymes by batch procedures [187]. Their use on columns is possible only when they are mixed with e.g. super-Cel-198, or by forming the gel in the presence of cellulose [189, 190]. Calcium phosphate has also been prepared in the form of brushite [191] or hydroxyapatite [192], which allows an adequate flow-rate of eluent.

The exchangers with a hydrophilic skeleton (ion-exchange cellulose etc.) are characterised by a high exchange capacity that is due to their extensive capacity to bind water [187]. The ion-exchange celluloses can be anion-exchangers:

$$-O-CH_2-CH_2-\overset{Et}{\underset{Et}{N}}H^+ \qquad \text{Diethylaminoethyl-cellulose (DEAE)}$$

$$-O-CH_2-CH_2-\overset{Et}{\underset{Et}{N^+}}-Et \qquad \text{Triethylaminoethyl-cellulose (TEAE)}$$

epichlorhydrin + triethanolamine (ECTEOLA),
or cation-exchangers:

$$-O-CH_2-C\overset{O}{\underset{}{\diagdown O^-}} \qquad \text{Carboxymethyl-cellulose (CM)}$$

$$-O-CH_2-S\overset{O}{\underset{O}{\diagdown O^-}} \qquad \text{Sulphomethyl-cellulose (SM)}$$

$$-O-CH_2-CH_2-S\overset{O}{\underset{O}{\diagdown O^-}} \qquad \text{Sulphoethyl-cellulose (SE)}$$

$$-O-P\overset{O}{\underset{O^-}{\diagdown O^-}} \qquad \text{Phosphorylated cellulose (P)}$$

Though the literature on the basic study of protein adsorption on the surface of ion-exchangers is rather scanty, the existence of multiple electrostatic bonds between the proteins and the exchangers was assumed long ago [187, 193–195]. Both proteins and ion-exchangers are polyelectrolytes and are therefore able to interact at several points if the internuclear distances are favourable. The elution of a substance from a certain adsorbent depends on the number of bonds which have been formed. For instance, for two proteins having the same surface charge density and adsorbed on the same ion-exchanger, stronger elution conditions will be needed for the protein with the larger molecule, as it will form more electrostatic bonds with the exchanger. Proteins can migrate along a column without being fixed if they lack a charge of the correct sign for forming multiple bonds with the exchanger. Other proteins may possess charges such that they can form electrostatic bonds with the adsorbent, and there will be a finite probability that these bonds can dissociate simultaneously; the molecule will then migrate to another place on the

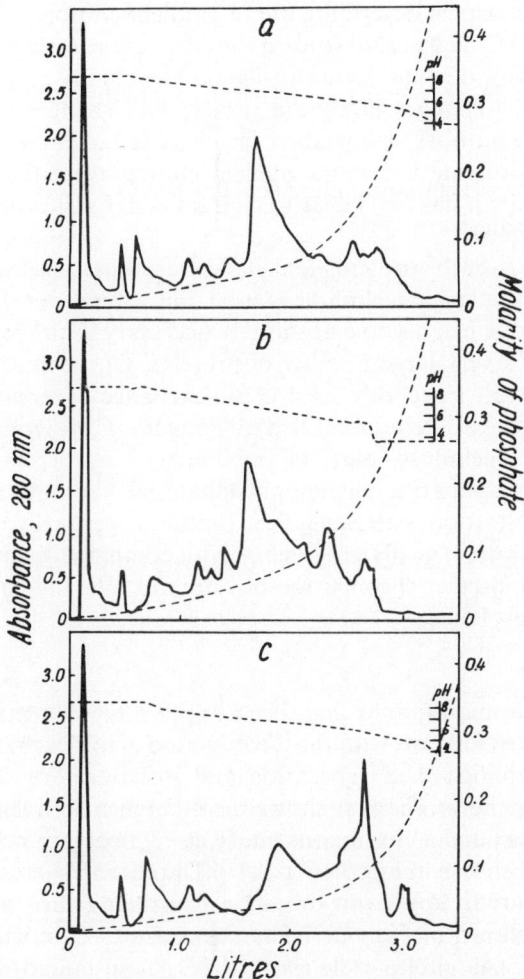

FIG. 7.95. Effect of shape of gradient on elution profile of serum. Column: 40 × 2·2 cm; 25 g of DEAE-cellulose. Starting buffer, 0·005M tris phosphate, pH = 8·6; limit buffer, 0·5M tris phosphate. Volume per cent of limit buffer in a nine-chambered 'Varigrad' gradient device: a: 0, 1·6, 2, 9, 9, 2, 20, 30, 100; b: 0, 0, 4, 0, 25, 0, 25, 40, 100; c: 0, 0, 4, 12, 0, 14, 0, 32, 100. Gradient volume, 9 × 375 ml; flow-rate 35 ml/hr.
(———) absorbance (protein concentration), (– – – –) phosphate concentration, (.....) pH gradient [196].

column, where the bonds will be re-established and the process repeated, until elution is complete. Hence the gradient elution technique is the best fitted to satisfy the consequent demands for continuous change of the ionic strength or pH of the eluent. Conditions for selective elution of a protein mixture are thus created. This is confirmed by the extraordinarily high number of papers describing use of gradient elution.

Peterson and Chiazze [196] studied the effect of the form of the gradient on the separation of some serum proteins. The results given in Fig. 7.95 clearly show the changes in elution profile with changes in the type of gradient. These authors also studied the effect of buffer composition and showed that succinate is a more efficient eluting agent than phosphate, probably because it has two negative charges at pH values at which phosphate has only one.

In the separation of proteins (enzymes, hormones, haemoglobins and plasma) not just a single technique is used, but several, yet the chromatographic technique has become absolutely necessary in the separation and purification of a very large number of proteins. Of the chromatographic methods, the most frequently used is ion-exchange chromatography, on DEAE-cellulose, CM-cellulose, DEAE-Sephadex, CM-Sephadex, TEAE-cellulose and P-cellulose. Next in popularity is adsorption chromatography on hydroxyapatite, calcium phosphate gel, kaolinite, and alumina, as well as gel-filtration with Sephadex. Elution is generally done with an ionic strength (saline) or pH gradient or with combined gradients.

Paper and thin-layer chromatography are much less often used for the separation of proteins.

7.3.8 Lipids

Gradient chromatography has been applied to separation of lipids, which took a step forward with the introduction of ion-exchange cellulose, which has simplified their separation and isolation. We shall mention only two papers here. The first shows the difference between the use of a stepwise gradient and a continuous one (Fig. 7.96), obtained by Wren [77] with the device shown in Fig. 7.16 (p. 147). The elution curves obtained for the same mixture of lipids and the same adsorbent (silicic acid) for both gradients are shown in Fig. 7.97. The peaks are C–G, biliary pigments; B, cholesterol; A, cholesterol ester; C–O, lipid-bound amino-acids; L, inositol phospholipid; L, M, phosphatidalcholine and phosphatidylcholine; H, phosphatidalethanolamine and phosphatidylethanolamine; H, phosphatidylserine; I–L, phosphatidylseride salt (?); N, sphingomyelin; A, triglyceride.

When the continuous gradient is used, the peaks are sharper and have smaller tails, and false peaks arising from abrupt changes in the eluent composition are avoided.

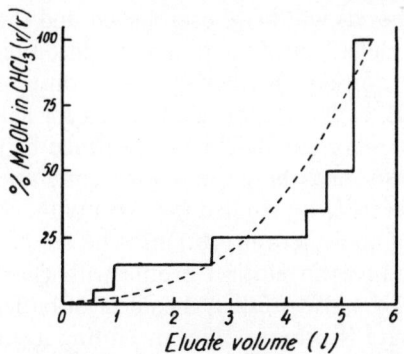

FIG. 7.96. Concentration continuous gradient (dotted line) produced by the apparatus illustrated in Fig. 7.16, and discontinuous gradient used for the separations illustrated in Fig. 7.97 [77].

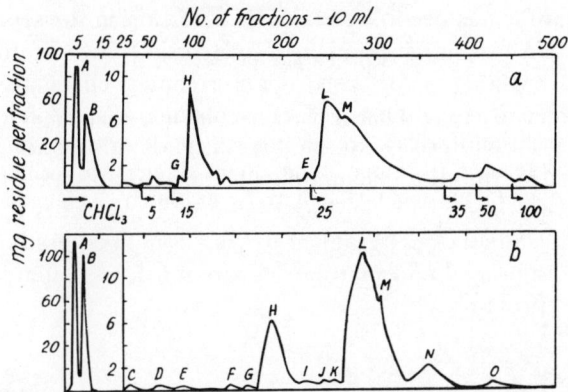

FIG. 7.97. Elution curves of equal quantities of lipid (1·23 g) from silicic acid columns (50 g) with the discontinuous gradient (a) and continuous gradient (b) illustrated in Fig. 7.96 [77].

The resolution of the chromatographic peaks evidently depends on the slope of the gradient. This was demonstrated in the linear gradient elution of a phospholipid mixture on a silicic acid column, by Billimoria et al. [85] whose results are shown in Fig. 7.98. In the case of peaks 1–5 resolution is poor. Rechromatography of the mixture containing these components, in the same conditions but with a gradient having a slope approximately 70 per cent of that used in the first separation, gave a clear separation of the five components (Fig. 7.99). The peaks are *1*, a complex phosphatidic acid; *2, 3, 4*, phosphatidylserine and phosphatidylserine galactoside, *5*, phosphatidylethanolamine, phosphatidylserine traces, compound X.

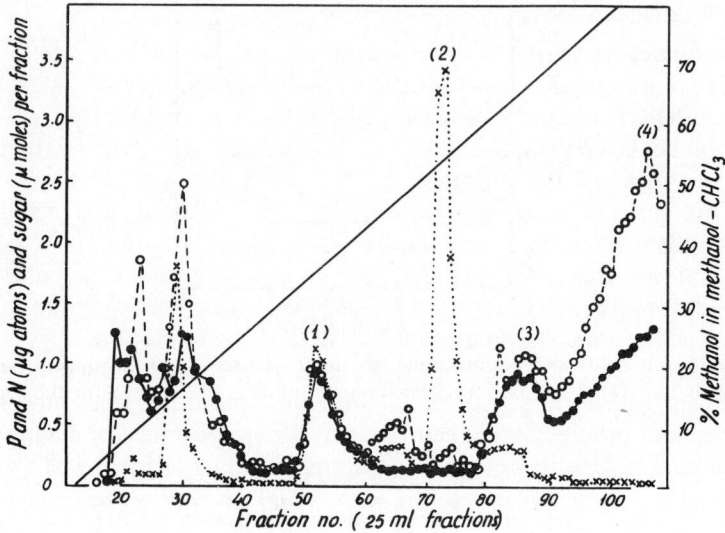

Fig. 7.98. Chromatography of butter phospholipids by a continuous linear gradient elution with methanol–CHCl$_3$ from a column of silicic acid. ●—phosphorus, ○—nitrogen, ×—sugar, (—) slope of the solvent gradient, 6—lecithin galactoside, 7—a sugar-containing phosphatidylcholine, 8—lecithin, 9—sphingomyelin [85].

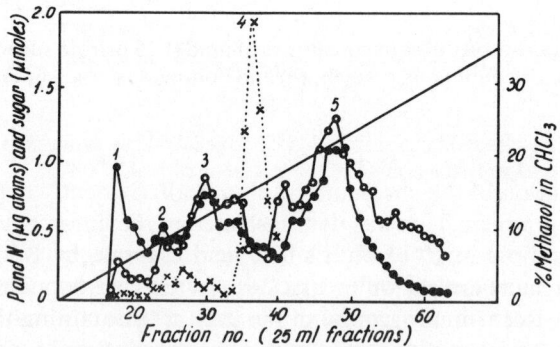

Fig. 7.99. Re-chromatography of peaks 1–5 from Fig. 7.98 with gradient of methanol increasing to 34 per cent after 70 fractions. ●—phosphorus, ○—nitrogen, ×—sugar, (—) slope of gradient. Reservoir diam. 83·5 mm and mixing vessel 73 mm. Concentration of methanol in CHCl$_3$ in mixing vessel, 57 per cent (v/v) [85].

7.3.9 NUCLEIC ACIDS

The numerous methods and techniques used for the study of nucleic acids have led to knowledge of many details of their chemical structure, physical behaviour and biological activity, and chromatography has also contributed largely to the study of nucleic acids and their structural elements. Among the chromatographic methods thus used, we may mention ion-exchange chromatography on cellulose and synthetic resins, adsorption chromatography on methylated albumin, calcium phosphate gel, charcoal, and kieselguhr, gel-filtration, and paper and thin-layer chromatography. In most of the papers on chromatography of nucleic acids, ionic strength gradients were used for elution.

Peterson and Kuff [197] studied the effect of the magnesium concentration in the saline gradient used for fractionation of the ribosomes from rat liver and mouse plasma cell tumour. The importance of magnesium ions in maintaining the integrity of the ribosomal structure is well known, so this study is of high interest. Figure 7.100 illustrates the results of a

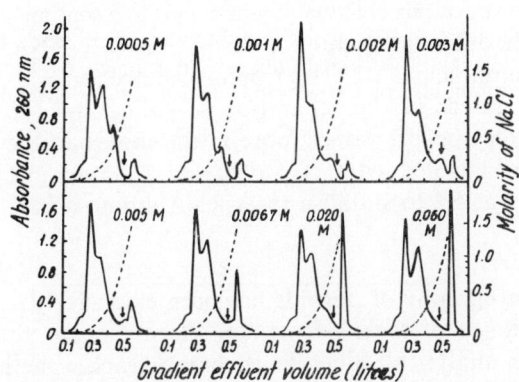

FIG. 7.100. The effect of increasing Mg^{2+} concentration in the salt gradient. The broken line represents the course of the gradient of NaCl [197].

series of experiments in which the magnesium concentration was initially 0.5mM, but increased at different rates during the chromatography, the Mg^{2+}/Na^+ ratio in the gradient portion of the chromatogram being higher for each successive experiment. The most satisfactory distribution of the ribonucleoprotein peaks appears with use of the lowest ratio. Optimum conditions evidently require the form of the magnesium gradient to be different from that of the sodium gradient. Cherayil and Bock [198] used two linear sodium chloride and magnesium chloride

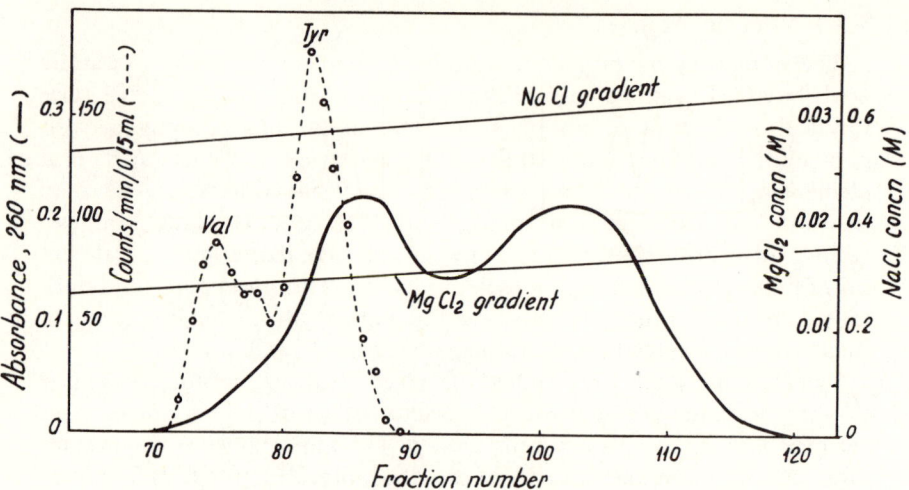

Fig. 7.101. Elution of s-RNA with NaCl and MgCl₂ gradient at pH = 5·2 [198].

gradients with different slopes for resolution of s-RNA on a DEAE-Sephadex column at pH 5·2. The results are given in Fig. 7.101. Resolution was very good for the separation of valine and tyrosine acceptor RNA. The latter appears rather more in the first than the last fractions. The s-RNA was labelled with ^{14}C-valine and ^{14}C-tyrosine, and the working pH was set at 5·2 to stabilise the s-RNA amino-acids.

7.3.10 STEROIDS

The chromatography of steroids has been extensively explored, owing to the importance of these substances. According to the desired end, preparative or analytical, different chromatographic methods are used. For preparations, liquid column techniques are preferred, and for analysis thin-layer chromatography, which has more and more replaced paper chromatography and gas chromatography for this purpose.

Owing to the complexity of the mixtures, gradient elution plays an important role in the liquid chromatography. We shall deal only with some aspects of major importance for the programming of the gradients, aspects examined in the work of Francois et al. [199] and of Johnson et al. [121].

Francois et al. used water on a silicic acid support (Merck, dried at 100°C for 3 hr) as the stationary phase (ratio 2·5:1 w/w), and dichloromethane (DCM) in petroleum ether (PE) as eluent. The following steroid solutions were prepared (100 μg/ml in anhydrous ethanol): Δ^4-pregnen-21-ol-

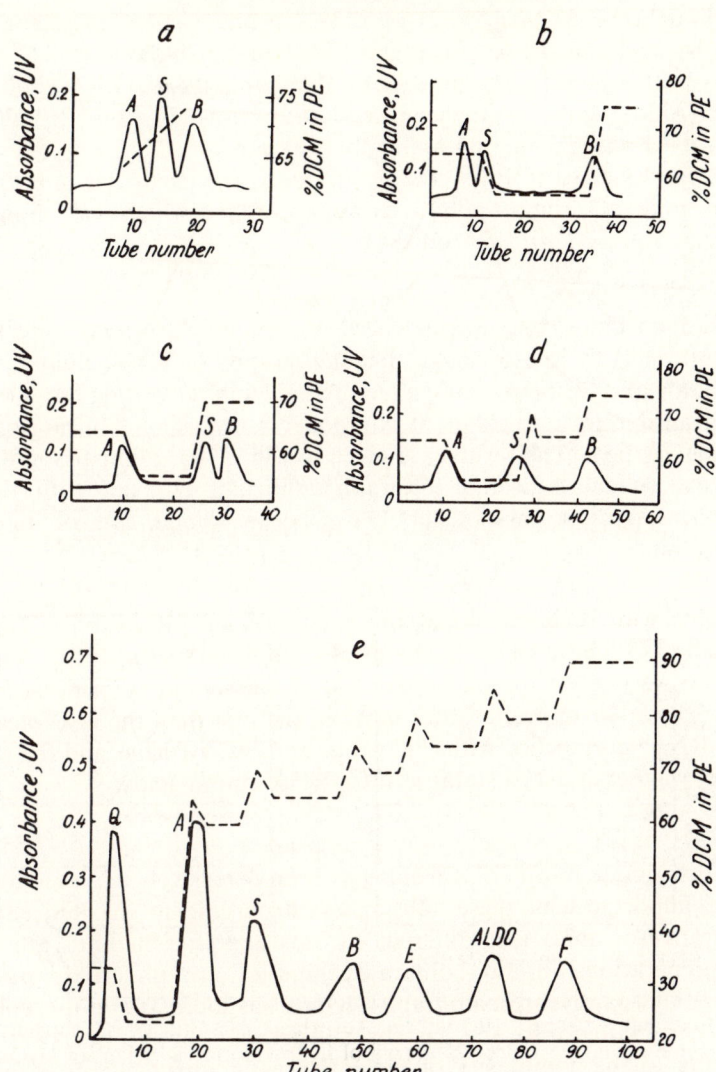

FIG. 7.102. Programmed gradient elution chromatography with the steroid analyser, using a gradient of dichloromethane (DCM) in petroleum ether (PE). *a*—Separation of compounds, A, S and B with a simple linear gradient; *b*—separation of compounds A, S and B with a programmed complex linear gradient; *c*—separation of compounds A, S and B with approx. same programmed complex linear gradient as in *b*; *d*—separation of compounds A, S and B with a more complex linear gradient than in *a* and *b*; *e*—separation of seven physiologically active adrenocortical hormones (compounds Q, A, S, B, E, aldo and F) with a programmed very complex linear gradient [199].

3,20-dione (Q), Δ^4-pregnen-21-ol-3,11-20-trione (A), Δ^4-pregnene-11β-21-diol-3,20-dione (B), Δ^4-pregnene-17α,21-diol-3,20-dione (S), Δ^4-pregnene-17α,21-diol-3–11,20-trione (E), Δ^4-pregnen-18-ol-11β-21-diol-3,20-dione (Aldo), and Δ^4-pregnene-11β,17α-21-triol-3,20-dione (F). These were mixed and chromatographically separated.

The results given in Fig. 7.102 show clearly the particular importance of the gradient programming (the apparatus for which was shown in Figs. 7.57 and 7.60, pp. 182 and 184).

7.3.11 ANTIBIOTICS

A sodium chloride gradient has also been used to fractionate some streptothricins by ion-exchange chromatography on CM-cellulose. Thus Shokholov and Reshetov [200] have succeeded in separating five streptothricin antibiotics into different components, by using a linear sodium chloride gradient. The results are given in Fig. 7.103. Comparison of different experiments is given by the graph in the right-hand part of the figure, and shows a close relationship between the components of different preparations.

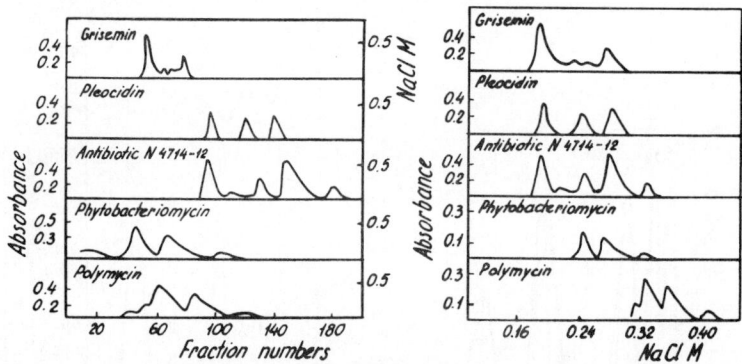

FIG. 7.103. Chromatography of five streptothricins on a carboxymethyl-cellulose (CMC) column. Column: 0·9 × 40 cm. Adsorbent: Na-form CMC, capacity 0·55 meq/g. Amount of substance: 20–30 mg of hydrochloride in 0·5 ml of H_2O. Rate 15–20 ml/hr. Fraction volume: 5·9 ml (grisemin), 4·8 ml (pleocidin), 5 ml (phytobacteriomycin), 4·6 ml (antibiotic No 4714-12), 5·2 ml (polymycin). Eluent: NaCl solution with linear concentration gradient [200].

7.3.12 INORGANIC IONS

Gradient elution has been as spectacularly successful in inorganic chemistry, when compared with conventional chromatography. To illustrate this we quote the results obtained by Nervik [32] for the separation

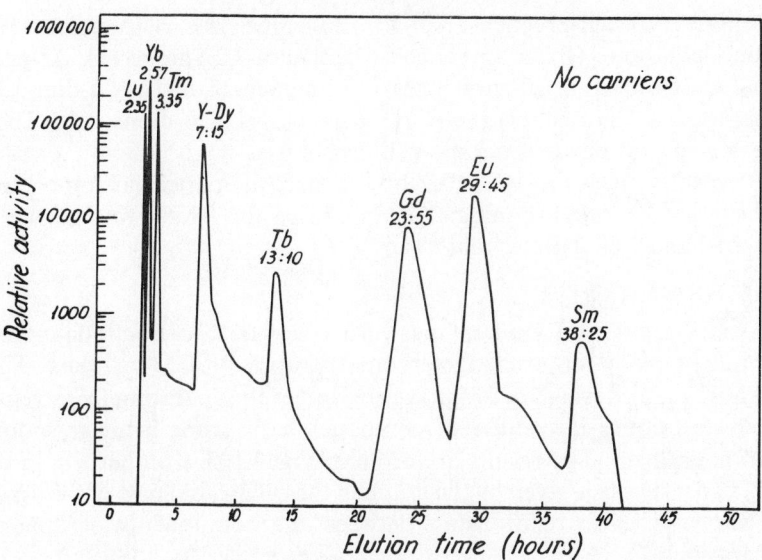

FIG. 7.104. Elution curve of a mixture of rare earths at a constant pH of 3·19 [32].

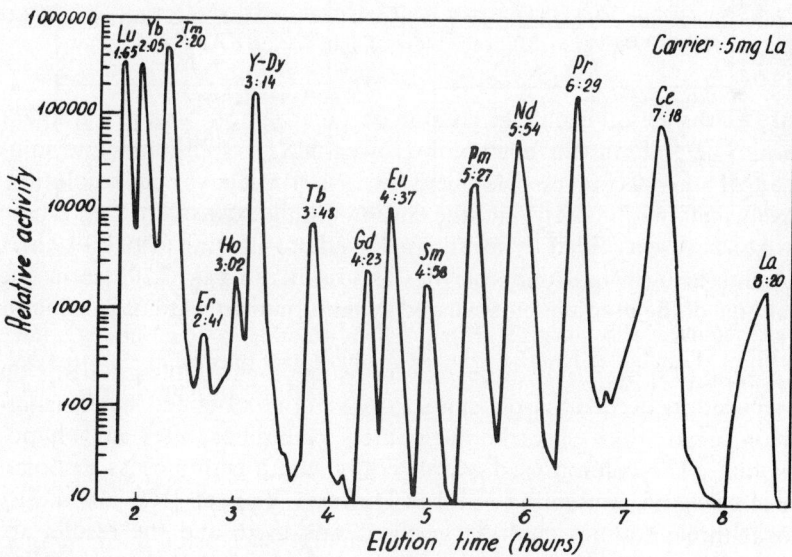

FIG. 7.105. Elution curve of a mixture of rare earths, for a continuous pH-gradient. Initial pH = 3·19 and an average rate of change of pH (dpH/dt) of 0·107/hr [32].

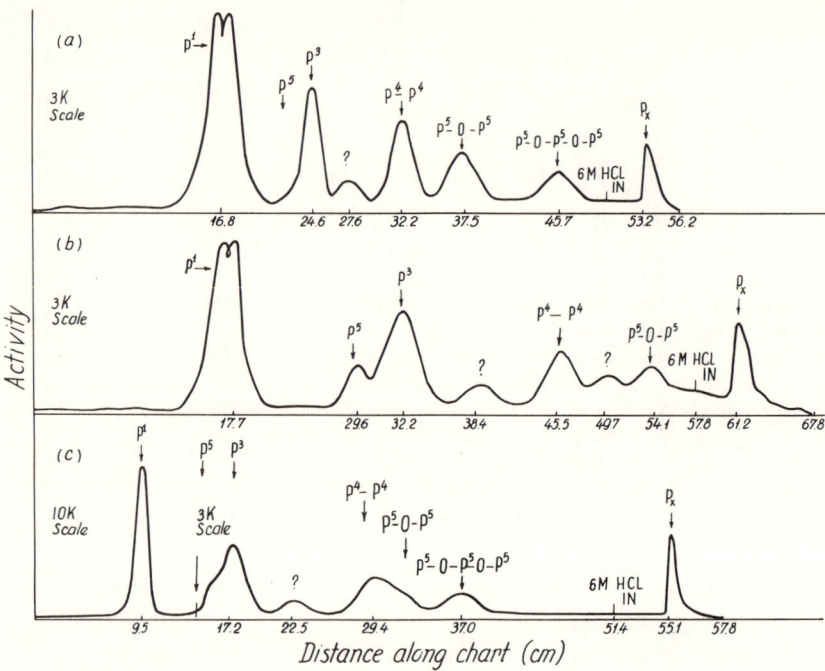

FIG. 7.106. Separation of the oxidation state of phosphorus oxyacids under different gradient-producing conditions. a—0·5M NaCl into 2·0 l of 0·025M NaCl, flow-rate 5·8 ml/min; b—0·3M NaCl into 2·0 l of 0·025M NaCl, flow-rate 6·0 ml/min; c—0·3M NaCl into 2·0 l of 0·1M NaCl, flow-rate 6·8 ml/min [73].

of rare earths by ion-exchange column chromatography and pH-gradient elution. The column was filled with Dowex-50X12. Sodium lactate solution at pH 3·19 was first used as eluent (Fig. 7.104) but gave poor resolution, the peaks having long tails and the elution taking excessively long. These drawbacks were avoided by using a pH gradient starting at pH 3·19 and increasing at 0·107 pH units per hr. The results in Fig. 7.105 show the advantage of the gradient elution, and the elution time was cut by a factor of about 8.

Anselmo [73] has studied the action of some logarithmic gradients on the separation of various phosphorus oxyanions obtained by thermal-neutron irradiation of orthophosphates, orthophosphites and hypophosphites. The column used was 43 cm in length with a cross-sectional area of 4·91 cm^2, containing 38 ml of Dowex 1-X8 resin (100–200 mesh). A logarithmic sodium chloride gradient was used, and the results are given in Fig. 7.106. The efficiency of separating such mixtures by saline gradient elution was confirmed by other workers [37, 63, 67].

REFERENCES

1. Lebreton, P., *Bull. Soc. Chim. France*, 2188 (1960).
2. Mikeš, O., *Chem. Listy* **54**, 576 (1960).
3. Dorfner, K., *Chemiker-Ztg. Chem. Apparatur* **87**, 871 (1963).
4. Snyder, L. R., *Chromatog. Rev.* **7**, 1 (1965).
5. Liteanu, C. and Gocan, S., *Bull. Soc. Chim. France*, 4527 (1970).
6. Roberts, E. J. and Mason, A., *Anal. Chem.* **28**, 1063 (1956).
7. Harpur, R. P., *Anal. Chem.* **31**, 2212 (1959).
8. Brusca, D. R. and Gavienowski, A. M., *J. Chromatog.* **14**, 502 (1967).
9. Anderson, N. G., Bond, H. E. and Canning, R. E., *Anal. Biochem.* **3**, 472 (1962).
10. Teekell, R. A., Boling, W. H., Lyke, W. A. and Chiriboga, J., *J. Chromatog.* **7**, 424 (1962).
11. Lerner, S. R., *Anal. Chem.* **35**, 1108 (1963).
12. Rombauts, W. A. and Raftery, M. A., *Anal. Chem.* **37**, 1611 (1965).
13. Busch, H., Hurlbert, R. B. and Potter, V. R., *J. Biol. Chem.* **196**, 717 (1952).
14. Bock, R. M. and Ling, N.-S., *Anal. Chem.* **26**, 1543 (1954).
15. Drake, B., *Arkiv. Kemi*, **8**, 1 (1955).
16. Hauton, J. C., *J. Chromatog.* **2**, 257 (1962).
17. Sorin, B. and Vargues, R., *Technicon Symposia, 1966. Automation in Analytical Chemistry*, Vol. II, p. 469, Mediad Incorporated, New York (1967).
18. Cherkin, A., Martinez, F. E. and Dunn, M. S., *J. Am. Chem. Soc.* **75**, 1244 (1953).
19. Svensson, H. and Forchheimer, A., *Acta Chem. Scand.* **16**, 2131 (1962).
20. Schulz, W. W. and Purdy, W. C., *Anal. Chem.* **35**, 2222 (1963).
21. Palo, V., Koman, V. and Harabe, Z., *Chem. Zvesti* **12**, 525 (1958).
22. Warner, H. R. and Lands, W. E. M., *J. Lipid Res.* **1**, 248 (1960).
23. O'Sullivan, D. G., *Analyst* **85**, 434 (1960).
24. Svensson, H., *Science Tools* **5**, 37 (1958).
25. Hadwiger, H. and Glur, P., *Experientia* **19**, 270 (1963).
26. Kocent, A., *J. Chromatog.* **6**, 324 (1961).
27. Bendich, A., Pahl, H. B., Korngold, G. C., Rosenkranz, H. S. and Fresco, J. R., *J. Am. Chem. Soc.* **80**, 3949 (1958).
28. Sakami, W., *Anal. Biochem.* **3**, 358 (1962).
29. Kesner, L., Muntwyler, E., Griffin, G. E. and Abrams, J., *Anal. Chem.* **34**, 1178 (1962).
30. Alm, R. S., Williams, R. J. P. and Tiselius, A., *Acta Chem. Scand.* **6**, 826 (1952).
31. Mader, C., *Anal. Chem.* **26**, 566 (1954).
32. Nervik, W. E., *J. Phys. Chem.* **59**, 690 (1955).
33. Donaldson, K. O., Tulane, V. J. and Marshall, L. M., *Anal. Chem.* **24**, 185 (1952).

34. Bannister, D. W., Phillips, C. S. G. and Williams, R. J. P., *Anal. Chem.* **26**, 1451 (1954).
35. Boman, H. G., *Nature* **173**, 447 (1954).
36. Boman, H. G., *Biochim. Biophys. Acta* **16**, 245 (1955).
37. Grande, J. A. and Beukenkamp, J., *Anal. Chem.* **28**, 1497 (1956).
38. Drake, B., *Arkiv. Kemi* **8**, 189 (1955).
39. Hurlbert, R. B., Potter, V. R., *J. Biol. Chem.* **209**, 1 (1954).
40. Busch, H. and Potter, V. R., *Cancer Res.* **12**, 660 (1952).
41. Busch, H. and Potter, V. R., *Cancer Res.* **13**, 168 (1953).
42. Hurlbert, R. B., Schmitz, H., Brumm, A. F. and Potter, V. R., *J. Biol. Chem.* **209**, 23 (1954).
43. Lerman, L. S., *Biochim. Biophys. Acta* **18**, 132 (1955).
44. Moore, S. and Stein, W. H., *J. Biol. Chem.* **211**, 893 (1954).
45. Williams, R. J. P., *Analyst* **77**, 905 (1952).
46. Schwab, H., Rieman, W., III and Vaughan, P. A., *Anal. Chem.* **29**, 1357 (1957).
47. Hirsch, J. and Ahrens, E. H., Jr., *J. Biol. Chem.* **233**, 311 (1958).
48. Alexa, J., *Chem. Listy* **53**, 761 (1959).
49. Hauton, J. C., *J. Chromatog.* **9**, 439 (1962).
50. Kesner, L., Muntwyler, E., Griffin, G. E. and Abrams, J., *Anal. Chem.* **35**, 83 (1963).
51. Campbell, P. N., Jacobs, S., Works, T. S. and Kaessman, T. R. E., *Chem. Ind. London* 117 (1955).
52. Chalkley, D. E. and Williams, R. J. P., *J. Chem. Soc.* 1718 (1954).
53. Clauser, H. and Li, C. H., *J. Am. Chem. Soc.* **76**, 4337 (1954).
54. Schmitz, H., Hurlbert, R. B. and Potter, V. R., *J. Biol. Chem.* **209**, 41 (1954).
55. Schmitz, H., Potter, V. R., Hurlbert, R. B. and White, D. M., *Cancer Res.* **14**, 66 (1954).
56. Sober, H. A. and Peterson, E. A., *J. Am. Chem. Soc.* **76**, 1711 (1954).
57. Sober, H. A., Gutter, F. J., Wickoff, M. M. and Peterson, E. A., *J. Am. Chem. Soc.* **78**, 756 (1956).
58. Sober, H. A. and Peterson, E. A., *Federation Proc.* **17**, 1116 (1958).
59. Thomson, A. R., *Biochem. J.* **61**, 253 (1955).
60. Vilkas, E. and Lederer, E., *Compt. Rend.* **240**, 1156 (1955).
61. Middleton, W. R., *Anal. Chem.* **39**, 1839 (1967).
62. De Bruyne, P., *J. Inorg. Nucl. Chem.* **32**, 348 (1970).
63. Ohashi, S., Tsuji, N., Veno, Y., Takeshita, M. and Muto, M., *J. Chromatog.* **50**, 349 (1970).
64. Hurlbert, R., in *Methods in Enzymology*, Vol. 3, S. P. Colowick and N. O. Kaplan (eds.), p. 785, Academic Press, New York (1957).
65. Gutter, F. J., Peterson, E. A. and Sober, H. A., *Arch. Biochem. Biophys.* **80**, 353 (1959).
66. Parr, C. W., *Biochem. J.* **56**, XXVII (1954).
67. Koguchi, K., Waki, H. and Ohashi, S., *J. Chromatog.* **25**, 398 (1966).
68. Molnár, F., Horváth, A. and Khalkin, V. A., *J. Chromatog.* **26**, 215 (1967).
69. Ohashi, S. and Koguchi, K., *J. Chromatog.* **27**, 214 (1967).
70. Barclay, R. W., *J. Chromatog.* **31**, 145 (1967).
71. Vestergaard, P., *J. Chromatog.* **3**, 560 (1960).
72. Wallach, D. F. H. and Nordby, G. L., *Biochim. Biophys. Acta* **70**, 188 (1963).
73. Anselmo, V. C., *U.S. At. En. Comm. Rept.* C00-1618-1, Avail. Dept. CFST.

74. Thomas, D. E. and Thomas, J. D. R., *Analyst* **94**, 1099 (1969).
75. Skelly, N. E., *Anal. Chem.* **33**, 271 (1961).
76. Miller, G. L., *Anal. Biochem.* **2**, 133 (1960).
77. Wren, J. J., *Nature* **184**, 816 (1959).
78. Nelson, G. J., *Anal. Biochem.* **5**, 116 (1963).
79. Nelson, G. J., *J. Am. Oil Chem. Soc.* **44**, 86 (1967).
80. Lakshmanan, T. K. and Lieberman, S., *Arch. Biochem. Biophys.* **53**, 258 (1954).
81. Piez, K. A., *Anal. Chem.* **28**, 1451 (1956).
82. Reiner, J. M. and Reiner, B., *Anal. Biochem.* **4**, 1 (1962).
83. Arcus, A. C., *J. Chromatog.* **3**, 411 (1960).
84. Choules, G. L., *Anal. Biochem.* **3**, 236 (1962).
85. Billimoria, J. D., Curtis, R. G. and Maclagan, N. F., *Biochem. J.* **78**, 185 (1961).
86. Peterson, E. A. and Sober, H. A., *Anal. Chem.* **31**, 857 (1959).
87. Chase, M. A., *Anal. Chem.* **35**, 1457 (1963).
88. Lakshmanan, T. K. and Lieberman, S., *Arch. Biochem. Biophys.* **45**, 235 (1953).
89. Liebl, V. and Mikeš, O., *Chem. Listy* **52**, 2153 (1958).
90. Liebl, V. and Mikeš, O., *Collection Czech. Chem. Commun.* **24**, 809 (1959).
91. Mikeš, O., Tomóšek, V. and Holeyšovský, V., *Chem. Listy* **53**, 609 (1959).
92. Kenyon, W. C., McCarley, J. E., Boucher, E. G., Robinson, A. E. and Wiebe, A. K., *Anal. Chem.* **27**, 1888 (1955).
93. Horton, B. F., *J. Chromatog.* **27**, 263 (1967).
94. Van Tamelen, E. E. and Taylor, C. W., *J. Am. Chem. Soc.* **79**, 5256 (1957).
95. Wren, J. J., *J. Chromatog.* **12**, 32 (1963).
96. Bader, H. and Morgan, H. E., *Biochim. Biophys. Acta* **57**, 562 (1962).
97. Hoban, N. and White, J. W., Jr., *Anal. Chem.* **30**, 1294 (1958).
98. Brown, G. L. and Watson, M., *Nature* **172**, 339 (1953).
99. Brown, G. L. and Watson, M., *Trans. Faraday Soc.* **50**, 294 (1954).
100. Kellie, A. E. and Wade, A. P., *Biochem. J.* **66**, 196 (1957).
101. Kellie, A. E., *Analyst* **82**, 722 (1957).
102. Lawson, G. J. and Purdie, J. W., *Mikrochim. Acta* 415 (1961).
103. Rosett, T., *J. Chromatog.* **18**, 498 (1965).
104. Snyder, L. R. and Warren, H. D., *J. Chromatog.* **15**, 344 (1964).
105. Desreux, V., *Rec. Trav. Chim.* **68**, 789 (1949).
106. Polson, A., *J. Chromatog.* **5**, 116 (1961).
107. Lindberg, B. and Wickberg, B., *Acta Chem. Scand.* **7**, 140 (1953).
108. Lerman, L. S., *Nature* **172**, 635 (1953).
109. Keller, P. J., Cohen, E. and Neurath, H., *J. Biol. Chem.* **233**, 344 (1958).
110. Alm, R. S., *Acta Chem. Scand.* **6**, 1186 (1952).
111. Katz, S., *Anal. Biochem.* **5**, 7 (1963).
112. Watts, P. R., *Chem. Ind. London* 76 (1963).
113. Allen, R. R. and Eggenberger, D. N., *Anal. Chem.* **27**, 476 (1955).
114. Bombaugh, K. J., King, R. N. and Cohen, A. J., *J. Chromatog.* **43**, 332 (1969).
115. Smith, M. A. and Stahmann, M. A., *J. Chromatog.* **9**, 528 (1962).
116. Schmidtmann, W., *Chemiker. Ztg. Chem. Apparatur* **89**, 231 (1965).
117. Burns, J. A., Curtis, C. F. and Kacser, H., *J. Chromatog.* **20**, 310 (1965).
118. Hagerman, D. D., *J. Chromatog.* **41**, 250 (1969).
119. Curtain, C. C., *J. Chromatog.* **7**, 24 (1962).
120. Keck, K., *Anal. Biochem.* **39**, 288 (1971).

121. Johnson, D. F., Lamontagne, N. S., Riggle, G. C. and Anderson, F. O., *Anal. Chem.* **43**, 1712 (1971).
122. Blattner, F. R. and Abelson, J. N., *Anal. Chem.* **38**, 1279 (1966).
123. Anderson, F. O., Crisp, L. R., Riggle, G. C., Vurek, G. G., Heftmann, E., Johnson, D. F., Francois, D. and Perrine, T. C., *Anal. Chem.* **33**, 1606 (1961).
124. Bailey, D. G., *J. Chromatog.* **50**, 137 (1970).
125. Lederer, M., *Nature* **172**, 727 (1953).
126. Muic, N. and Meniga, A., *Arh. Hig. Rada Toksik.* **15**, 341 (1964).
127. Franks, F., *Analyst* **81**, 384 (1956).
128. Franks, F., *Analyst* **81**, 390 (1956).
129. Curtain, C. C., *Nature* **191**, 1269 (1961).
130. De Wachter, R., *J. Chromatog.* **36**, 109 (1968).
131. Koman, V. and Palo, V., *Chem. Zvesti* **12**, 513 (1958).
132. Wieland, Th. and Determann, H., *Experientia* **18**, 431 (1962).
133. Rybika, S. M., *Chem. Ind. London* 308 (1962).
134. Niederwieser, A. and Honegger, C. G., in *Advances in Chromatog.* Vol. 2, p. 123, Giddings, G. C. and Keller, R. A. (eds.), Arnold, London (1966).
135. Luzzatto, L. and Okoye, C. N., *Biochem. Biophys. Res. Commun.* **29**, 705 (1967).
136. Stickland, R. G., *Anal. Biochem.* **10**, 108 (1965).
137. Niederwieser, A. and Honegger, C. G., *Helv. Chim. Acta* **48**, 893 (1965).
138. Keulemans, A. I. M., *Gas Chromatography*, Chap. 4, Reinhold, New York (1957).
139. Hagdahl, L., Williams, R. J. P. and Tiselius, A., *Arkiv. Kemi* **4**, 193 (1952).
140. Freiling, E. C., *J. Am. Chem. Soc.* **77**, 2067 (1955).
141. Tiselius, A., *Endeavour* **11**, 5 (1952).
142. Williams, R. J. P., Hagdahl, C. and Tiselius, A., *Arkiv. Kemi* **7**, 4 (1954).
143. Snyder, L. R., *J. Chromatog.* **16**, 55 (1964).
144. Snyder, L. R., *Advan. Anal. Chem. Instr.* **3**, 251 (1964).
145. Snyder, L. R., *J. Chromatog.* **8**, 178 (1962).
146. Snyder, L. R., *J. Chromatog.* **13**, 415 (1964).
147. Snyder, L. R., *J. Chromatog.* **6**, 22 (1961).
148. Snyder, L. R., *J. Chromatog.* **11**, 195 (1963).
149. Snyder, L. R., *J. Chromatog.* **12**, 488 (1963).
150. Snyder, L. R. and Saunders, D. L., *J. Chromatog. Sci.* **7**, 195 (1969).
151. Jermyn, M. A., *Australian J. Chem.* **10**, 55 (1957).
152. Barker, S. A., Bourne, E. J. and Theander, O., *J. Chem. Soc.* 4276 (1955).
153. Rony, P. R., *Sepn. Sci.* **3**, 425 (1968).
154. Carles, J. and Abravanel, G., *J. Chromatog.* **56**, 231 (1971).
155. Martin, A. J. P., *Ann. Rev. Biochem.* **19**, 517 (1950).
156. Lederer, E. and Lederer, M., *Chromatography*, 2nd Ed., p. 105, Elsevier, Amsterdam (1957).
157. Marshall, L. M., Donaldson, K. O. and Friedberg, F., *Anal. Chem.* **24**, 773 (1952).
158. Giddings, J. C., *Anal. Chem.* **39**, 1027 (1967).
159. Horváth, C. G. and Lipsky, S. R., *Anal. Chem.* **39**, 1893 (1967).
160. Freiling, E. C., *J. Phys. Chem.* **61**, 543 (1957).
161. Mayer, S. W. and Tompkins, E. R., *J. Am. Chem. Soc.* **69**, 2866 (1947).
162. Martin, A. J. P. and Synge, R. L. M., *Biochem. J.* **35**, 1358 (1941).
163. Maslova, G. B., Nazarov, P. P. and Chumutov, K. V., *Ionoobmen. Sorbenty v Prom.*, p. 103, Akad, Nauk SSSR, Inst. Fiz. Kim. (1963).

164. Inczédy, J., *Magy. Kém. Lapja* **24**, 232 (1969).
165. Inczédy, J., *Magy. Kém. Lapja* **24**, 488 (1968).
166. Massart, D. L. and Bossaert, W., *J. Chromatog.* **32**, 195 (1968).
167. Kraus, K. A. and Moore, F. L., *J. Am. Chem. Soc.* **71**, 3855 (1949).
168. Kraus, K. A. and Moore, F. L., *J. Am. Chem. Soc.* **73**, 9, 13, 2900 (1951).
169. Deelstra, H., *Ph.D. Thesis*, Ghent University.
170. Galeffi, C., Ciasca-Rendina, M. A., Delle Monache, E. M., Del Fresno, A. V. and Bettòlo, G. B. M., *J. Chromatog.* **45**, 407 (1969).
171. Snyder, L. R., in *Principles of Adsorption Chromatography*, Dekker, New York (1968).
172. Snyder, L. R., *Anal. Chem.* **39**, 698 (1967).
173. Snyder, L. R., *J. Chromatog. Sci.* **8**, 692 (1970).
174. Snyder, L. R. and Saunders, D. L., *J. Chromatog.* **44**, 1 (1969).
175. Snyder, L. R., *Anal. Chem.* **39**, 705 (1967).
176. Hartley, R. D. and Lawson, G. J., *J. Chromatog.* **4**, 410 (1960).
177. Hauton, J. C., *J. Chromatog.* **7**, 341 (1962).
178. Sjövall, J., *Clin. Chim. Acta* **4**, 652 (1959).
179. Jermyn, M. A., *Australian J. Biol. Sci.* **15**, 787 (1962).
180. Barker, S. A., Bourne, E. J. and Theander, O., *J. Chem. Soc.* 4276 (1955).
181. Kimura, M., *J. Chromatog.* **41**, 462 (1969).
182. French, D., Robyt, J. F., Weintraub, M. and Knock, P., *J. Chromatog.* **24**, 68 (1966).
183. Logie, D., *Analyst* **82**, 563 (1957).
184. Popl, M., Mostecky, J. and Havel, Z., *J. Chromatog.* **53**, 233 (1970).
185. Calzolari, C., Favretto, L. and Stancher, B., *J. Chromatog.* **47**, 209 (1970).
186. Starcher, B. C., Wenger, L. Y. and Johnson, L. D., *J. Chromatog.* **54**, 425 (1971).
187. Sober, H. A. and Peterson, E. A., in *Amino Acids, Proteins and Cancer Biochemistry*, p. 61, Academic Press, New York, (1960).
188. Swingle, S. M. and Tiselius, A., *Biochem. J.* **48**, 171 (1951).
189. Price, V. E. and Greenfield, R. E., *J. Biol. Chem.* **209**, 363 (1954).
190. Black, S. and Wright, N. G., *J. Biol. Chem.* **213**, 51 (1955).
191. Tiselius, A., *Arkiv. Kemi* **7**, 443 (1954).
192. Tiselius, A., Hjertén, S. and Levin, O., *Arch. Biochem. Biophys.* **65**, 132 (1956).
193. Boardman, N. K. and Portridge, S. M., *Biochem. J.* **59**, 543 (1955).
194. Moore, S. and Stein, W. H., *Advan. Protein Chem.* **11**, 191 (1956).
195. Peterson, E. A. and Sober, H. A., *J. Am. Chem. Soc.* **78**, 751 (1956).
196. Peterson, E. A. and Chiazze, E. A., *Arch. Biochem. Biophys.* **99**, 136 (1962).
197. Peterson, E. A. and Kuff, E. L., *Biochemistry* **8**, 2916 (1969).
198. Cherayil, J. D. and Bock, R. M., *Biochemistry* **4**, 1174 (1965).
199. Francois, D., Johnson, D. F. and Heftmann, E., *Anal. Chem.* **35**, 2019 (1963).
200. Shokholov, A. S. and Reshetov, P. D., *J. Chromatog.* **14**, 495 (1964).

CHAPTER 8

STATIONARY-PHASE GRADIENTS

Though there are many ways of changing the mobile phase (see Section 7.1), things are quite different for the stationary phase. This is because there are more solvents than adsorbents, and also because it is much easier to achieve mobile-phase gradients.

However, a stationary-phase gradient is a special way of obtaining rapid and optimal separation on a thin layer. Depending on the choice of start-line and flow-direction of eluent, three types of gradient are possible, orthogonal, parallel and antiparallel. The stationary phase gradient may also be discontinuous or continuous, and be based on composition, impregnation, activity and pH (see Section 6.2).

8.1 Apparatus

8.1.1 DISCONTINUOUS GRADIENTS

Berger et al. [1] and Abbott and Thomson [2] have separated compartments of different lengths in a Desaga spreader by introduction of close-fitting partitions of PTFE or aluminium sheet into the cylinder. The compartments are filled simultaneously with suspensions of different adsorbents. Elution can take place either perpendicularly to the direction of spreading or parallel to it; the two can also be combined.

Liebmann and Schumann [3] similarly created a Desaga spreader with five compartments, which was used to give a discontinuous parallel gradient or stationary phase.

Other procedures for obtaining a discontinuous gradient, such as the spraying of a solution or suspension on the partly covered plate, or the partial immersion of the plate, will be described in the practical part of the chapter.

8.1.2 CONTINUOUS GRADIENTS

Stahl et al. [4–9] worked out techniques and apparatus for obtaining continuous stationary-phase gradients. Stahl [5] described a device for filling columns, consisting of a chamber divided into two compartments, A_1 and A_2, by a diagonal partition B (Fig. 8.1). Both compartments are filled with two adsorbents, for instance basic and acidic alumina, of the

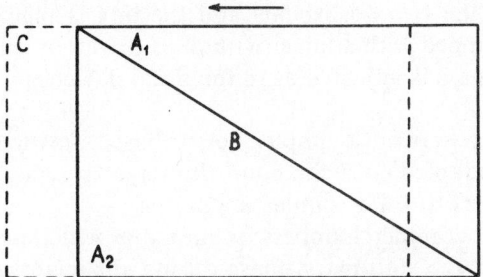

FIG. 8.1 Chamber with two compartments for obtaining gradients of column packing, from [5].

same grain size. The chamber has a sliding partition C at the base, and when this is opened, the two adsorbents will fall, mixed in different proportions, into a mixing chamber, after which they are continuously transported into the chromatographic column.

Recently, Kamm and Mes [10] made an apparatus for preparation of discs with a gradient of acrylamide gels for use in electrophoresis.

The problem of obtaining thin layers with a gradient was solved by Stahl [6] by constructing a mechanically driven device, based on the same principle of mixing two different adsorbents. It consisted of a box with a diagonal partition (Fig. 8.2). Two compartments A and B contain

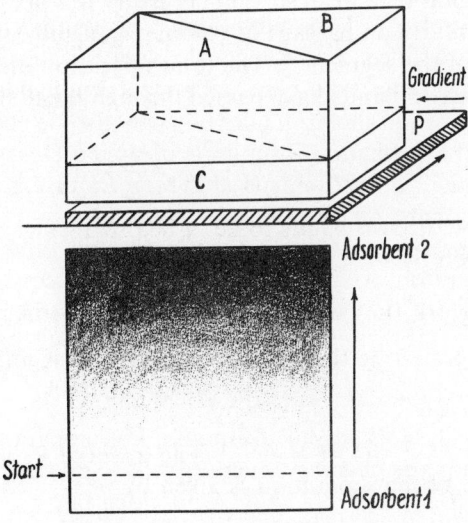

FIG. 8.2 Diagram of apparatus for obtaining layers with composition gradient, from [6].

suspensions of the two adsorbents and the box is placed on a mixing chamber C equipped with a mixer with discs driven by an electric motor. The layer thickness is adjusted as in the usual devices for spreading thin layers.

Warren [11] described a simpler device based on the same principle, which can be adapted to a Shandon thin-layer spreader and Unoplan leveller, as well as to other similar apparatus.

The chromatographic chambers by means of which activity gradients are obtained on the stationary phase during chromatography in a programmed vapour atmosphere, will be described in the next chapter.

8.2 Resolution

To establish the equations giving the resolution in stationary-phase gradient chromatography, Snyder and Saunders [12] considered only the the case of a parallel gradient. The calculation is the same as for a mobile phase gradient. The adsorbent is divided into segments, $1, 2, \ldots, i, \ldots, n$, of relative lengths $\bar{L}_i = L_i/L$, with corresponding average α^* values (for a pair of components in adjacent bands A and B) $\alpha_1^*, \alpha_2^*, \ldots, \alpha_i^*, \ldots, \alpha_n^*$. On passage of a volume V_1 of solvent which is constant in composition the two are carried to the end of segment 1 of the bed bands i.e.

$$\frac{V_1}{V^\circ} = \bar{L}_1 \alpha_1^* \tag{8.1}$$

where V° is the total volume of solvent necessary to soak the whole length of the bed L. Similarly on passage of volume V_2 of solvent, the two bands pass to the end of the segment 2. The total migration distance of solvent front L_f, after the two bands have passed through the first i bed segments, is given by

$$L_f/L = \sum^i (V_i/V^\circ) + \sum^i \bar{L}_i \tag{8.2}$$

The condition for the two bands to be in bed segment j at the end of the separation is given by

$$\sum^{j-i} [(V_i/V^\circ) + \bar{L}_i] < 1 < \sum^j [(V_i/V^\circ) + \bar{L}_i] \tag{8.3}$$

The distance migrated by the two bands in the jth segment is given by

$$\bar{L}_{j'} = (\Delta L/L)/(1 + \alpha_j^*) \tag{8.4}$$

where $\Delta L/L = 1 - \sum^{j-1} [(V_i/V^\circ) + \bar{L}_i]$. The average R_f value of the two bands at the end of the separation is given by

$$\left(\sum^{j-1} L_i \right) + \bar{L}_j \bar{L}_{j'} = \bar{L} \tag{8.5}$$

as in the case of the elution gradient.

The resolution R_s, and the separation function F may be calculated [12] essentially in the same way as in the case of gradient elution of a column [13] [see Eq. (7.204)]:

$$F_g = (R_s^2)_g = \frac{N[(\alpha_A^*/\alpha_B^*) - 1]^2 E_n^2}{16} \tag{8.6}$$

where N is the total number of theoretical plates in the bed of length L, and E_n^2 is given by

$$E_n^2 = \frac{\sum\limits^{j} \left(\prod\limits^{i<m<j} \right) G_m \bar{L}_i \alpha_i^*/(1 + \alpha_i^*)}{\sum\limits^{j} \left(\prod\limits^{i<m<j} G_m^2 \right) \bar{L}_i} \tag{8.7}$$

in which the compression factor G_m of the bands is given by

$$G_m = \alpha_{m+1}^*(1 + \alpha_m^*)/\alpha_m^*(1 + \alpha_{m+1}^*) \tag{8.8}$$

By means of Eq. (8.7) we can calculate the effective number of theoretical plates NE_n^2 in the stationary-phase gradient in thin-layer chromatography as a function of the adsorptivity of the sample α_1^* and of the adsorptivity gradient of the adsorbent, i.e. according to the position on the layer. In Fig. 8.3 are plotted the results of such a calculation for the function

$$\log \alpha^* = \log \alpha_1^* + b(x/L) \tag{8.9}$$

where x is the distance along the plate from the starting point.

On examining Fig. 8.3 we notice that at higher values of b the resolution of the two components is poorer. The resolution also varies with the position of the bands on the plate.

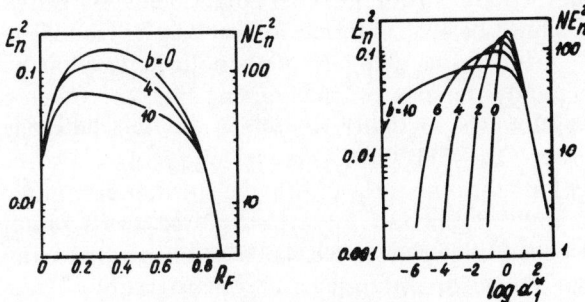

FIG. 8.3. Resolution on a gradient-layer as a function of gradient steepness b from Eq. (8.9) (logarithmic programme with V_i small) [12].

Comparing the results from Fig. 8.3 with those for elution gradients on thin layers (Figs. 7.87 and 7.88), we notice better resolution in the case of the stationary-phase gradient. This is only valid, however, for values of $b > 4$. In the case of $b < 4$, the resolutions for both gradients become practically equal.

8.3 Applications

Here we shall give some applications of stationary-phase gradients, classified according to the type of gradient.

8.3.1 COMPOSITION GRADIENT

Parallel discontinuous gradient (successive). A plate with two layers was used to separate Fe^{3+}, Ni^{2+} and Co^{2+} ions. The first portion of the plate was covered with an aqueous suspension of MN 300 cellulose (3 g), dimethylglyoxime (0·10 g) and distilled water (35 ml), and the second with a mixture of cellulose and Dowex 50 WX2 (NH_4^+ form) cation-exchanger. The cation mixture was added to the first portion of the plate, and developed upwards with an eluent of distilled water (70 ml), ethanol (20 ml), tartaric acid (5 g) and ammonia (added till its smell persisted). The nickel was retained on the starting line, as its dimethylglyoximate, while the iron and cobalt were carried to the second layer. Part of the cobalt remained at the boundary between the two layers, and the remainder was distributed in several spots corresponding to the various cobaltammine complexes, which could be detected by spraying with aqueous sodium sulphide solution. The iron, probably in the form of an anionic tartrate complex, was practically not retained by the resin and migrated with the solvent front.

In separation of a mixture of iodide, T_4 (thyroxine), DIT (di-iodotyrosine) and MIT (monoiodotyrosine), the most difficult problem is the separation of the iodide from the other components. Berger et al. [1] have used for this a layer of silver chloride, followed by a second layer of Dowex 1 X2 (OH^- form) resin, with $3N$ methanolic potassium hydroxide as eluent. The iodide remains on the starting line, the other components migrating towards the second layer where they will be separated in the following order: T_4, DIT, MIT. This technique may also be used with good results for other mixtures of organic substances containing iodine.

Liebmann and Schuhmann [3] have suggested layers with five zones of different adsorbent materials in parallel and antiparallel arrangements in one- and two-dimensional thin-layer chromatography.

Parallel continuous gradient. To test the effect of the composition gradient (kieselguhr–silica gel) Warren [11] has separated a mixture of

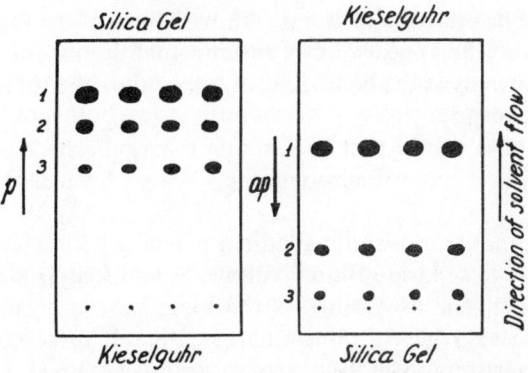

FIG. 8.4 Chromatograms of a dye mixture. Solvent system: benzene. Development distance: 18 cm. *1*—Butter Yellow, *2*—Sudan Red G, *3*—indophenol; *p*—parallel gradient, *ap*—antiparallel gradient [11].

three dyes, using a parallel and an antiparallel gradient. Results are given in Fig. 8.4.

Orthogonal gradient. This stationary-phase gradient opens new possibilities in thin-layer chromatography, both for separation of mixtures and for distribution of the zones on the chromatogram.

Stahl [5] has applied the orthogonal composition gradient of the stationary phase (cellulose–silica gel) in thin-layer chromatography to the separation of some amino-acids. A mixture of butanol–acetic acid–water (5:1:4 v/v) was used as eluent. Results are given in Fig. 8.5.

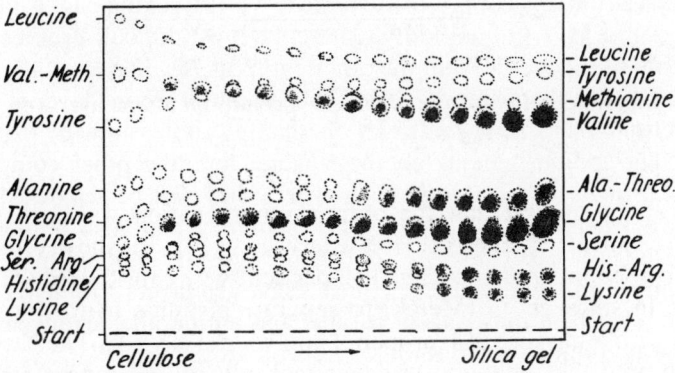

FIG. 8.5. Separation of a mixture of amino-acids on a thin layer with a composite gradient (cellulose/silica gel) [5].

It is advisable that this technique should be used to separate an unknown mixture. The appearance of the chromatogram will allow conclusions to be drawn as to the behaviour of one or the other of the substances forming the layer [5].

Figure 8.5 also shows that the colour reaction between amino-acids and ninhydrin is far more intense on the silica gel layer than on the cellulose layer.

The orthogonal composition gradient (silica gel–kieselguhr) was used by Warren [11] to separate Butter Yellow, Sudan Red G and indophenol by means of benzene as eluent. As the kieselguhr concentration in the silica gel increases, the R_f values increase for all three components in different proportions. The best separation is obtained with approximately 75 per cent kieselguhr.

8.3.2 Impregnation Gradient

Orthogonal gradient. The optimal degree of impregnation of a thin layer may easily be established, as shown by Stahl and Vollmann [7] by using an orthogonal gradient on a chromatographic plate. Figure 8.6 shows the separation of some terpene alcohols on a silica gel layer impregnated with silver nitrate.

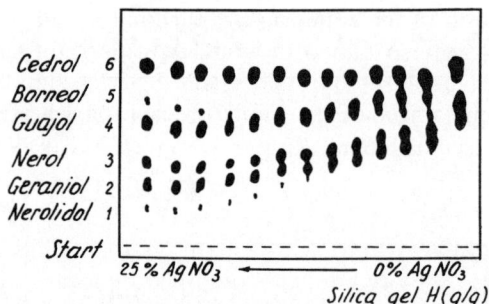

Fig. 8.6. Separation of a mixture of terpene alcohols on a thin layer with impregnation mixture [7].

This study showed that the optimal impregnation was with 2·5 per cent silver nitrate in silica gel H. Niederwieser [14] also showed the effect of an orthogonal gradient, using 0–2 per cent w/w cupric sulphate pentahydrate in silica gel G (Merck) to separate histidine hydrochloride by elution with a mixture of 1-propanol and water (7:3 v/v). The chromatogram showed that retention was maximal only at low concentrations of the complexation agent.

Rozumek [15], using an orthogonal silver nitrate gradient, showed that the best separation of the steroid alkaloids (tomatidenol, solasodine and soladulcidine) was obtained in the part of the plate containing 4·1–6·3 per cent silver nitrate in silica gel G. A content of 1–1·5 per cent silver nitrate in silica gel G is sufficient for the separation of steroid sapogenins (yamogenin, diosgenin and tigogenin).

8.3.3 ACTIVITY GRADIENT

Honegger [16] achieved an antidiagonal activity gradient on a thin layer in the following way: a plate with a thin layer which had been soaked in a 25 per cent solution of water in acetone (the right hand side of the plate being covered up to the diagonal line) for 5 sec, was activated at 110° for 20 min. The results are given in Fig. 8.7.

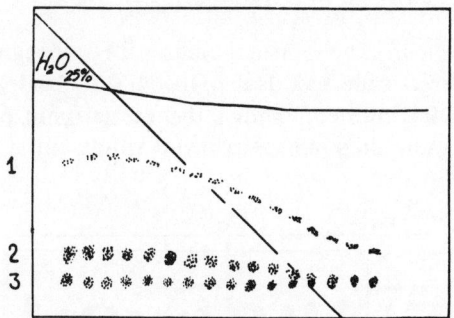

FIG. 8.7. Separation of a mixture of squalene (*1*), cholesteryl stearate (*2*) and cholesterol (*3*) using an activity gradient on silica gel G. Solvent: heptane. The left side of the plate was activated [16].

Thin-layer chromatography with activity gradients has been explored extensively of late. Activity gradients obtained by means of programmed vapour phase composition will be dealt with in the next chapter.

8.3.4 pH GRADIENTS

Orthogonal gradient. A full study was undertaken by Stahl and Dumont [9, 17] on the effect of an orthogonal pH gradient on the separation of such substances as: mono- and polybasic acids, ampholytes, organic bases, alkaloids and pH indicators. To obtain the pH gradient [9] silica gel G (Merck) suspensions in sulphuric acid or sodium hydroxide were used. The plates covered with the mixture of the two suspensions (by the device shown in Fig. 8.2) were activated for 45 min at 110°. The pH gradient covers the following ranges, depending on the concentration of

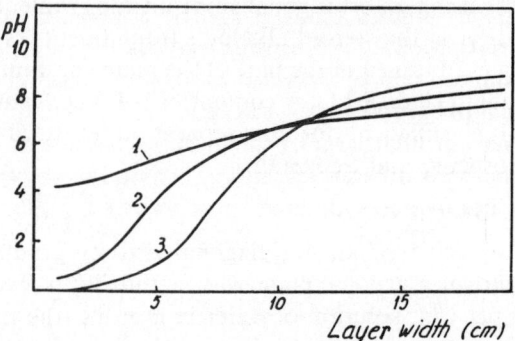

FIG. 8.8. pH values of the layer as a function of plate width (orthogonal pH-layer gradient) for three concentrations of NaOH and H_2SO_4 solutions used for preparing the silica gel suspensions. (1) $0.1N$; (2) $0.3N$; (3) $0.5N$ [9].

the acid and base, as shown: 0·1N, pH = 4·3–7·7; 0·3N, pH = 0·5–8·3; 0·5N, pH = 0·2–8·9. Figure 8.8 shows the variation of pH with distance along the layer for the three concentrations used.

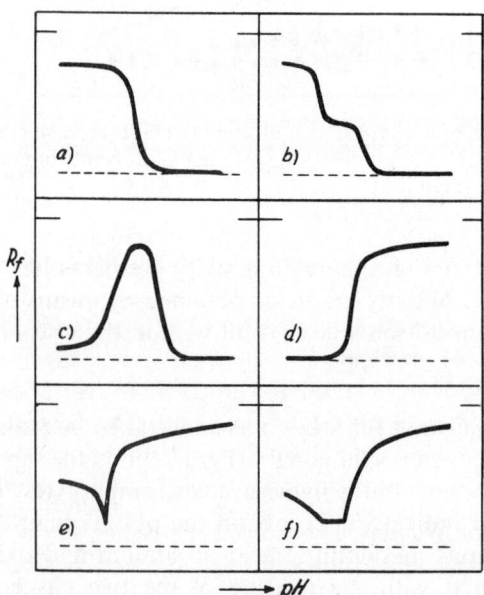

FIG. 8.9. Typical curve shapes obtained in pH orthogonal gradient thin-layer chromatography. ---- Start line [17].

From the various studies made it can be concluded that all curves showing the variation of R_f with orthogonal pH gradient for the abovementioned substances, can be classified in six types, given in Fig. 8.9.

The orthogonal pH gradient has also been used by Dumont and Kraus [18] for thin-layer chromatography on silica gel of caffeine, theophylline and theobromine with different eluents. This study showed that the R_f values of these components differed most at pH 6·2–8·5, as shown in Fig. 8.10.

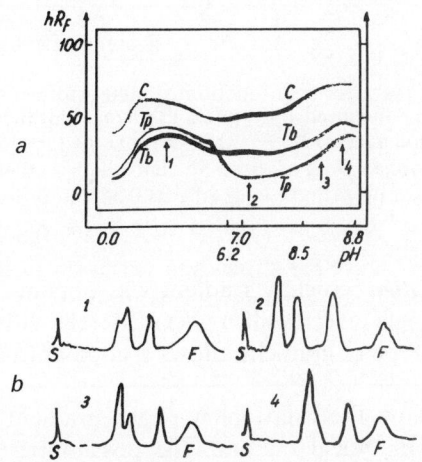

FIG. 8.10. Chromatogram (*a*) and densitogram (*b*) of a mixture of caffeine (*c*), theophylline (*Tp*) and theobromine (*Tb*) on a thin layer (silica gel) with orthogonal pH gradient. Solvent chloroform–methanol (90:90; v/v). Densitograms correspond to the positions shown by the arrows. *S*—start line, *F*—solvent front [18].

Warren [11] also used an orthogonal pH gradient on a layer of silica gel to separate Butter Yellow, Sudan Red and indophenol by elution with benzene. The best separation was obtained in the alkaline region.

An orthogonal pH gradient on a silica gel layer (suspensions prepared with 0·1N phosphoric acid and 0·05N potassium hydroxide) also proved useful in the separation of simple indoles [4, 19]. The mixtures which could otherwise only be separated by two-dimensional chromatography could be separated into 14 components on a single layer with a pH gradient.

Parallel gradient. Warren [11] showed the possibility of using parallel and antiparallel pH gradients on silica gel layers to separate certain dyes. Niederwieser [14] showed that when a parallel impregnation on gradient is used, tailed spots are obtained. This inconvenience can be avoided by using an antiparallel gradient (Fig. 8.11).

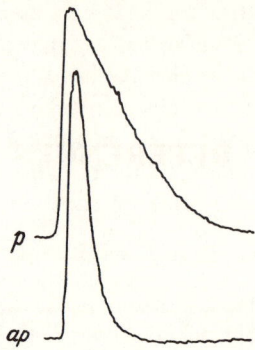

FIG. 8.11. Spot shape influenced by type of gradient. Parallel (p) and antiparallel (ap) impregnation gradients of 0·23 $\xrightarrow{\text{linear}}$ 2·3 per cent w/w tartaric acid on silica gel G (Merck). Substance applied: 2 μg of eosin. Solvent: diethyl ether, migration distance 10 cm. R_f-value of eosin in (p) and (ap) gradients 0·35 and 0·70 respectively [14].

Antidiagonal gradient. Such a gradient was obtained by Niederwieser [14] by impregnating a layer of silica gel G (Merck) with 0–1 per cent w/w tartaric acid. This type of gradient allows a good antiparallel gradient to be found quickly.

These results show that stationary-phase gradient chromatography represents a new analytical method, the possibilities of which are far from exhausted. It allows a better use of the entire layer, i.e. a better distribution of the zones and therefore an increase of the separation efficiency. It also permits the optimal separation conditions for an unknown mixture to be established very rapidly.

REFERENCES

1. Berger, J. A., Meyniel, G., Petit, J. and Blanquet, P., *Bull. Soc. Chim. France* 2662 (1963).
2. Abbott, D. C. and Thomson, J., *Chem. Ind. (London)* 310 (1965).
3. Liebmann, R. and Schuhmann, H., *Chem. Tech.* **19**, 693 (1967).
4. Stahl, E., *Chem.-Ing. Techn.* **36**, 941 (1964).
5. Stahl, E., *Z. Anal. Chem.* **221**, 3 (1966).
6. Stahl, E., *German Pat. No.* 1175912 (1964) (Appl. 1961).
7. Stahl, E. and Vollmann, H., *Talanta* **12**, 525 (1965).
8. Stahl, E. and Pfeifle, J., *Naturwiss.* **52**, 620 (1965).
9. Stahl, E. and Dumont, E., *Talanta* **16**, 657 (1969).
10. Kamm, L. and Mes, J., *J. Chromatog.* **62**, 383 (1971).
11. Warren, B., *J. Chromatog.* **20**, 603 (1965).
12. Snyder, L. R. and Saunders, D. L., *J. Chromatog.* **44**, 1 (1969).
13. Snyder, L. R. and Saunders, D. L., *J. Chromatog. Sci.* **7**, 195 (1969).
14. Niederwieser, A., *Chromatographia* **2**, 23 (1969).
15. Rozumek, K.-E., *J. Chromatog.* **40**, 97 (1969).
16. Honegger, C. G., *Helv. Chim. Acta* **47**, 2384 (1964).
17. Stahl, E. and Dumont, E., *J. Chromatog. Sci.* **7**, 517 (1969).
18. Dumont, E. and Kraus, Lj., *J. Chromatog.* **48**, 106 (1970).
19. Stahl, E. and Kaldewey, H., *Z. Physiol. Chem.* **323**, 182 (1961).

CHAPTER 9

ENVIRONMENTAL GRADIENTS

Though chromatography is essentially due to repeated partition equilibria of the components between two phases, the environment in which it takes place can also influence the equilibria. We have already shown (see Chapter 5) how the environmental conditions (temperature, flow-rate of the eluent, etc.) can influence the resolution in conventional chromatography. The optimization of the chromatographic process was also discussed. The more complex the system to be analysed, the greater are the chances that its chromatography in a constant environment may lead to unsatisfactory results. It is very unlikely that the same conditions are the best for all the components of the mixture. Thus, introduction of an environment programmed to change during the chromatographic process, i.e. a gradient, gives chromatography a new variable.

Such a gradient—temperature programming—has become frequently used in gas chromatography, but in liquid chromatography developments have been slower, largely because mobile-phase gradients are easier to work with and give good results.

It is convenient to classify these environmental gradients according to the parameter that is programmed.

9.1 Temperature Gradients

9.1.1 Closed Columns

9.1.1.1 *Apparatus*

There exist two types of device for achieving a temperature gradient. The first type varies the temperature as a function of time, i.e. the temperature of the whole column is the same at any instant ($dT/dx = 0$, where T is the temperature and x the length) but varies with time ($dT/dt \neq 0$; $t =$ time). The second type gives a parallel or antiparallel temperature gradient, i.e. the temperature along the column varies ($dT/dx \neq 0$), but the temperature at a certain point in the column remains constant ($dT/dt = 0$).

Programmed temperature. Scott and Lawrence [1] described an apparatus (Fig. 9.1) for study of elution with a temperature programme under axial

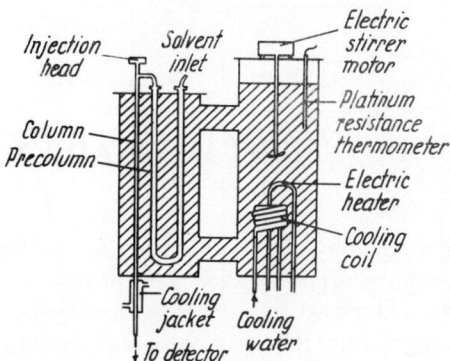

FIG. 9.1. Apparatus for gradient elution by axial equilibrium [1].

equilibrium conditions ($dT/dx = 0$, $dT/dt \neq 0$). This apparatus consists of two communicating cylinders, one of which contains the thermostat system consisting of heating and cooling coils, a thermometer with platinum resistance, working as a temperature transducer, and a stirrer operated from outside by an electric motor. The second cylinder contains two pre-columns and the column proper. In Fig. 9.1 only one of the pre-columns is shown. The liquid filling the two cylinders is chosen according to the temperature range in which the column works. For instance it can be filled with ethylene glycol. The heating coil and the resistance thermometer are connected to a temperature programming unit. The initial temperature was chosen to be 28° so that the fluctuation of the ambient temperature did not matter. The column (300 × 2 mm, stainless steel) and the pre-column (600 × 2 mm, stainless steel) were packed [2] with 'Gasil' silica gel (300/350 BS mesh). The volume of the pre-column must be large enough to provide the entire quantity of eluent needed for an analysis. Solvent is delivered by a non-pulsating pump, able to work up to 1000 psig.

Maggs and Young [3] built a simpler device consisting of a glass tube (4 mm inner diameter), the temperature of which could be varied from −2° to 80° by means of a water-jacket surrounding the column. The water-jacket was connected to a thermostatically controlled water-bath. The water in the thermostat contained 30 per cent ethylene glycol as antifreeze.

Antiparallel temperature gradient. In a recent paper Liteanu and Gocan [4] have described an apparatus for achieving a parallel or antiparallel temperature gradient. It (Fig. 9.2a) is composed of a copper tube (1) having an inner diameter approximately equal to the outer diameter

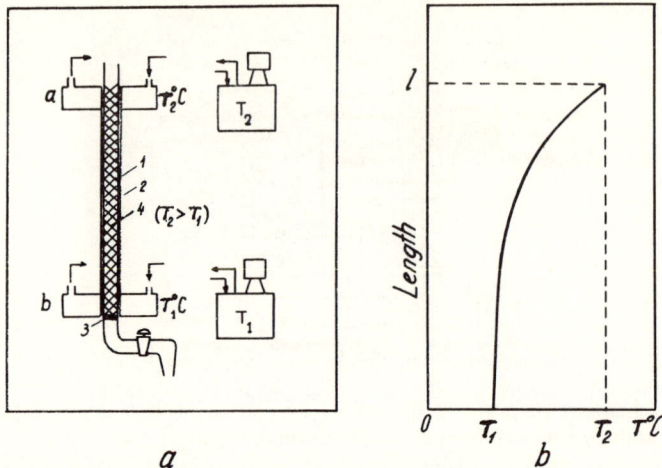

FIG. 9.2 (a) Device for producing a parallel or antiparallel temperature gradient for closed column; (b) the temperature variation along the column [4].

of the glass tube (2). The metallic tube is fitted at both ends with circular chambers, a and b, through which water flows from two thermostats, at temperature T_1 in the lower chamber, and at temperature T_2 at the upper chamber. Now if $T_2 > T_1$ the temperature gradient is antiparallel, and if $T_2 < T_1$, it is parallel. The glass tube (2) has a fitted disc (3) at the bottom, which supports the ion-exchanger (4).

From the theory of thermal conductivity, for a metal bar with temperatures T_1 and T_2 at the two ends ($T_2 < T_1$) the temperature decreases along the bar according to an exponential law having the form $T_1 = T_2 e^{-al}$, where l is the distance, and a is a constant equal to $\sqrt{\alpha p/\lambda A}$, where α is the surface conductivity of the bar, p the perimeter of the bar, λ the thermal conductivity of the bar and A the cross-sectional area of the bar. The profile of temperature variation along the column is represented in Fig. 9.2b. The temperature gradient will also be of exponential form.

Orthogonal temperature gradient. To achieve an orthogonal temperature gradient, Thompson et al. [5, 6] used the apparatus shown in Fig. 9.3. There are two variants of the form of the column: the coiled capillary system [5, 6] and the rectilinear channel system [6].

The coiled capillary system consists of two aluminium cylinders, with the capillary tube coiled between them. The cylinder is solid in its upper part (except the central aperture), and is electrically heated. The lower cylinder is milled out to allow flow of the cooling water. The apparatus was built in two different sizes, the smaller having cylinders 8 in. in diameter and 3 in. in height [6], and the larger with cylinders about 14 in.

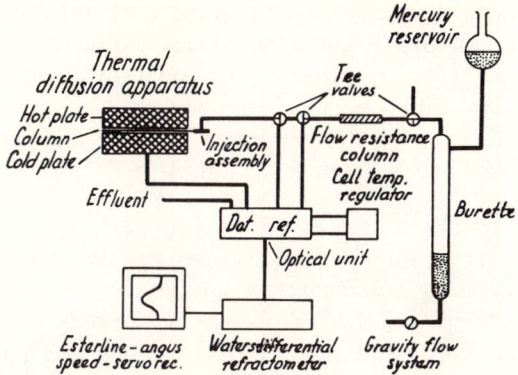

FIG. 9.3. Schematic diagram of the coiled capillary system and auxiliary equipment for producing an orthogonal temperature gradient [6].

in diameter and 2 in. in height [5]. The temperature of the two cylinders is measured with a thermocouple. The eluent is passed through the capillary column by a gravity flow apparatus, and the effluent is led to a refractometric detector. Capillaries of different forms, lengths and materials have been tried. The most successful columns were of 'Teflon' of 0·56 mm bore (0·25 mm wall thickness) and lengths 20·5 and 54·6 m [5, 6].

The rectilinear channel system is shown in Fig. 9.4. It consists of two tubes (made of 304 SS stainless steel) having a cross-sectional area of 2·54 cm^2 (wall thickness 0·27 cm) and 3·05 m in length and closed at each end by a massive block of the same material. These blocks are provided with apertures for the circulation of liquids for heating (oil) and for cooling (water), as well as for the injector at one end and for the removal of eluent at the other end.

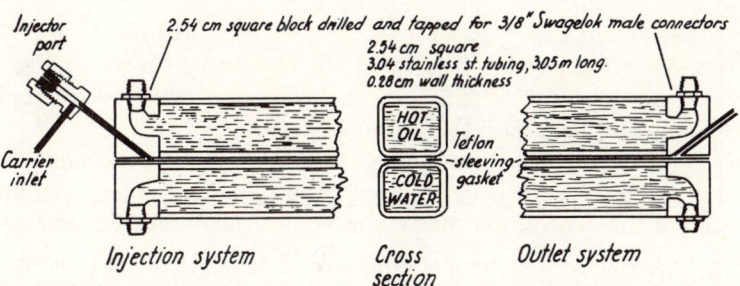

FIG. 9.4. Channel system for producing an orthogonal temperature gradient [6].

The column, which is a rectilinear channel, is obtained by grinding and polishing one face of each of the tubes and then setting these faces together, with a 'Teflon' strip between them at the edges. In this way a rectilinear channel is achieved with a 12 × 0·25 mm section.

Heating liquid (oil) flows in the upper tube in the opposite direction to the cooling water flowing in the lower tube.

9.1.1.2 *Practical uses*

The effect of programmed temperature was compared with simple temperature increase in liquid chromatography, by Hesse and Engelhardt [7]. The results plotted in Fig. 9.5 show that programming the temperature improves the separation and makes it faster. Maggs and Young [3], using heptane as non-polar solvent (containing 0·2 per cent n-propanol as moderator), showed that in the separation of squalane, methyl oleate

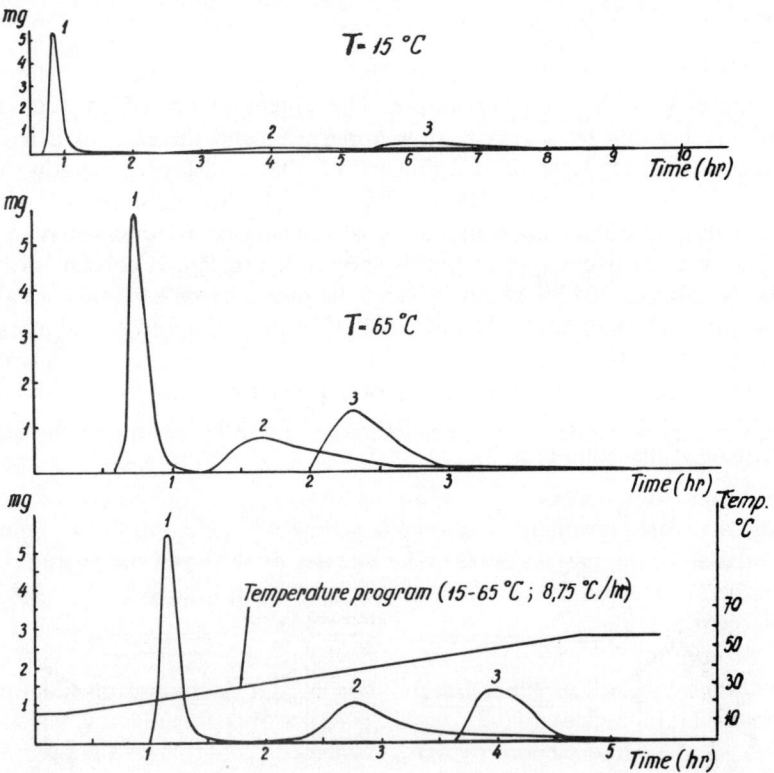

Fig. 9.5. Separation of a mixture of *p*-methoxyazobenzene (1), Sudan III (2) and *p*-aminoazobenzene (3) on a column (basic Al_2O_3) in isothermal conditions (*a* and *b*) and with a temperature programme (*c*) [7].

and dinonyl phthalate on a column (30 × 0·4 cm) filled with silicic acid, a temperature programme could successfully replace an elution gradient.

This technique (programmed temperature, heptane/isopropanol as eluent) was further developed by Scott and Lawrence [1]. Silica gel was used for the stationary phase, and the apparatus was that in Fig. 9.1. At the beginning the moderator is in equilibrium with the stationary phase. The polarity of the mobile phase increases during the chromatographic development, owing to the temperature programme of the column (increasing temperature) as well as of the pre-column. It therefore results that any increase in the temperature of the system will increase the concentration of moderator in the mobile phase. This conclusion was checked experimentally, and the results are given in Fig. 9.6, which shows

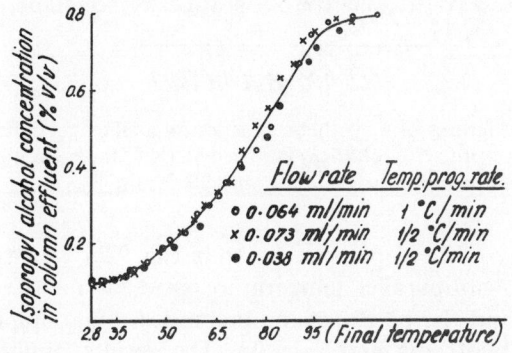

FIG. 9.6. Effect of temperature and flow-rate on the isopropyl alcohol concentration in the column effluent [1].

that a constant rate of increase of either the temperature or the eluent flow-rate gives the same effect on the concentration of isopropyl alcohol in the eluent in the column, within the limits of experimental error. In this way a polarity gradient is obtained on the whole length of the column, (and also of the pre-column), and evidently the effect of the programmed temperature is equivalent to that of a mobile-phase gradient (polarity gradient).

Programmed temperature was tested by Scott and Lawrence [1] on a synthetic mixture of squalane, methyl palmitate, dinonyl phthalate and tristearin, in approximately equal proportions. Results are shown in Fig. 9.7. This figure demonstrates the effect of programmed temperature, as compared to that of isothermal conditions. When the rate of change of the programmed temperature increases, the peaks become more symmetrical and the elution time is reduced.

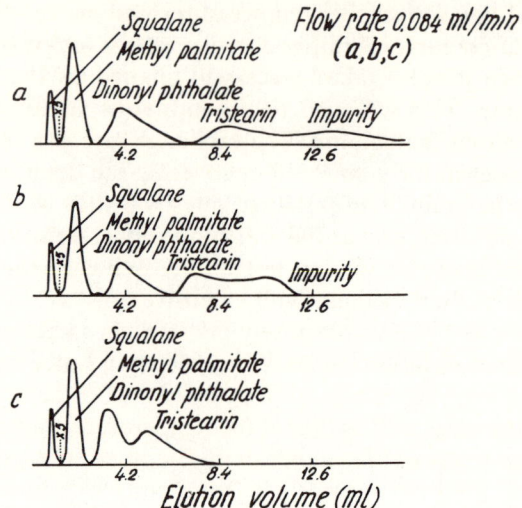

FIG. 9.7. Chromatograms of a synthetic mixture on a silica gel column moderated with 0·1 per cent isopropyl alcohol. (*a*) Isothermal. (*b*) Temperature programmed at 0·25°C/min. (*c*) Temperature programmed at 0·75°C/min [1].

By making use of the apparatus shown in Fig. 9.2*a*, we have studied [4] the effect of an antiparallel temperature gradient on the form of the elution curves and the height of the theoretical plate, for ion-exchange chromatography of zinc and calcium. The results obtained with the gradient are compared in Table 9.1 with those for simple increase in temperature. The effect of temperature on H was discussed in Section 5.2.5 and it was shown that it is very complex. In this case it has a beneficial effect by decreasing H, but the antiparallel temperature gradient gives a bigger effect than a simple temperature increase does.

TABLE 9.1. The values of the plate-height H(cm) for a column (30 × 1 cm) of Dowex 50 X8 (H^+ form), eluent 0·5N HCl, charge 5 ml of $ZnCl_2$ solution (17 mg/ml) and 2 ml of $CoCl_2$ solution (19·5 mg/ml) [4]

Ion	Temperature (°C)			Temperature grad. (°C/cm)	
	20	60	80	$\frac{60-40}{30} = 0.66$	$\frac{80-60}{30} = 0.66$
Zn^{2+}	2·40	2·12	2·06	1·92	1·75
Co^{2+}	5·66	3·40	1·90	1·88	1·80

The effect of the antiparallel temperature gradient on H and on the shape of the elution curves is rather complex, as the temperature at which the ion-exchange process takes place will change, as the ions migrate along the column. This will lead to a continuous variation of the ion-exchange rate along the column. The exchange rate is a function of the diffusion processes in the solution film covering the grain, and especially of diffusion in the grain. The exchange rates will thus be different in the frontal part of the zone, and at the rear of the zone. By suitably choosing the temperature gradient, we can favourably influence the shape of the asymmetric curves, and of those with extensive tails.

In the paper cited [4] we purposely chose a chromatographic system giving very extensive elution curves [H very large in isothermal conditions (20°)], and thus were better able to draw attention to the action of higher initial temperatures, as well as that of the temperature gradient (Fig. 9.8a). The higher the temperature at which the temperature gradient starts, the smaller the retention volume, the total elution volume, and consequently the time required for the analysis (Fig. 9.8b) since the antiparallel temperature gradient leads to a narrowing of the zones, as well as an increase in their symmetry.

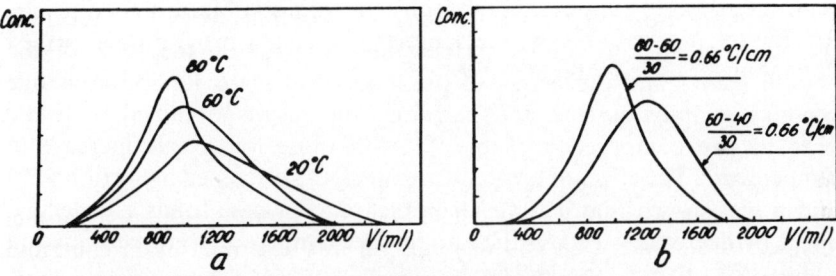

FIG. 9.8. Influence of temperature (a) and temperature gradient (b) on the shape of elution curves. Column (30 × 1 cm) filled with Dowex 50-X8, in H^+ form; 19·5 mg of $CoCl_2$/ml, mobile phase 0·5M HCl [4].

The efficiency of the orthogonal temperature gradient was tested by Thomson et al. [5, 6] in the fractionation of some polystyrene samples. The samples were of polystyrenes with molecular weights of 3525 and 4·11 × 10^5. Isothermal chromatographic measurements using the polystyrene–toluene system were carried out at the two extreme temperatures which created the orthogonal gradient, as well with the gradient itself. The results obtained are given in Fig. 9.9. It can be seen that isothermal chromatography does not give any fractionation but the orthogonal

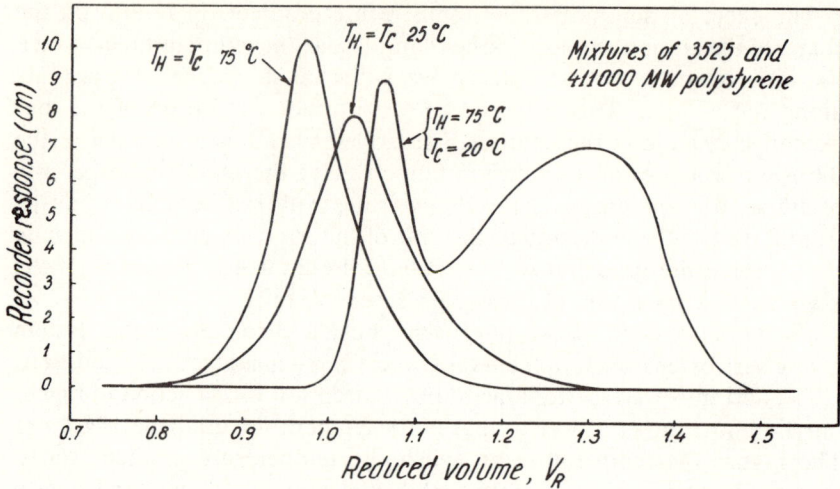

FIG. 9.9. Comparison of elution peaks for a polystyrene sample at temperature extremes with those produced in a thermal gradient [5].

temperature gradient does. Thompson et al. [6] also obtained good results for fractionation of polystyrene samples by using a programmed orthogonal temperature gradient.

9.1.2 OPEN COLUMNS

9.1.2.1 *Apparatus*

Parallel temperature gradient. The first device for developing a parallel temperature gradient on an open column was described by Liteanu and Gocan [9] in 1961. This device was improved [10, 11] and two variants developed (Fig. 9.10).

The glass tube (E) is specially made with double walls like a Dewar vessel, the lower part having a single wall. The temperature gradient along the length of the paper or thin layer (s) is achieved by plunging the lower part of the glass tube (e) into a vessel with water at constant temperature (T_i), maintained by passing water from a thermostat (T) through the spiral (S_p). The higher temperature (T_s) is achieved by passing warm water from another thermostat through a glass spiral (S) (Fig. 9.10a), or by means of an electric resistance (R) fed from an autotransformer (AT) (Fig. 9.10b). The glass tube is closed at the top by a plug (b) of plastic material, a good fit being ensured by use of a piece of filter paper (a). The temperatures are measured by thermometers (T_s and T_i). Later,

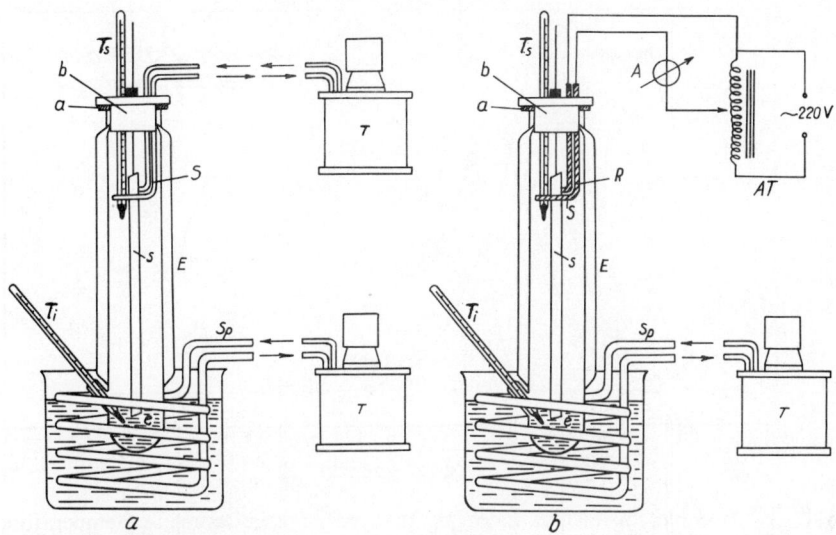

FIG. 9.10. Chromatographic device for producing a parallel temperature gradient. Top thermostated with water (a) or electrically (b) [10, 11].

Liteanu and Gocan [12] made a chromatographic chamber of the usual dimensions (Fig. 9.11) suitable for use with temperature gradients.

The chromatographic chamber has double walls made of 'plexiglass'. The top is provided with two removable lids (Fig. 9.11, a, b). The eluent (E) is in the vessel (B), on the bottom of which there is also set the spiral (S_p) through which water is passed from the thermostat (T_s) to maintain the temperature T_1 constant (Fig. 9.11e).

The higher temperature, measured by thermometer T_2, is achieved either by means of a thermostatic system, made of a contact thermometer (T_c), an electromagnetic relay (RE) and a heater (R) (Fig. 9.11c), or (Fig. 9.11, b, d), by means of a spiral (S'_p) through which water flows from a thermostat (T_A). The shape of the parallel temperature gradient along the central axis of the chamber is given in Fig. 9.12 for three values of the higher temperature. Inside the chromatographic chamber, there is also a vapour pressure gradient, leading to a downward diffusion of the vapours towards the surface of the eluent, where they condense.

Turina et al. [13] described an adapter for solvent evaporation on the upper side of the plate (Fig. 9.13). This device consists of a chromatographic chamber in the lid of which is inserted a glass tube having inside it an electrical resistance (R), fed by an autotransformer. The non-coated face of the chromatographic plate rests against the heating tube, and during

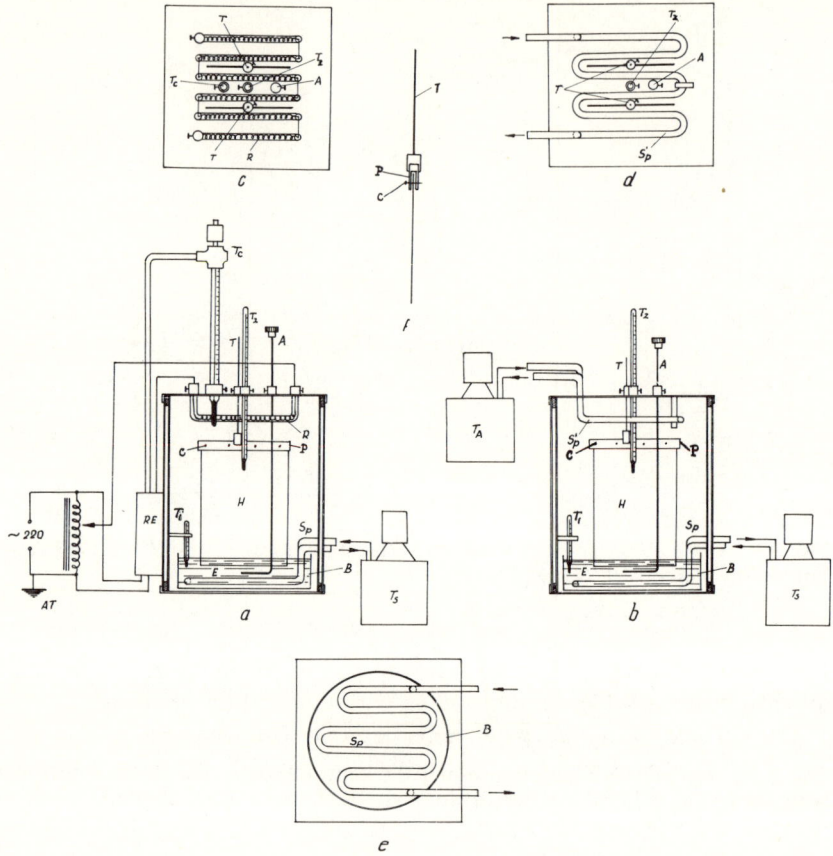

Fig. 9.11. 'Plexiglass' chromatographic chamber of large capacity with parallel temperature gradient [12]. (a) Electrical temperature control at top; (b) Thermostatic water-circulation for temperature control at top; (c) lid for tank in arrangement a; (d) lid for tank in arrangement b; (e) temperature control system for eluent; (f) method of fixing plate or paper. T_c—contact thermometer; R—electrical resistance; RE—electromagnetic relay; AT—autotransformer; B—container for eluent; S_p—glass spiral; A—stirrer; H—paper or glass plate; T and P—means of fixing H; C—'plexiglass' nail.

the chromatography, owing to the heating of the plate, part of the eluent evaporates into the atmosphere of the chromatographic chamber.

Antiparallel temperature gradient. Drapon and Guilbot [14] described a chromatographic chamber with a mobile-phase evaporation gradient caused by a temperature gradient. The chamber is cylindrical (60 cm in

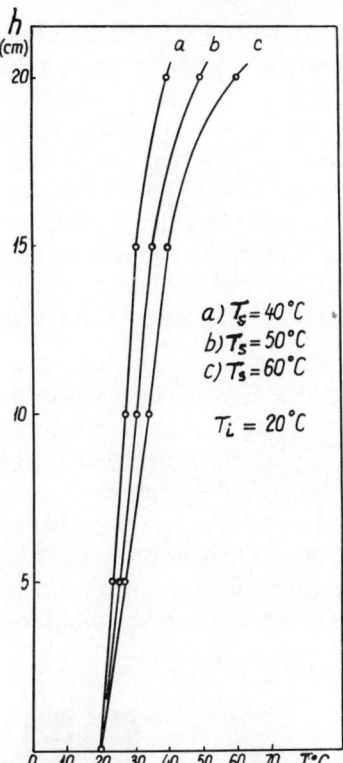

FIG. 9.12. Temperature variation along the central axis of the chromatographic chamber in Fig. 9.11 [12].

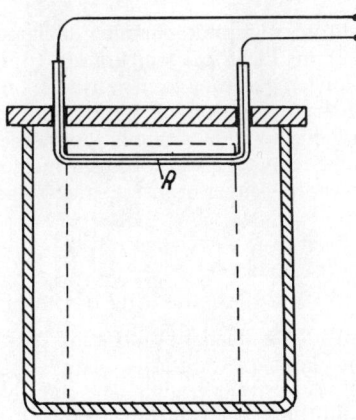

FIG. 9.13. Adapter for warming top of plate [13].

height and 20 cm in diameter) and is made of glass. It is closed by a polished glass plate, provided with an aperture allowing admission of the solvent into the reservoir, without significant disturbance of the atmosphere. The atmosphere in the chamber is rapidly homogenised by an electric fan with four stainless-steel blades, the axle of which passes through the side of the tank 25 cm above the base.

The chromatographic tank is used in a room thermostatically controlled at $26 \pm 0.5°$. To effect a temperature gradient, the lower part of the tank is placed in a caisson maintained at a temperature lower than that of the upper part of the chamber. Development is made downwards.

Orthogonal temperature gradient. In the case of stationary-phase gradients we discussed the orthogonal gradient set in the plane of the thin layer and perpendicular to the flow direction of the eluent. An orthogonal temperature gradient (as related to the electric field) in the plane of the plate has been applied only in electrophoresis, by Hollmén and Kulonen [15], for the investigation of the collagen–gelatin transition.

There is another type of temperature gradient, however, which is perpendicular not only to the flow of the eluent, but also to the plane of the chromatographic plate (gradient orthogonal to the plane of the plate).

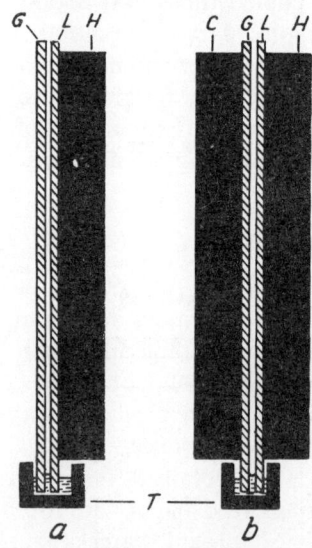

FIG. 9.14. Two arrangements for partial evaporation of mobile phase (orthogonal temperature gradient on plate). *L*—thin-layer plate, *G*—cover plate, *H*—heating block, *C*—cooling block, *T*—solvent trough. The right-hand arrangement allows full thermostating to establish a reproducible temperature gradient between the thin-layer and the cooling plate [18].

Turina *et al.* [16] used an electric heater attached to the back of the entire surface of the chromatographic plate, which was placed in a chamber for ascending chromatography. Part of the solvent evaporates from the plate during the migration, and a better resolution is finally obtained.

Stahl [17] has used a sort of temperature programming (or more precisely, a programmed orthogonal temperature gradient) in thin-layer chromatography. This orthogonal temperature gradient is obtained by introducing the chromatographic plate, which has the temperature of the surrounding medium, into a chromatographic chamber cooled to $-20°$. Approximately 30 min are necessary for the plate to cool to that temperature.

A temperature gradient orthogonal to the plane of the plate was also applied in the case of sandwich-type chromatographic chambers. Niederwieser [18] described two types of arrangement, which are shown in Fig. 9.14. The two sandwich-chambers are in a vertical position. The first consists of the plate with the thin layer (L), with the thermostated heating block (H) on the non-coated side. Facing the thin layer and separated from it by a distance of 1 mm there is the cover plate (G). The two plates are lowered into a small tank (T) filled with solvent (Fig. 9.14a). If necessary, the cover plate can be cooled through a cooling block (C) (Fig. 9.14b). During the chromatography, part of the mobile phase vaporises into the gaseous phase adjacent to the layer and may be removed by condensation on the cooler covering plate. The fraction of solvent removed is reproducibly adjustable by changing the temperature of the two plates (G and L). It also can be adjusted by lowering the pressure.

9.1.2.2 *Variation of R_f Values as a Function of the Temperature Gradient*

Parallel temperature gradient. A series of papers [9–12, 19–37] has dealt with the variation of R_f values with temperature gradient for a variety of substances and eluents, the apparatus being that in Figs. 9.10 and 9.11. Figure 9.15 gives the results obtained [12] for the separation of nickel(II) and manganese(II) with n-butanol saturated with $4M$ hydrochloric acid, on Whatman No. 4 paper.

Similar results [22] were obtained for separation of histidine and aspartic acid on Whatman No. 1 paper with n-butanol–acetic acid–water (4:1:5 v/v), n-butanol–acetic acid–water (9:1:1 v/v) and n-butanol–ethanol–water (4:1:1 v/v). For small temperature gradients, the differences between the R_f values were not enough to give separation, but the R_f values became sufficiently differentiated at temperature gradients $>2°/cm$ (Fig. 9.16). Isothermal chromatography gave no separation of these

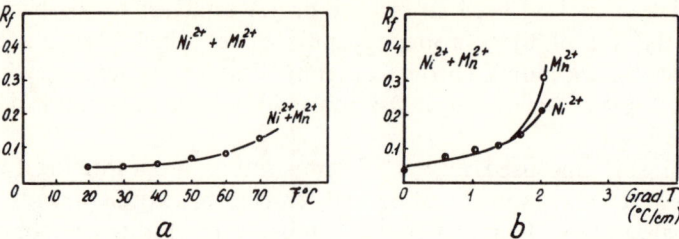

FIG. 9.15. Plot of R_f values vs. temperature (a) and temperature gradient (b) for Ni^{2+} and Mn^{2+} (chlorides) [12].

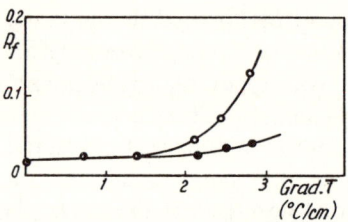

FIG. 9.16. Plot of R_f values vs. temperature gradient for mixture of histidine (—●—●—) and aspartic acid (—○—○—). Whatman No. 4 paper, eluent n-butanol + acetic acetic + water (90:10:10 v/v) [22].

components at 20° for an eluent travel of 10 cm, and even at 60° (the higher temperature used for the 2·81°/cm gradient) gave no separation, but only a slight increase of the R_f values.

The efficiency of the temperature gradient was also examined for chromatography on a thin cellulose layer [37]. Isothermal conditions and a parallel temperature gradient were compared for chromatography of a mixture of leucine (1), phenylamine (2), alanine (3), glutamic acid (4), arginine (5) and cystine (6) with the organic phase of n-butanol–acetic acid–water (40:10:50 v/v) as eluent. To avoid the formation of double spots the plates (with sample on) had to be equilibrated for 1–2 hr with the vapours of the development system. The results obtained are given in Fig. 9.17, and show that some substances can be separated by use of a temperature gradient even though in isothermal chromatography they have practically the same R_f values.

The chromatographic separation of the mixtures $Ni^{2+} + Mn^{2+}$, $Ce^{3+} + La^{3+}$, and $Hf^{4+} + ZrO^{2+}$, in a parallel temperature gradient has been studied as a function of the temperature range of the thermal gradient [32]. Whatman No. 1 paper was used with n-butanol saturated

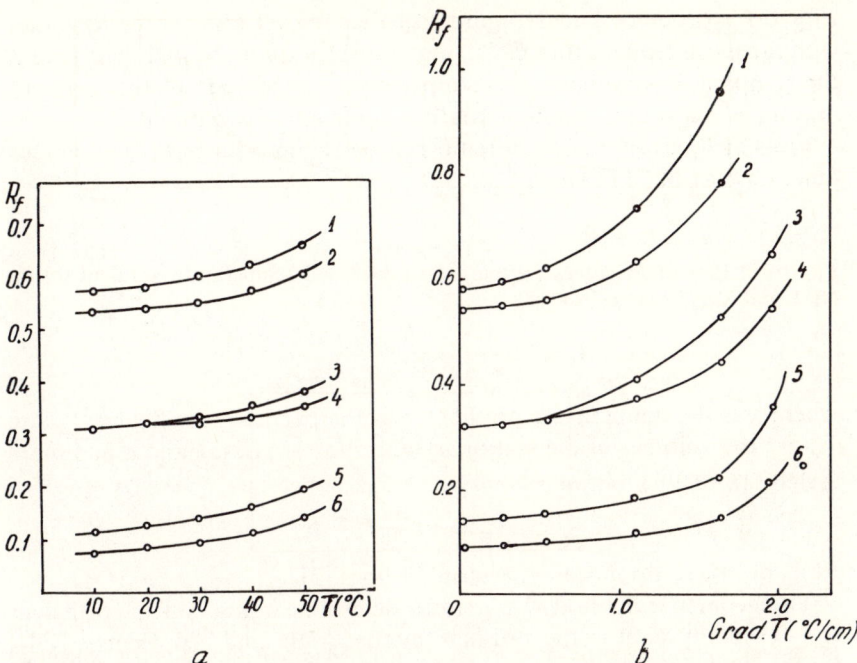

FIG. 9.17. Plot of R_f values in isothermal conditions (a) and in temperature gradient on cellulose thin-layer (b), for mixture of amino-acids; 1—leucine, 2—phenylalanine, 3—alanine, 4—glutamic acid, 5—arginine, 6—cystine [37].

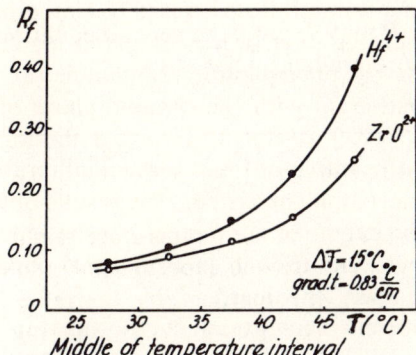

FIG. 9.18. Plot of R_f values for Hf^{4+} and ZrO^{2+} vs. the middle of the temperature interval, at temperature gradient 0·83°C/cm [32].

with 4M hydrochloric acid. Figure 9.18 gives the variation of the R_f values with the mean temperature in the gradient. This study showed that besides the temperature gradient, the temperature of the start of the gradient may also be a source of differentiation in the chromatography.

Plots of R_f as a function of temperature gradient suggest [21] that this function may be of the form:

$$y = \frac{1}{a - bx} \tag{9.1}$$

i.e.

$$R_f = \frac{1}{a - bx} = \frac{1}{1 + \alpha V_S/V_M} \tag{9.2}$$

where x is the temperature gradient, α is the partition coefficient, V_S and V_M are the volumes of the stationary and mobile phases, and a and b are constants. From (9.2), we obtain:

$$\alpha V_S/V_M = (a - 1) - bx \tag{9.3}$$

Plots of $\alpha V_S/V_M$ do indeed give straight lines [21].

The geometric method was used to determine a and b. The coefficient $(a - 1)$ in Eq. (9.3) is the ordinate at the origin and represents $\alpha V_S/V_M$ for $x = 0$ (isothermal conditions) and can be written as $\alpha_0 k_0$ (where $k = V_S/V_M$). The temperature gradient for $\alpha k = 0$, i.e. $R_f = 1$, can be denoted by x_1 (limiting temperature gradient).

The coefficient b is the slope of the line, and will be equal to $\alpha_0 k_0/x_1$. Equation (9.3) can therefore be written as

$$\alpha k = \alpha_0 k_0 - \frac{\alpha_0 k_0 x}{x_1} \tag{9.4}$$

Substitution of (9.4) in (9.2) gives the general equation for the variation of R_f with temperature gradient:

$$R_f = \frac{1}{1 + \alpha_0 k_0 - \frac{\alpha_0 k_0 x}{x_1}} = \frac{1}{1 + \alpha_0 k_0 (1 - \gamma)} \tag{9.5}$$

where $\gamma = x/x_1$ is the degree of the temperature gradient. This empirical equation describes well enough the experimental results [21].

In temperature gradient chromatography, the value R_f is a function of the time for development (see Section 2.1.2), differing from that of isothermal chromatography which, except for the first 20 min or so, does not depend on time (see Section 2.1.1). This finding was experimentally checked [33] for nickel, lead and bismuth, as seen in Fig. 9.19.

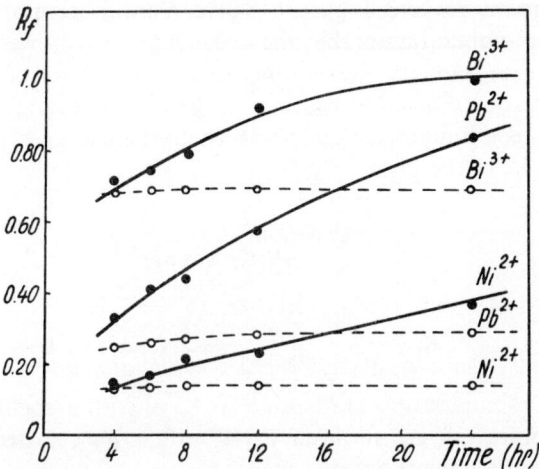

FIG. 9.19. Plot of R_f values vs. time for Bi^{3+}, Pb^{2+} and Ni^{2+} in isothermal conditions at 20°C (----) and in a temperature gradient of 1·30°C/cm (———) [33].

The increase of the R_f values with a temperature gradient is as much due to the extra eluent volume ΔV as to the change of the partition coefficient α. When a parallel temperature gradient is used, a parallel vapour pressure gradient is also created, favouring the vaporisation of a quantity of eluent ΔV which will condense in the lower part of the chromatographic chamber, on the surface of the eluent at the lowest temperature in the chamber. For eluent to migrate to the same height as in isothermal conditions, the quantity of eluent $V + \Delta V$ must pass through the column (V being the volume of eluent necessary in isothermal conditions). These facts were indirectly checked by kinetic studies on eluent migration [26] (see Section 1.2) as well as by direct measurements [33].

As is well known, in closed column chromatography [38] and open column chromatography (on paper) [39], the distance travelled by the concentration maximum of the migrating component, z'_z, is dependent on the volume of the mobile phase V' passing through the column, i.e.

$$z'_z = \frac{V'}{A_M + \alpha A_S} \tag{9.6}$$

where A_S and A_M are the cross-sectional areas of the stationary and of the mobile phase.

The distance z_f traversed by the eluent front is proportional to the volume of the mobile phase V in the column:

$$z_f = \frac{V}{A_M} \tag{9.7}$$

By definition:

$$R'_f = \frac{z'_z}{z_f} = \frac{V'A_M}{V(A_M + \alpha A_S)} \tag{9.8}$$

In the absence of a temperature gradient, $V' = V$ and

$$R'_f = \frac{A_M}{A_M + \alpha A_S} = \frac{1}{1 + \alpha k} = R_f \tag{9.9}$$

In a temperature gradient $V' > V$ ($V' = V + \Delta V$), so that from (9.8) and (9.9) it results that:

$$(R_f)_{grad} = (R_f)_{iso}\frac{V + \Delta V}{V} = \frac{1}{1 + \alpha_0 k_0} \cdot \frac{V + \Delta V}{V} \tag{9.10}$$

Writing (9.10) for two components on the same chromatogram, we obtain:

$$(R_{f_1})_{grad} = (R_{f_1})_{iso}\frac{V + \Delta V}{V} \tag{9.11}$$

and

$$(R_{f_2})_{grad} = (R_{f_2})_{iso}\frac{V + \Delta V}{V} \tag{9.12}$$

Taking the difference in the R_f values, we obtain:

$$(\Delta R_f)_{grad} = (\Delta R_f)_{iso}\frac{V + \Delta V}{V} \tag{9.13}$$

Equations (9.10) and (9.13) show that $(R_f)_{grad}$ and $(\Delta R_f)_{grad}$ are functions of α, k, V and ΔV. It must be stressed that ΔV is only the volume of eluent that has evaporated on the portion of column above the zone, and has thus participated in the chromatographic partition process [33].

The study carried out by Stewart and Gierke [40] showed that slight changes occur in the concentration profile of the mobile phase in the middle region of the column, which is the region where the chromatographic process for the final resolution of the components takes place. Comparison of the concentration profile of the steady-state evaporation system with the profile observed in a completely non-evaporating system

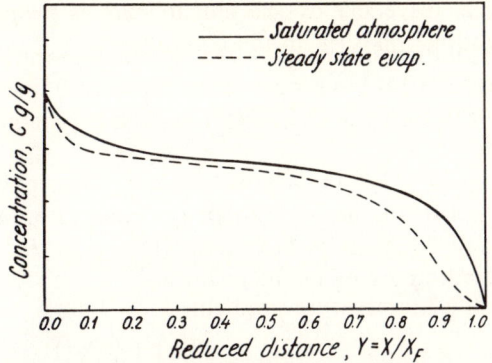

FIG. 9.20. Concentration of mobile phase in thin-layer (g/g) as a function of reduced distance for non-evaporation and steady evaporation conditions. The system is water–Whatman No. 3 MM paper [40].

for the region corresponding to R_f 0.2–0.8, shows a decrease of 5–20 per cent in the concentration when evaporation takes place (Fig. 9.20). The consequence of this is that the cross-sectional area A_M of the mobile phase, which influences the R_f value since it represents the molar fraction of solute in the mobile phase, may be compared for temperature gradient conditions with the one for a saturated atmosphere. That is to say, the capacity in terms of concentration of the component that can be chromatographed is not much reduced by evaporation.

We may draw the conclusion that the changes in the values of k and V in a temperature gradient make a minor contribution to the change in $(R_f)_{\text{grad}}$ and $(\Delta R_f)_{\text{grad}}$ values and that the important variation in these values (see Figs. 9.15–9.19) is due to ΔV and α. (See also Section 9.1.2.3.)

Antiparallel temperature gradient. Drapon and Guilbot [14], using the chromatographic chamber described (Section 9.1.2.1) for the chromatographic separation of some glucosides, have obtained a better resolution of the zones, good reproducibility of the R_f values, and an improvement in the accuracy of the absorbance measurements made on the effluent.

Orthogonal temperature gradient on plates. Turina et al. [16] achieved a gradient effect on plates covered with silica gel H (0.25 mm thick) by heating the plates during the chromatography. The solvent partially evaporated from the plate, increasing the concentration in the centre of the zone and thus improving detection and resolution. Dyes were used as test substances and were eluted with toluene.

9.1.2.3 Resolution

To find the resolution in open column chromatography with a parallel temperature gradient [41], we start from Eqs. (4.57) and (4.58) and the resultant Eq. (4.76) which give the resolution on open columns in isothermal chromatography.

By substituting in (4.76) the values of $(R_f)_{grad}$ and $(\Delta R_f)_{grad}$ given by Eqs. (9.10) and (9.13), and assuming that the chromatographic zones are close together and equal in surface area $[(H_1)_{grad} = (H_2)_{grad} = (H)]$, we can write the equation giving us the resolution:

$$(R_s)_{grad} = \frac{\sqrt{L}}{4\sqrt{(H)_{grad}}} \left(\frac{\Delta R_f}{\sqrt{R_f}}\right)_{iso} \frac{V + \Delta V}{V} \qquad (9.14)$$

where L is the distance from start line to eluent front.

The temperature gradient acts on all the terms determining the resolution, except the length of the column (if we ignore expansion).

Effect of a parallel temperature gradient on the height equivalent to a theoretical plate $(H)_{grad}$. The problem of plate-height H in open column chromatography (paper and thin layer) was discussed in Section 4.1.2. A temperature gradient influences the local non-equilibrium term in Eq. (4.32), by diminishing the values of θ and k owing to the evaporation of eluent in the column. A much better equilibrium is achieved in the temperature gradient than in isothermal chromatography, in the first place because of the temperature increase while the zone advances, and in the second, owing to the fact that the rate of travel of the zone decreases because of the stronger evaporation as the zone migrates (Fig. 9.21). Initially, the shape of a spot is circular, and at the end of the chromatographic operation it becomes elliptical, with the small axis in the direction

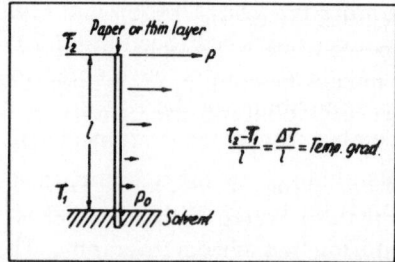

FIG. 9.21. Diagrammatic representation of vapour tension along an open column (paper or thin-layer plate) in a chromatographic chamber with parallel temperature gradient [41].

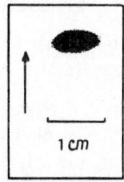

FIG. 9.22. Shape of the spot for Mn^{2+} at the end of development. Whatman No. 3 paper; eluent, n-butanol satd. with $4M$ HCl; temperature gradient 0·25°C/cm [41].

of flow (Fig. 9.22). This is because evaporation in the frontal part of the zone is more intensive, and therefore the rate of travel of the eluent, as well as of the zone, is smaller. Consequently, in a temperature gradient the height of the theoretical plate decreases, and hence a better resolution is obtained [see Eq. (9.14)].

Change of $(R_f)_{grad}$ and $(\Delta R_f)_{grad}$ from the variation of the partition coefficient. The partition coefficient also varies in a temperature gradient, owing to two factors: the temperature effect and dehomogenisation of the eluent.

The modification of the partition coefficient, and of the R_f value with temperature change was discussed in Section 3.1.3 and 5.3.4 and it was shown that for substances having $\Delta G° < 0$, the value of R_f will increase with rise in temperature, and decrease if $\Delta G° > 0$. These variations will evidently differ for the different components, so that the ratio $(\Delta R_f)_{iso}/\sqrt{(R_f)_{iso}}$ may increase and the resolution improve [see Eq. (9.14)].

Change in the partition coefficient with dehomogenisation of the eluent is known to occur in isothermal chromatography (see Section 1.4). When a temperature gradient exists, the dehomogenisation is accentuated because of the supplementary evaporation. The eluent, being usually a mixture, is thus submitted to a more marked dehomogenisation than in isothermal conditions [19, 25] (see Fig. 1.14).

Through the dehomogenisation brought about by the action of the temperature gradient in causing continuous evaporation, it is possible to achieve mobile-phase, pH, concentration, polarity or ionic-strength gradients. The action of the temperature gradient is combined with the effect of these gradients obtained by the dehomogenisation of the eluent, and thus the resolution is further improved.

The increase in resolution caused by the supplementary volume of eluent ΔV passing through the column and evaporating. At the start, the atmosphere in the chromatographic chamber will be unsaturated, but it becomes saturated through the evaporation of the eluent on the column. Later on, if there is a parallel temperature gradient, there will also be a vapour pressure gradient, which creates an antiparallel diffusion of the vapours condensing on the eluent surface (cold wall). The evaporation process will hence continue as in the case of a distillation process and the quantity of eluent ΔV leaving the column will steadily increase. This quantity ΔV depends on the evaporation rate, which in its turn depends on the size of the temperature gradient, on the surface area exposed to evaporation, and on the degree of saturation of the chamber, as well as on the duration of the chromatographic process [33, 34]. By ΔV we understand, as already shown, only that fraction of the eluent volume that has evaporated from the portion of the column above the zone, and has thus taken part in the chromatographic partition process [33]. As the duration of the chromatographic process can be of any length, we may obtain any value for ΔV, and therefore a higher resolution, as can be seen from Eq. (9.14).

For the change in flow-rate of the eluent front under a temperature gradient, the equation $v_f = dz_f/dt = A/(A + Bt)^2$ has been established [26] where A and B are constants depending on the working conditions. The flow-rate of the eluent decreases continuously owing to evaporation from the column, and becomes nil after not too long a time, as can be seen from Table 9.2.

This does not mean that the chromatographic process has stopped. In fact it continues, but there is a dynamic equilibrium between the eluent entering the column, and that evaporating from the column and the condensates. Evidently ΔV increases throughout and if we leave the process to continue for a fairly long time, we shall risk all the components reaching R_f values $= 1$. The process will therefore have to be interrupted when the resolution of the components is optimal, as in the case of continuous chromatography.

For instance, consider a component having an R_f value $= 0.10$. In a column in which the eluent front has travelled $z_f = 15$ cm, the component

TABLE 9.2. The migration distance and flow-rate of eluent as a function of time [34]. Eluent: n-butanol saturated with $4M$ HCl. Paper: Schleicher–Schüll 2040 bM. Temperature gradient: $2°/\text{cm}$, $T_i = 20°C$, $A = 2.28$ min/cm, $B = 0.058$ cm^{-1}

t (min)	z_f (cm)	v_t (cm/s)	t (min)	z_f (cm)	v_f (cm/s)
1	0.43	0.95	1000	16.58	0.00066
10	3.49	0.70	2000	16.90	0.00016
100	12.37	0.05	2500	16.97	0.00010
500	15.98	0.0026	∞	17.24†	0

† Calculated.

has travelled $z_z = z_f R_f = 1.5$ cm. If a temperature gradient is chosen so that when ΔV is big enough $(R_f)_{grad} = 0.80$, the component will have travelled $(z_z)_{grad} = z_f(R_f)_{grad} = 12$ cm. To have the component travel 12 cm in isothermal conditions, it would have been necessary for the eluent front to travel $z'_f = 12/0.1 = 120$ cm, i.e. $z'_f = z_f + z_f = 15 + 105$ cm. In this example a supplementary column 105 cm in length is achieved, so $L'/L = 8$, and as the resolution is proportional to $\sqrt{L}$, we obtain $(R_s)_{grad} = \sqrt{8}(R_f)_{iso}$.

If we consider two components, the R_f values of which vary with the temperature gradient according to Eq. (9.5), as seen in Fig. 9.23a, the chromatographic process need be continued only up to the limiting temperature gradient for the faster moving component [42]. This is due to the fact that for this value the resolution of the two components is maximal, as seen in Fig. 9.23b.

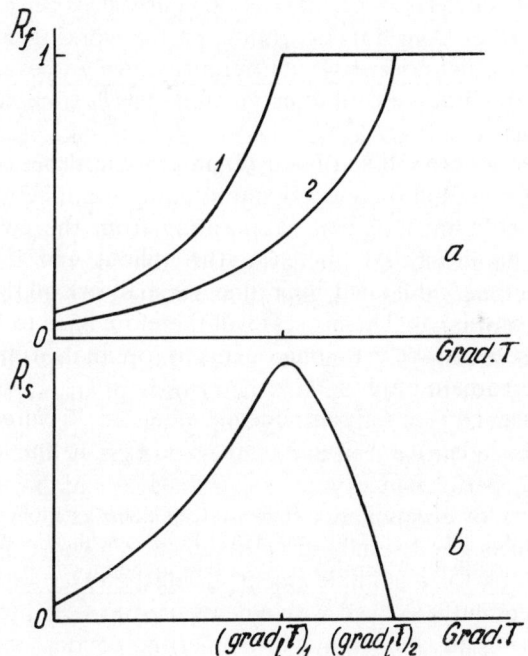

FIG. 9.23. Variation of resolution on open column as a function of temperature gradient [42].

For the separation of the amino-acids for which the variation of R_f was shown in Fig. 9.17, the resolution has also been calculated from densitometry of the spots [43]. The results obtained are given in Fig. 9.24 which

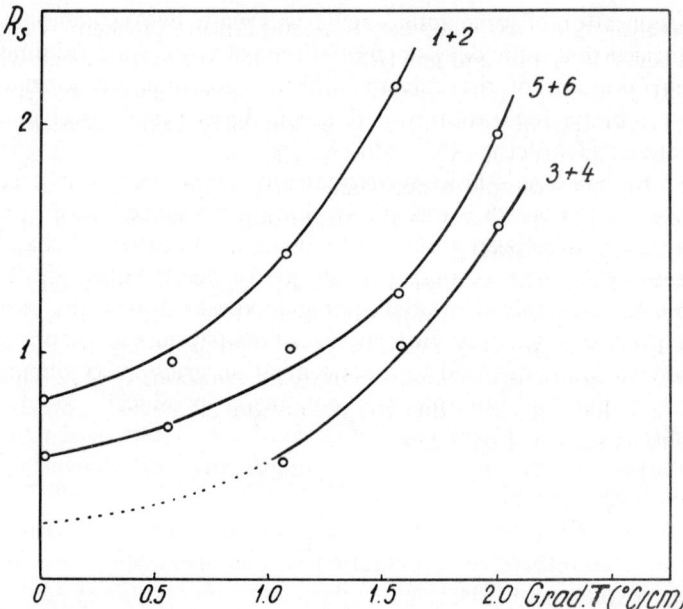

Fig. 9.24. Variation of resolution as a function of temperature gradient in separation of amino-acids presented in Fig. 9.17, on a cellulose thin layer; the dotted portion is extrapolated [43].

shows that the resolution increases with the increase in temperature gradient, reaching values >1, and hence complete separation of all the components is obtained.

It was also shown [43] that the variation of the resolution with the temperature gradient (Fig. 9.23) is described by an empirical equation having the form $(R_s)_{grad} = 1/[a - b(\text{temp.grad.})]$.

From all this it can be seen that temperature gradient chromatography offers specific performances and is particularly suitable in the case of some mixtures of components that are very difficult to separate. The technique allows good results to be obtained in a short time, as it saves the need to look for a suitable eluent. In all the cases studied by us, we have most frequently worked with eluents used to separate substances of the class concerned, and even if in isothermal conditions the R_f values were practically equal, a good separation was obtained with a temperature gradient.

9.1.2.4 Comparison between the Parallel Temperature Gradient and other Chromatographic Techniques

Chromatographic technique in an 'unsaturated N-chamber'. The first indication of the way to obtain a better resolution in unsaturated chambers for the separation of some amino-acids, was made by Arx and Neher [44]. They noticed that if the development of the plate was started immediately after introduction of the solvent, without previous saturation of the chamber, a better separation was obtained. These results were confirmed by Jones and Heathcote [45].

This problem was studied systematically by de Zeeuw in a series of papers [46–50] in which it was shown that improved separations of some hypnotic substances were obtained by using an unsaturated chamber. At the beginning of the process in an unsaturated space, the quantity of vapour available for adsorption is small, but it increases during the chromatographic process. The adsorption process of the vapours in the unsaturated chamber will therefore lead to a concentration gradient of the adsorbed vapours, a parallel polarity gradient being produced, giving better separations [47]. De Zeeuw et al. [50] also obtained improved separations in thin-layer chromatography when using a single-component eluent in unsaturated chambers.

It is shown below that though for multiple-component eluents, used in unsaturated chambers, an explanation can be given [48], there is yet no complete explanation for single-component eluents. Geiss [51] showed that both single- and multi-component eluents give improved separations owing to the same effect, i.e. the increase of the total quantity of eluent carried through the thin layer [40, 52]. That the mechanism stated by de Zeeuw for a multi-component eluent [48] to form an antiparallel activity gradient (decrease in activity of the layer in the direction of migration) is the decisive factor in obtaining improved separation in unsaturated chambers, has become doubtful [52–55].

During migration in unsaturated N-chambers, the eluent evaporates at a higher rate than in saturated spaces. The evaporated eluent will be replaced by the migration of more eluent from the reservoir [50].

Unlike the case for unsaturated N-chambers, in chambers with a small volume (S-chambers), ideally unsaturated, the R_f values, and hence ΔR_f, for single-component solvents are relatively higher than in saturated chambers (if all other conditions are constant). This is not due to evaporation in the S-chamber (the volume of the chamber is extremely small, and the quantity of evaporated eluent is negligible), but to the absence of pre-adsorption of vapours of the dry layer [51].

Geiss [51] also made some remarks on the statement by de Zeeuw et al. [50], namely that more solvent will be carried through the cross-section of the lower part of the plate than through the upper part, which should lead to a bigger migration of the lower spots, and therefore a worsening of the separation, a fact that was not confirmed experimentally. It is known that even in unsaturated N-chambers the medium next to the eluent becomes saturated in approximately 15 min [51]. Consequently, the solvent will evaporate only from the frontal part of the eluent, where the unsaturation is the highest, so the whole quantity of eluent that evaporates passes through the whole layer, except the portion nearest to the front where it evaporates (Fig. 9.25a). It follows that owing to this supplementary volume of eluent ΔV, a supplementary length of layer $\Delta L = \Delta V/A_M$ will be achieved, thus leading to an increase of the resolution.

Evidently, this supplementary evaporation which takes place in the unsaturated N-chamber, also takes place in a chamber with a parallel temperature gradient (see Section 9.1.2.3). There also exists the possibility of extending the duration of the evaporation, and of achieving any desired supplementary volume ΔV necessary for an improved resolution. Further, the chromatographic operation can begin in unsaturated conditions, and be continued in saturated conditions in a temperature gradient.

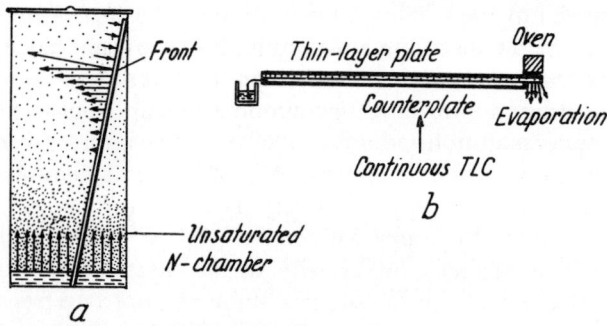

FIG. 9.25. Similarities of 'unsaturated N-chamber' (a) and continuous thin-layer chromatography (b). In each system integral solvent flow is increased by solvent evaporation from the upper parts of the layer [51].

Continuous chromatographic technique. As seen in Fig. 9.25b, the quantity of eluent ΔV can be forcibly evaporated by means of a heating system [51]. Brenner and Niederwieser [56] also described a horizontal apparatus for continuous development, the plate being placed with the thin layer upwards, and nearly entirely covered, leaving only a portion free for evaporation. In this case, the increase in resolution is due exclusively to the supplementary volume of eluent ΔV that has passed through the layer. A supplementary length ΔL is hence achieved and an increased resolution as in the case of the parallel temperature gradient. However, the action of the parallel temperature gradient, as shown before, is more complex, and besides the supplementary volume ΔV, the partition coefficients are also modified.

One-dimensional multiple chromatographic technique. This technique was discussed in Section 2.2. The supplementary volume ΔV is achieved by repeated irrigation of the same column, which is dried between the repetitions. What is discontinuously achieved in this technique, is actually achieved continuously in temperature gradient chromatography. The greater efficiency of the gradient technique is explained by the multiple action of the parallel temperature gradient, (Section 9.1.2.3), which besides causing the supplementary volume of eluent to be used, increases the dehomogenisation of the initial eluent, which is equivalent to an elution gradient. Thus a better resolution is obtained in a shorter time and with a reduced number of operations [57, 58].

9.2 Vapour Gradient

The vapour gradient technique has made its appearance fairly recently. The part played by the vapours in chromatographic chambers on separation on thin layers is already known. Apparatus has been designed for the control and programming of these vapours. By means of the vapour gradient the migration rates of the components can be adjusted so as to give maximum resolution.

9.2.1 APPARATUS

Discontinuous parallel vapour gradient. The vapour-gradient chamber designed by de Zeeuw [59–62] is shown in Fig. 9.26. The device consists of three parts made of chromium-plated brass. The lowest plate (*A*) carries the reservoir (*B*) of solvent, and a chamber with 21 troughs (*C*). When in use, the troughs are filled with mixtures of polar and non-polar solvents (*S*) according to a certain programme forming a composite gradient. The vapours of these solvent mixtures with a polarity gradient will equilibrate with the adsorbent on the chromatographic plate (*D*) which is placed with the thin layer (*E*) over the troughs. The eluent in the reservoir is connected to the thin layer of adsorbent by means of a filter paper strip (*F*). The plate (*A*) forming the base is provided on the inner face with tubes (*G*)

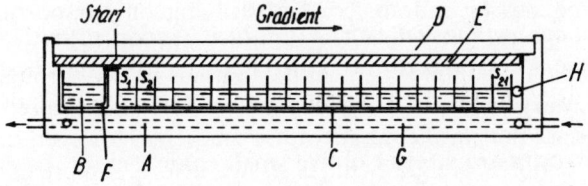

FIG. 9.26. Diagram of the vapour-programming chamber (longitudinal section) from [60].

carrying water from a thermostat. The development may continue for an unlimited period of time, as the eluent is evaporated at the end of the plate by means of a tube (H) insulated with asbestos and through which warm water runs. A 0·5 cm breadth of adsorbent is removed from three sides of the plates (20 × 20 cm). During the development the plate lies on two Teflon distance pieces 0·5 mm thick, placed at opposite ends. In this way the space between the plate and the grooved chamber is small enough to prevent vapour currents, without the layer touching the grooves.

Van Dijk [63] described a chromatographic installation for circular thin-layer plates, with parallel vapour gradient. This differs from the normal circular chromatographic technique in that the eluent migrates centripetally, the components are placed by dropper in a circle at the outer edge of the plate, the elution is continuous, and the components are collected separately.

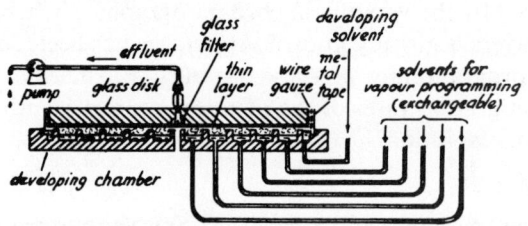

Fig. 9.27. Side view of the centripetal thin-layer chromatography apparatus [63].

Van Dijk's apparatus [63] is shown in Fig. 9.27. The vapour gradient is achieved by use of a number of concentric circular troughs, the eluent being fed to the innermost trough from a reservoir bottle, and the solvents for vapour programming being in the others. The chromatographic plate is a glass disc (23 cm in diameter and 1 cm thick) with a central aperture, and covered on the thin-layer side with a glass frit. To cover it with adsorbent by means of a layer-spreading device, a surface is previously cut out into which the glass disc is introduced. The effluent can be siphoned or pumped away to a detector or collector.

Discontinuous orthogonal vapour gradient. Sandroni and Schlitt [64] adapted some accessories to the 'Vario KS-Chamber' described by Geiss and Schlitt [65], to create a discontinuous orthogonal vapour gradient (Fig. 9.28).

The accessories are a series of five small chambers for the eluents, and a tank with five rectangular compartments. For other types of gradients, chambers of appropriate shape should be used.

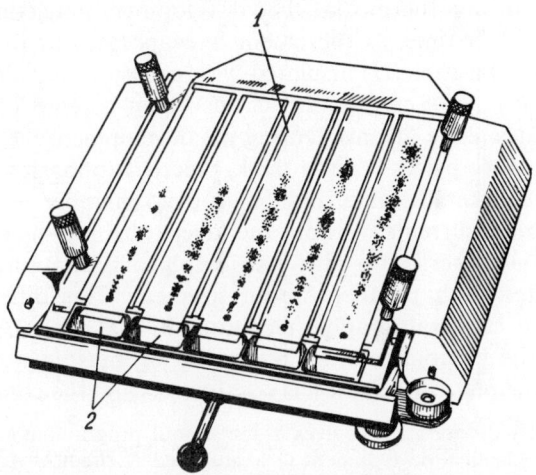

FIG. 9.28. Modified 'Vario- KS-chamber' (made by joining a number of the vapour-programming chambers presented in Fig. 9.26) for obtaining an orthogonal solvent and vapour gradient. *1*—Barriers, *2*—solvent tanks [64].

9.2.2 APPLICATIONS

To obtain good results, de Zeeuw [59, 60] recommended the use of chambers of the type described in Fig. 9.24. He also showed that the distance between the sides of the grooves and the plate, the number of grooves, and the temperature conditions are important factors that must be taken into account. To obtain spots without tails (as the spots of the components have a definite size, the upper part of the spot will tend to migrate more rapidly than the lower part) it will be necessary to intercalate some troughs with low polarity solvents between those containing solvent mixtures of higher polarity. The number of decelerating troughs (low polarity) and accelerating troughs (high polarity) must be chosen so as to yield optimal resolution.

Parallel vapour gradient. To illustrate the improvement of separation in a parallel vapour gradient as compared to classical development, we have chosen an example from de Zeeuw's paper [60] that we give in Fig. 9.29. The plates were prepared with silica gel GF 254 (Merck), 30 g in 60 ml of distilled water. This suspension was enough to spread a 0.25 mm layer on five 20×20 cm plates. The plates were then dried in air for 15 min, activated at 110° for 30 min, cooled, and stored in a desiccator. It was shown that the reproducibility was good if a series of factors involved in the separation process, such as temperature of the chamber, cooling

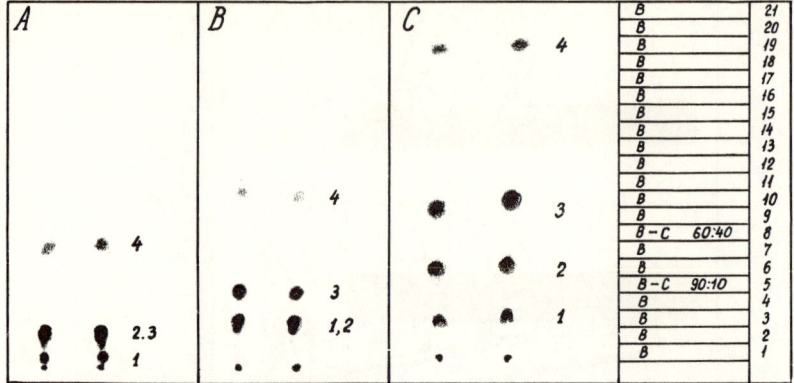

Fig. 9.29. Improved separation of dyes in the vapour programming chamber (VP) as compared to classical development in a saturated N-chamber. (A) N-chamber, solvent benzene, temp. 21°C, rel. humidity 26 per cent, development 36 min, saturation 45 min. (B) N-chamber, solvent benzene–chloroform (80:20 v/v), temp. 22°C, rel. humidity 30 per cent, development 39 min, saturation 45 min. (C) VP-chamber, solvent benzene, temp. 21·7°C, rel. humidity 29 per cent, development 110 min, saturation 110 min, strips 0·3 mm, cooling 19°C. Code: B—benzene, C—chloroform, 1—indophenol, 2—p-nitroaniline, 3—Sudan Red G, 4—Butter Yellow. Plates silica gel GF 254, load 20 μg [60].

temperature, relative humidity, saturation, thickness of the strips and of the layers, were kept constant. Small variations of these parameters had a large influence on the effect of the vapour gradient, and therefore on the separation.

This technique offers new possibilities of separating some classes of substances for which the classical method on thin layers does not offer sufficient resolution.

This method has also been used for preparative purposes [61]. The samples were put as a narrow strip, on a 40 × 20 cm plate 2·5 cm from the lower edge of the plate, and at least 2·5 cm from the sides. Silica gel PF 254 + 366 (Merck) (70 g) and distilled water (175 ml) were used to prepare two plates (40 × 20 cm) with a layer 1 mm thick. The plates were then dried in air for 3 hr and activated for 45 min at 110°, cooled and put into a desiccator. The results obtained for the separation of some sulphonamides are given in Fig. 9.30.

Discontinuous orthogonal vapour gradient. Besides the vapour gradient resulting from the combination of different organic solvents, discussed above (Section 9.2), the vapour gradient obtained by means of water

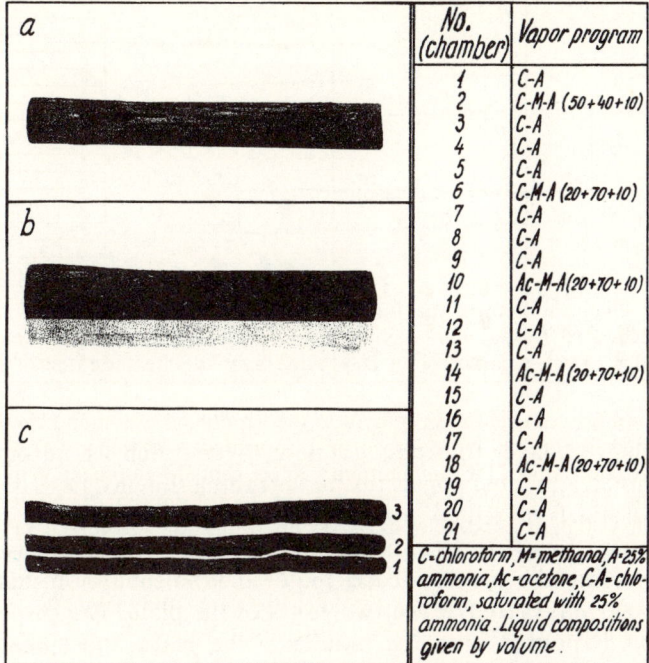

FIG. 9.30. Separation of a mixture of sulfadiazine (*1*), sulfamerazine (*2*) and sulfamethazine (*3*) on a silica-gel layer; (*a*) conventional development at 20·9°C in saturated chamber, eluent chloroform + methanol + 25 per cent ammonia (50:40:5); (*b*) as for *a*, development repeated twice, (*c*) vapour-programmed chamber, eluent: chloroform + methanol + 25 per cent ammonia (75:20:5) [61].

vapour (expressed as relative humidity) has also been studied. The influence of atmospheric moisture in thin-layer chromatography has already been dealt with in Chapters 3 and 5. The orthogonal gradient of relative humidity may be used as a working technique for rapidly finding the best humidity conditions in the chromatographic chamber (see Fig. 5.15).

9.3 Layer Thickness (Section) Gradient

An antiparallel layer thickness gradient has been used only a few times [66–68]. The use of this gradient is advantageous for the separation of mixtures containing some components in traces. If the correct eluents are chosen, the components in high concentration will be adsorbed in the first parts of the layer, where the adsorption capacity is the highest (the layer is the thickest) and thus the trace components will no longer be masked.

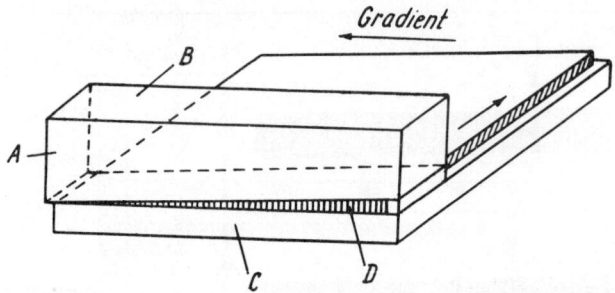

FIG. 9.31. Diagram of apparatus for preparing wedge-layer plates [66]. *a*. 'Perspex' plate attached to Desaga spreader. *b*. Plan of the apparatus, A—Perspex plate, B—spreader, C—glass carrier plate, D—wedge layer. *c*. Schematic view of apparatus.

To achieve a layer thickness gradient, Abbott and Thomson [66, 67] have adapted a manual apparatus for spreading thin layers. A diagram of this apparatus is presented in Fig. 9.31.

Bazan and Joel [68] described a simpler device which consists of two wedges having the profile of the required gradient (for instance from 125 to 1000 μm) placed at the two edges of the plate. The suspension of adsorbent is poured onto the middle of the plate, and spread with a perfectly straight glass rod till the surface becomes smooth. The plates are then dried, activated and kept in a desiccator till they are used.

By use of an antiparallel thickness gradient of a silica gel G layer, some very small quantities of free fatty acids in rather large and complex lipid samples have been separated by a single chromatographic operation and determined densitometrically [68]. The method may also be used for preparative work. It has been used [67] for the separation of Dinoseb (2-s-butyl-4,6-dinitrophenol) extracted from vegetable materials. A 1:1 alumina G and kieselguhr C mixture was used as adsorbent.

9.4 Eluent Flow-velocity Gradient

The use of the velocity gradient to reduce the duration of the analysis in liquid–solid column chromatography was studied by Scott and Lawrence [2]. The results obtained are given in Fig. 9.32. The results obtained in experiments *a* and *b* show that the velocity gradient gives a better separation and shortens the analysis time by a factor of approximately two.

Snyder [69] showed that the resolution per unit time with a velocity gradient is approximately the same as that with a temperature gradient in liquid–solid adsorption chromatography.

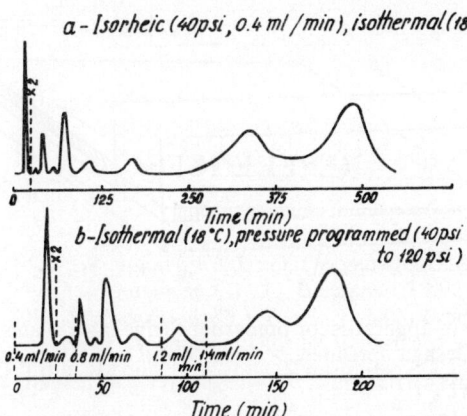

FIG. 9.32. Chromatograms of high-boiling fractions of essential oil. Column length 1 m, column diameter 4 mm, packing 200–250 BS mesh silica gel, mobile phase 0.15 per cent isopropyl alcohol in heptane; (a) with constant pressure, (b) with programmed pressure [2].

REFERENCES

1. Scott, P. P. W. and Lawrence, J. G., *J. Chromatog. Sci.* **8**, 619 (1970).
2. Scott, R. P. W. and Lawrence, J. G., *J. Chromatog. Sci.* **7**, 65 (1969).
3. Maggs, R. J. and Young, T. E., *Gas Chromatography*, 1968, p. 217, C. L. A. Harbourne (ed.), Institute of Petroleum, London (1969).
4. Liteanu, C., Gocan, S., Hodișan, T., Nașcu, H. and Măruțoiu, C., *Rev. Roumaine Chim.* **17**, 497 (1972).
5. Thompson, C. H., Myers, M. N. and Giddings, J. C., *Sepn. Sci.* **2**, 797 (1967).
6. Thompson, G. H., Myers, M. N. and Giddings, J. C., *Anal. Chem.* **41**, 1219 (1969).
7. Hesse, G. and Engelhardt, H., *J. Chromatog.* **21**, 228 (1966).
8. Giddings, J. C., *J. Chem. Physics* **49**, 81 (1968).
9. Liteanu, C. and Gocan, S., *Stud. Univ. Babeș-Bolyai* **6**, (2) 99 (1961).
10. Liteanu, C. and Gocan, S., *Stud. Univ. Babeș-Bolyai* **7**, (1) 99 (1962).
11. Liteanu, C. and Gocan, S., *Stud. Univ. Babeș-Bolyai* **11**, (2) 71 (1966).
12. Liteanu, C. and Gocan, S., *Bull. Soc. Chim. France* 1416 (1969).
13. Turina, S., Marjanovič-Krajova, V. and Šoljič, *Anal. Chem.* **40**, 471 (1968).
14. Drapron, R. and Guilbot, A., in *Chromatographie et Méthodes des Séparation Immediate*, Vol. II, p. 27, G. Parissakis (ed.), Publication de l'Union de Chimists Hellénes, Athens (1966).
15. Hollmén, T. and Kulonen, E., *J. Chromatog.* **21**, 454 (1966).
16. Turina, S., Šoljič, Z., Marjanović, V., *J. Chromatog.* **39**, 81 (1969).
17. Stahl, E., *Z. Anal. Chem.* **221**, 3 (1966).
18. Niederwieser, A. *Chromatographia* **2**, 262 (1969).
19. Liteanu, C. and Gocan, S., *Rev. Chim. Acad. RPR* **7**, 1041 (1962).
20. Liteanu, C. and Gocan, S., *Rev. Roumaine Chim.* **9**, 651 (1964).
21. Gocan, S. and Liteanu, C., *Rev. Roumaine Chim.* **9**, 715 (1964).
22. Liteanu, C. and Gocan, S., *Rev. Roumaine Chim.* **10**, 1051 (1965).
23. Liteanu, C. and Gocan, S., *Stud. Univ. Babeș-Bolyai, Chem.* **11**, (2) 71 (1966).
24. Liteanu, C., Gocan, S. and Onișor, M., *Stud. Univ. Babeș-Bolyai, Chem.* **11**, (2) 79 (1966).
25. Liteanu, C. and Gocan, S., *Bull. Soc. Chim. France* 3836 (1967).
26. Gocan, S. and Liteanu, C., *Bull. Soc. Chim. France* 1409 (1969).
27. Liteanu, C. and Gocan, S., *Stud. Univ. Babeș-Bolyai, Chem.* **13**, 135 (1968).
28. Liteanu, C., Mărgineanu, F. and Macarovici, D., *Rev. Roumaine Chim.* **12**, 503 (1967).
29. Liteanu, C. and Dulămiță, N., *Rev. Roumaine Chim.* **13**, 437 (1968).
30. Liteanu, C. and Gocan, S., *Talanta* **17**, 1115 (1970).
31. Liteanu, C., Gocan, S. and Hodișan, T., *Rev. Roumaine Chim.* **15**, 1751 (1970).
32. Gocan, S., Hodișan, T., Liteanu, M. and Liteanu, C., *Stud. Univ. Babeș-Bolyai, Chem.* **16**, (1) 37 (1971).
33. Gocan, S., Ciupală-Costea, C., Hodișan, T. and Liteanu, C., *Rev. Roumaine Chim.* **16**, 1069 (1971).

References

34. Liteanu, C. and Gocan, S., *Bull. Soc. Chim. France*, 4527 (1970).
35. Hodişan, T., Gocan, S. and Liteanu, C., *Stud. Univ. Babeş-Bolyai, Chem.* **17**, (2) 63 (1972).
36. Liteanu, C. and Constantinescu, A., *Rev. Roumaine Chim.* **18**, 155 (1973).
37. Hodişan, T. and Liteanu, C., *Stud. Univ. Babeş-Bolyai, Chem.* **17**, (2) 73 (1972).
38. Martin, A. J. P. and Synge, R. L. M., *Biochem. J.* **35**, 1358 (1941).
39. Consden, R., Gordon, A. H. and Martin, A. J. P., *Biochem. J.* **38**, 244 (1944).
40. Stewart, G. H. and Gierke, T. D., *J. Chromatog. Sci.* **8**, 129 (1970).
41. Gocan, S. and Liteanu, C., *Rev. Roumaine Chim.* **17**, 661 (1972).
42. Liteanu, C., Hodişan, T., Sîrbu, N. and Gocan, S., *Rev. Roumaine Chim.* in press.
43. Liteanu, C. and Hodişan, T., *Rev. Roumaine Chim.* **17**, 1085 (1972).
44. Von Arx, E. and Neher, R., *J. Chromatog.* **12**, 329 (1963).
45. Jones, K. and Heathcote, J. G., *J. Chromatog.* **24**, 106 (1966).
46. De Zeeuw, R. A., *J. Chromatog.* **32**, 43 (1968).
47. De Zeeuw, R. A., *J. Chromatog.* **33**, 222 (1968).
48. De Zeeuw, R. A., *Anal. Chem.* **40**, 915 (1968).
49. De Zeeuw, R. A. and Wijsbeek, J., *Pharm. Weekblad* **104**, 901 (1969).
50. De Zeeuw, R. A., Compaan, H., Ritter, J. J., Dhont, J. H., Vinkenborg, C. and Labadie, R. P., *J. Chromatog.* **47**, 382 (1970).
51. Geiss, F., *J. Chromatog.* **53**, 620 (1970).
52. Geiss, F., Sandroni, S. and Schlitt, H., *J. Chromatog.* **44**, 290 (1969).
53. Snyder, L. R. and Saunders, D. L., *J. Chromatog.* **44**, 1 (1969).
54. Niederwieser, A., *Chromatographie* **2**, 23 (1969).
55. Niederwieser, A., *Chromatographia* **2**, 519 (1969).
56. Brenner, M. and Niederwieser, A., *Experientia* **17**, 237 (1971).
57. Gocan, S., Liteanu, C., Hodişan, T., Naşcu, H. and Bîndeanu, St., *Rev. Roumaine Chim.* **17**, 669 (1972).
58. Hodişan, T., Bîndeanu, St. and Liteanu, C., *Stud. Univ. Babeş-Bolyai, Chem.* **17**, (2) 119 (1972).
59. De Zeeuw, R. A., *J. Pharm. Pharmac.* **20**, 54 S (1968).
60. De Zeeuw, R. A., *Anal. Chem.* **40**, 2134 (1968).
61. De Zeeuw, R. A. and Wijsbzek, J., *Anal. Chem.* **42**, 90 (1970).
62. De Zeeuw, R. A., *Pharm. Weekblad* **104**, 1173 (1969).
63. Van Dijk, J. H., in *Chromatographie sur colonnes*, p. 234, E. Kováts (ed.), Lausanne (1970).
64. Sandroni, S. and Schlitt, H., *J. Chromatog.* **52**, 169 (1970).
65. Geiss, F. and Schlitt, H., *Chromatographia* **1**, 392 (1968).
66. Abbott, D. C. and Thomson, J., *Chem. Ind. London* 481 (1964).
67. Abbott, D. C. and Thomson, J., *Analyst* **89**, 613 (1964).
68. Bazán, N. G., Jr. and Joel, C. D., *J. Lipid Res.* **11**, 42 (1970).
69. Snyder, L. R., *J. Chromatog. Sci.* **8**, 692 (1970).

CHAPTER 10

COMBINED GRADIENTS

The effects of two or several gradients may be combined to obtain a better resolution. By combined gradients we understand the combination of two or more gradients of the same or of different phases. We gave a classification of these in Section 6.3.

10.1 Combined Mobile-phase and Temperature Gradient

This mobile-phase and temperature combined gradient (antiparallel) was first applied by Baker and Williams [1] to the chromatographic fractionation of high polymers. Their apparatus consists of a column (A), the temperature jacket (B), the mixing vessel (C), and the solvent receiver (D) (Fig. 10.1). The mobile-phase gradient is achieved with ethyl methyl ketone running from the receiver (D) into the mixing chamber (C), ($V_m = 150$ ml) through the capillary (O), where it is mixed with ethanol by means of the magnetic stirrer (P). The flow-rate is 5 ml/hr. The column,

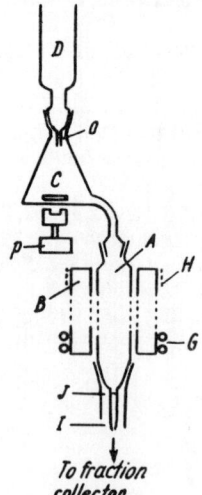

FIG. 10.1. Diagram of apparatus for the fractionation of polymers by elution and temperature gradients on a column [1].

made of a glass tube 35 cm in length and 24 mm in outer diameter, is filled with glass balls of average diameter 0·1 mm, the balls being added mixed with solvent of the same composition as the eluent in vessel C. A sintered-glass disc J is fused onto the bottom of the column. The capillary is used to control the flow-rate of the eluent through the column, as the glass-ball bed allows too quick a flow.

The jacket is an aluminium cylinder (B) 30 cm in length, with outer diameter 50 mm and inner diameter 25 mm. The electrical resistance (H) heats the upper part of the column to 60–65°. The lower part of the column is cooled by copper-tube spirals (G) through which cold water (10°) runs.

The polymer solution (300 mg dissolved in 10 ml of ethyl methyl ketone) is added to 30 g of the glass balls, and the solvent is evaporated with a hot-air blower. The glass balls loaded with polymer are introduced into the upper part of the column, where they occupy approximately 4 cm of its length, a distance similar to that occupied by the spiral of the heating resistance.

In each experiment fifty 10-ml fractions were collected. Each fraction was transferred quantitatively into a 250-ml beaker with methyl ethyl ketone and the polymer precipitated by addition of an excess of methyl alcohol. After decantation, the precipitate was dried. The polymer recovered from the fractions represented 90–95 per cent of the quantity processed.

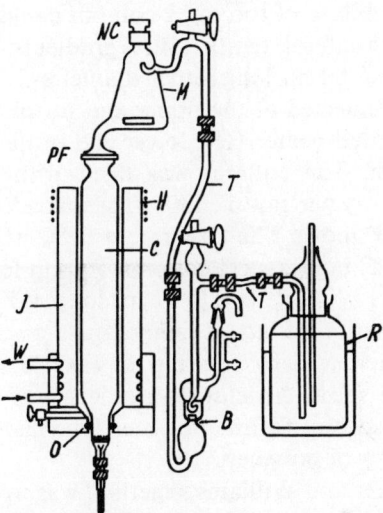

FIG. 10.2. Diagram of chromatographic system for the fractionation of polymers by elution and temperature gradients on a column where all ground-glass joint connections have been eliminated [2].

Schneider et al. [2] suggested an improved device (Fig. 10.2) which avoided losses of eluent due to dripping through the joints and release of air-bubbles into the eluent during heating.

The joint between mixing vessel and column was made of glass pipe fittings with a Teflon gasket (PF). The mixing chamber (M) was filled through its upper part with heated air-free solvent, and was then closed by a nylon plug (NC). The receiver (R) worked on the principle of the Mariotte bottle, maintaining a constant level of the liquid in the boiling vessel (B), where the solvent was boiled free from air. B, M and R were connected by a polyethylene tube (T). The lower part was cooled by circulating refrigerated cooling water (W) through grooves cut in the aluminium jacket (J). An O-ring (O) allowed the filling of the space between the jacket and the column (C) with mineral oil to give better heat transfer. The temperature adjustment was improved by connecting the heating resistance (H) coiled round the upper end to a 'Variac' fed through an electronic voltage-regulator and by controlling the temperature of the cooling-bath with a sensitive mercury thermoregulator.

A series of such devices is described in the literature [3–10], based on the same principles as those described above, and differing only in constructional detail.

For pilot plant fractionation of polymers, Cantow et al. [11] built an apparatus based on the same principle as those used for analysis. To avoid an outsize system when using large quantities of polymer, six columns were used in parallel. Use of too big a column could lead to preferential flow phenomena or lateral temperature gradients. The columns were made of aluminium (100 cm long, inner diameter 2·5 cm, outer diameter 3·8 cm) and were connected at the upper end to the mixing vessel, fitted with a thermoregulated heater. The lower end of the columns was set in a thermostatic bath. The column was filled with glass balls (0·1 mm diameter) and the polymer sample (30–40 g) was added according to the procedure described above. The flow-rate was 80 ml/hr.

Slonaker et al. [12] built a steel column (500 cm long, 15 cm diameter) filled with small bits of steel wire (0·77 cm long, 0·7 mm diameter), thus preventing the formation of lateral temperature gradients. The temperature gradient was achieved by winding an electric resistance round the column in a graded spiral. An elution-gradient was formed by mixing a solvent with a non-solvent for the polymer. The installation allowed the fractionation of 800 g of polymer.

The theory of Baker and Williams's method was worked out by Schultz et al. [13]. To explain the fractionation mechanism by the use of the two gradients (mobile-phase and temperature gradients) the existence of two stages was postulated, namely the formation of a gel phase around the support particles, followed by chromatographic separation between the two phases. The first phase, i.e. the gel phase, appears when the eluent (made of a solvent in the receiver and a non-solvent for the polymer in the mixing chamber, see Fig. 10.1) is able to dissolve the first quantity of polymer. The decrease of the temperature in the lower part of the column determines the precipitation of the solvent as a gel. In the second stage, an equilibrium is established at each level of the column, determining the partition of each molecular special between the two phases. The polymer migrates towards the bottom of the column through a series of dissolutions and precipitations determined by the solvent gradient and temperature, i.e. each molecular species passes several times from the gel phase into solution and back again, till it leaves the column. The influence of the temperature on the dependence between the length of the polymer chain and the solubility, makes each migrating species spread out from the point of entrance of the solvent precipitant mixture (x_0) at a characteristic rate. The antiparallel temperature gradient induces the fronts of these zones to migrate more slowly. Consequently there is a compression effect, consisting of continuous narrowing of the band of each species as it moves down the column (Fig. 10.3), thus increasing the fractionation efficiency.

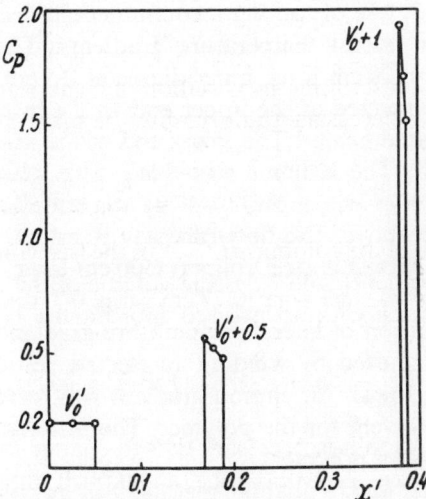

FIG. 10.3. The displacement through the column (spatial partition of C_P as a function of x') of a polymer having $P = 1 \cdot 000$. Elution of this polymer begins at the volume V'_0 [13].

The actual fractionation process depends on the establishment of a thermodynamic equilibrium between sol and gel, determined by the diffusion process of the macromolecules between the two phases. The migration of the species from one phase into the other may be disturbed by adsorption of polymer on the support or by hydrodynamic forces. These phenomena have the effect of causing continuous retention of some species or their release after the front of these fractions has passed the appropriate level on the column. The compression effect is thus limited, and the elution curve of a given molecular species is Gaussian in form.

From this model of the fractionation, Schultz et al. [13, 14] have inferred the general transport equation:

$$q\frac{\partial C_P}{\partial V} = -\frac{\partial (C_P \cdot g)}{\partial x} \tag{10.1}$$

where C_P represents the total concentration of a molecular species of degree of polymerization P, in an infinitesimal volume $q\,dx$, where q is the free cross-sectional area of the column, V the volume of eluent passed in time t, and g the fraction of the species P in the sol phase at the height x on the column, i.e. $g = m_{sol}/(m_{sol} + m_{gel})$, and m_{sol}/m_{gel} represents the partition ratio of the mass of a polymer between the sol phase (m_{sol}) and the gel phase (m_{gel}), which is a function of the degree of polymerisation P, temperature T, and volume fraction of the packing γ and of the volume ratio of solution to gel phase φ.

$$g = f(P, T, \gamma, \varphi) \tag{10.2}$$

To solve Eq. (10.1) describing the variation of C_P at point x with volume V of eluent used, it is necessary to normalise the main parameters of the column:

$$x' = x/L;\ V' = V/V_m \quad \text{and} \quad V'_t = V_t/V_m \tag{10.3}$$

where L is the length of the column, $qL = V_t$ is the free volume of the whole column (filled with eluent), and V_m is the volume of the mixing chamber. Equation (10.1), written with normalised coordinates, becomes:

$$V'_t \frac{\partial C_P}{\partial V'} = -\frac{\partial}{\partial V'}[C_P \cdot f(P, T, \gamma, \varphi)] \tag{10.4}$$

To solve this equation it is necessary that to change the coordinates to $f(P, x', V', \varphi)$ in (10.2), thus obtaining an explicit function instead of Eq. (10.4):

$$V'_t \frac{\partial C_P}{\partial V'} = -\frac{\partial}{\partial V'}[C_P \cdot f(P, x', V', \varphi)] \tag{10.5}$$

For the partition ratio m_{sol}/m_{gel} the authors' previously-inferred relationship [15, 16] was used:

$$m_{sol}/m_{gel} = \varphi \exp[P\varepsilon(\gamma, T)] \qquad (10.6)$$

Because of the temperature gradient, T is a function depending on x' while γ depends on V' and x'. By introducing into Eq. (10.6) a linear expression for the function $\varepsilon(\gamma, T)$ and by an exponential change of V' for γ, the following expression is obtained for g:

$$g = \frac{\varphi \exp\{P[a - bx' - c\exp(V'_t x' - V')]\}}{1 + \varphi \exp\{P[a - bx' - c\exp(V'_t x' - V')]\}} \qquad (10.7)$$

where $a = A' + c(T_t - T) - B\gamma_B$; $b = c(T_t - T_b)$ and $c = B(\gamma_A - \gamma_B)$, T_t and T_b are the temperatures at the top and the bottom of the column, T is an arbitrarily chosen reference temperature, γ_A and γ_B are the initial and final fractions of non-solvent in the elution mixture (for two interconnected vessels and V_m = constant, these vary exponentially, see Chapter 7). Hence, the constants a, b and c may be determined experimentally, while φ may be replaced by a constant value [14].

For chromatographic fractionation, parameters a, b and c must satisfy a series of limit conditions. A very large molecular species P will leave the column after a definite time t and after passage through the column of a volume of liquid V'_e. For $P'_{max} = \infty$, this condition is written $(a - b)/c = \exp(V'_t - V'_e)$. The moment at which the molecules pass from the gel phase into the sol phase is a characteristic of each species. The large molecules evidently start to migrate through the column later, after the passage of a volume of eluent mixture equal to V'_a. For $P_{max} = \infty$, this condition, i.e. of passing from gel into sol, can be expressed as $a/c = \exp(-V_a)$. The species with small molecular weights will also fractionate, a fact evidenced by the relation $c - a \sim \ln(0.1\varphi)/P_{min}$, in which P_{min} represents the degree of polymerisation of the smallest molecular species which can be fractionated. The shortest polymer chains are concentrated in the sol phase and therefore migrate quicker through the column.

Equation (10.5) may be solved numerically [14] by means of Eq. (10.7). Figure 10.3 shows the concentration C_P plotted vs. the normalised co-ordinate x' for three values of V'_0.

The compression effect mentioned above, coupled with the effect of the solvent gradient, determines the efficiency of the method, characterised by the separation coefficient σ_0 given by the relation:

$$\sigma_0 = \frac{1}{\Delta V'_P} \cdot \frac{dV'}{dP} = \frac{1}{\Delta V_P} \cdot \frac{dV}{dP} \qquad (10.8)$$

where V_P is the volume throughput when the species P appears at the outlet of the column ($x' = 1$).

Flowers et al. [8] have checked for poly-α-olefins (copolymer 1-octadecene + 1-dodecene), the contribution of the temperature gradient in the separation process. They draw the conclusion that the solvent gradient plays the main role in the fractionation, while the temperature gradient, inherent in the chromatographic precipitation, may in fact decrease the efficiency of the fractionation.

On the other hand, Guillet et al. [17], after a study on the fractionation of polyethylene, concluded that both the temperature gradient and the solvent gradient are necessary for a better separation of the fractions by chromatographic precipitation. It was also shown that a linear solvent gradient leads to a greater efficiency of polymer fractionation than an exponential gradient does. The effect of the temperature gradient is shown by an increase of the number of 'equilibrium plates' in the fractionation. Figure 10.4 compares the integral distribution curves for the same polymer

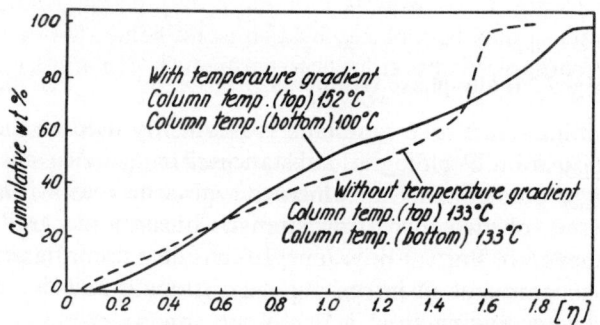

FIG. 10.4. Integral weight distribution curves of a polymer (commercial polyethylene) showing the effect of a temperature gradient on efficiency. $[\eta]$ = inherent viscosity ($[\eta] = 2\cdot50 \times 10^{-3} M^{0\cdot536}$; M = molecular weight) [17].

with and without a temperature gradient. The integral curve without a temperature gradient is distorted in the region of large molecular weights, in a similar way to the curves obtained by fractionation with a solvent gradient. It hence results that both gradients are necessary to obtain better fractionations through column chromatographic methods.

Mobile-phase gradients (pH and ionic strength) combined with a stepwise temperature gradient of the column were used by Moore and Stein [18] to separate a synthetic mixture of 46 amino-acids and related compounds. The column (150 × 0·9 cm) was filled with Dowex 50-X4 resin.

The amino-acid solution (2 mg/1–5 ml) was brought to pH 2·2 and introduced into the column, which was then filled with a buffer at pH = 3.1. Water at a temperature of 30° was allowed to flow through the jacket round the column and fractions were collected at a maximum rate of 8 ml/hr. When the serine had left the column (after passage of around 210 ml of buffered eluent at pH 3·1) the temperature of the water heating the column was raised to 50° and elution continued with the same buffer till the elution volume was 340 ml. A device was then switched in which gave a pH and ionic strength gradient. It consisted of two interconnected vessels: a mixing chamber filled with 0·2N sodium citrate at pH 3·1, and a receiver filled with a buffer solution of 2N sodium citrate + sodium acetate at pH 5.1. The temperature was kept at 50° till the elution volume reached 700 ml. To separate arginine and tryptophan, the temperature of the column was raised to 75°, and the gradient elution continued till the arginine, the last component, was eluted (volume 850 ml).

This technique of continuous or stepwise pH gradient combined with temperature programming is frequently used in the chromatography of amino-acids.

10.2 Combined Mobile-phase Gradients

The combination of two gradients is frequently used in the purification and separation of biological substances. For instance, an increasing ionic strength gradient can be combined with a decreasing pH gradient (see Fig. 7.95), or two increasing strength linear ionic gradients with different slopes (see Fig. 7.101). Cohn [19] has also used the combination of two linear gradients of increasing ionic strength and decreasing pH, to separate some nucleosides.

REFERENCES

1. Baker, C. A. and Williams, R. J. P., *J. Chem. Soc.* 2352 (1956).
2. Schneider, N. S., Holmes, L. G., Mijal, C. F. and Loconti, J. D., *J. Polymer Sci.* **37**, 551 (1959).
3. Weakley, T. J. R., Williams, R. J. P. and Wilson, J. D., *J. Chem. Soc.* 3963 (1960).
4. Jungnickel, J. L. and Weiss, F. T., *J. Polymer Sci.* **49**, 437 (1961).
5. Schulz, W. W. and Purdy, W. C., *Anal. Chem.* **35**, 2222 (1963).
6. Chapiro, A., Cordier, P., Jozefowicz, J. and Sebban-Danon, J., *J. Polymer Sci.*, *C* **4**, 491 (1963).
7. Hansen, C. M. and Sather, G. A., *J. Appl. Polymer Sci.* **8**, 2479 (1964).
8. Flowers, D. L., Hewett, W. A. and Mullineaux, R. D., *J. Polymer Sci. A-1* **2**, 2305 (1964).
9. Polacek, J., Schultz, L. and Kössler, I., *J. Polymer Sci.*, *C* **16**, 1327 (1967).
10. Caplan, R. S., *J. Polymer Sci.* **35**, 409 (1959).
11. Cantow, M. J. R., Porter, R. S. and Johnson, J. F., *J. Polymer Sci. C* **1**, 187 (1963).
12. Slonaker, D. F., Combs, R. L., Guillet, J. E. and Coover, H. W., Jr., *J. Polymer Sci. A-2* **4**, 523 (1966).
13. Schultz, G. V., Berger, K. C. and Scholz, A. G. R., *Ber. Bunsenges. Physik. Chem.* **69**, 856 (1956).
14. Schultz, G. V., Deusen, P. and Scholtz, A. G. R., *Makromol. Chem.* **69**, 47 (1963).
15. Schultz, G. V. and Nordt, E., *J. Prakt. Chem.* **155**, 115 (1940).
16. Schultz, G. V., *Z. Physik. Chem. Abt. B* **46**, 137 (1940); **47**, 155 (1940).
17. Guiller, J. E., Combs, R. L., Slonaker, D. F. and Coover, H. W., Jr., *J. Polymer Sci.* **47**, 307 (1960).
18. Moore, S. and Stein, W. H., *J. Biol. Chem.* **211**, 893 (1954).
19. Cohn, W. E., *J. Biol. Chem.* **235**, 1488 (1960).

APPENDIX I

COMMERCIAL SOURCES OF EQUIPMENT FOR GRADIENT LIQUID CHROMATOGRAPHY

No.	Manufacturer	Address
1.	Ace Glass Inc.	Vineland, New Jersey, 08360 (Box 688), USA
2.	Applied Research Laboratories Ltd.	Wingate Road, Luton, England
3.	Becker Delft N.V.	Vulcansweg, 259, Delft, The Netherlands
4.	Beckman Instruments, Inc. Spinco Div.	1117 California Avenue, Palo Alto, CA 94304, USA
5.	Brinkmann Instruments	Cantiague Road, Westbury, New York 11590, USA
6.	Buchler Instruments Div., Emerson Electric Company	407 West Vine St., Hatfield, Pa, USA
7.	Camag	Hamburgerstrasse 24, 4132 Muttenz, Switzerland
8.	Carlo Erba, S.p.A. Scientific Instruments Div.	20100 Milano, Italy
9.	Chromatec, Inc.	30 Main St., Ashland, Mass. 01721, USA
10.	Chromatronix Inc.	2743 Ninth St., Berkeley, California 94710, USA
11.	Desaga GmbH	69 Heidelberg 1, West Germany
12.	E. I. Du Pont de Nemours & Co. (Inc.)	Wilmington, Del. 19898, USA
13.	Glenco Scientific Inc.	3121 White Oak Dr., Houston, Texas 77077, USA
14.	Hoefer Scientific Instruments	520 Bryant St., San Francisco, California 94107, USA
15.	Hupe-Busch	D 7501 Grötzigen (Karlsruhe), Gutenbergstrasse 6, West Germany
16.	Instrument Research Assoc. Inc.	2352 John Glenn Dr., Atlanta, Georgia 30341, USA

Appendix I

No.	Manufacturer	Address
17.	Instrumentation Specialities Co.	4700 Superior, Lincoln, NE 68504, USA
18.	Isolab, Inc.	Drawer 4350, Akron, Ohio 44321, USA
19.	Japan Electron Optics Laboratory Co., Ltd.	1418 Nagakami, Akishima, Tokyo, Japan
20.	Kontes Glass Co.	Spruce St., Vineland, New Jersey, 08360, USA
21.	Laboratory Glass Apparatus	1200 Fourth St., Berkeley, California 94710, USA
22.	LKB-Produkter AB	S-161 25 Bromma, Stockholm, Sweden
23.	Metaloglass Inc.	466 Blue Hill Ave., Boston, Massachusetts 02121, USA
24.	Perkin-Elmer Corp.	702-A Main Ave., Norwalk, Conn. 06852, USA
25.	Pharmacia Fine Chemicals AB	P.O. Box 175, S-751, 04 Uppsala 1, Sweden
26.	Phoenix Precision Instrument	3803-05 North Fifth Street, Philadelphia, Penn. 19140, USA
27.	Pye Unicam Ltd.	York Street, Cambridge, CB1 2PX, England
28.	Quartz General Corp.	12440 Euline St., El Monta, California 91732, USA
29.	Quickfit Instrumentation James A. Jobling Co.	Stone, Staffordshire, England
30.	Sage Instruments Inc. Div. of Orion Research Inc.	11 Blackstone St., Cambridge, Massachusetts 02139, USA
31.	Siemens AG	Reinbruckenstrasse 50, 75 Karlomhe 21, West Germany
32.	Sigmamotor Inc.	14 Elizabeth St., Middleport, New York 14105, USA
33.	Technicon Industrial Systems	511 Benedict Ave., Tarrytown, New York 10591, USA
34.	Varian Associates Inc.	Steinhauserstrasse 6300, Zug, Switzerland
35.	Waters Associates	61 Fountain Street, Framingham, Massachusetts 01701, USA

APPENDIX II

THE CALCULATION OF THE POSITION OF THE PEAKS IN ION-EXCHANGE CHROMATOGRAPHY AND ELUTION WITH A MOBILE-PHASE GRADIENT, AFTER THE METHOD WORKED OUT BY OHASHI AND KOGUCHI
([69] in Chapter 7)

We start from the fundamental equation for the position of the elution peak:

$$K_{dr}V_b = \int_0^{V_R} [1 - c \exp(-V/V_m)]^n \, dV \tag{II.1}$$

where V_R represents the retention volume, K_{dr} the distribution coefficient of a given ion at the concentration C_r of eluent in the receiver, defined as the ratio between the quantity of ion per unit volume of resin and the quantity of the ion per unit volume of solution, n is a constant given by

$$K_d = aC^{-n} \tag{II.2}$$

and is the charge of the ion in an ideal system, V_b is the partial volume of resin, dV the increase in the eluent volume, and $c = (C_r - C_{0m})/C_r$ in which C_{0m} represents the initial concentration in the mixing chamber.

In deducing Eq. (II.1) the following suppositions were made: (a) the plate-theory is used for all ion-exchange processes; (b) the solution volume in the connecting tube between the mixing vessel and the column, the dead volume and the interstitial volume of the resin bed, are negligible compared to the retention volume; (c) the eluent concentration and the distribution coefficient of the ion are related by Eq. (II.2).

To solve equation (II.1) the following changes of variables are made:

$$V/V_m = t; \quad V_R/V_m = x; \quad K_{dr}V_b/V_m = y; \quad dV = V_m \, dt$$

so that Eq. (II.1) becomes:

$$y = \int_0^x (1 - ce^{-t})^n \, dt \tag{II.3}$$

A new change of variables is made:

$$ce^{-t} = u \quad \text{and} \quad dt = -(1/u)\,du$$

so that Eq. (II.3) becomes

$$\begin{aligned}
y &= \int_c^{ce^{-x}} -[(1-u)^n/u]\,du = -\int_c^{ce^{-x}} (1/u)\,du \\
&\quad + \int_c^{ce^{-x}} [1-(1-u)^n](1/u)\,du \\
&= x - \int_0^c [1-(1-u)^n](1/u)\,du + \int_0^{ce^{-x}} [1-(1-u)^n](1/u)\,du \\
&= x - f_n(c) + f_n(ce^{-x}) = G(x)
\end{aligned} \qquad (II.4)$$

The second and third terms can be expressed as

$$f_n(z) = \int_0^z [1-(1-u)^n]\cdot(1/u)\,du \qquad (II.5)$$

Following a new change of variables:

$$1 - u = v; \quad du = -dv$$

Equation (II.5) may be represented by

$$f_n(z) = \int_{1-z}^1 [(1-v^n)/(1-v)]\,dv \qquad (II.6)$$

The authors show that a function such as $f_n(z)$ is very complicated and is very difficult to solve for any value of n. Nevertheless, Eq. (II.6) may be integrated when n is n_0, $n_0 + 1/2$, $n_0 + 1/4$ or $n_0 + 3/4$, where n_0 is a whole number. Results are given below:

$n = n_0$:

$$f_n(z) = 1 + \frac{1}{2} + \frac{1}{3} + \cdots + \frac{1}{n_0} \\
- \left[1 - z + \frac{(1-z)^2}{2} + \frac{(1-z)^3}{3} + \cdots + \frac{(1-z)^{n_0}}{n_0}\right] \qquad (II.7)$$

$n = n_0 + 1/2$:

$$f_n(z) = 2 + \frac{2}{5} + \frac{2}{5} + \cdots + \frac{2}{2n_0 + 1} - 2\log 2$$

$$- 2(1 - z)^{1/2} \left[1 + \frac{1 - z}{3} + \frac{(1 - z)^2}{5} + \cdots + \frac{(1 - z)^{n_0}}{2n_0 + 1} \right]$$

$$+ 2\log[1 + (1 - z)^{1/2}] \tag{II.8}$$

$n = n_0 + 1/4$:

$$f_n(z) = 4 + \frac{4}{5} + \frac{4}{9} + \cdots + \frac{4}{4n_0 + 1} - 3\log 2 - \frac{\pi}{2}$$

$$- 4(1 - z)^{1/4} \left[1 + \frac{1 - z}{5} + \frac{(1 - z)^2}{9} + \cdots + \frac{(1 - z)^{n_0}}{4n_0 + 1} \right]$$

$$+ \log[1 + (1 - z)^{1/2}] + 2\log[1 + (1 - z)^{1/4}]$$

$$+ 2\tan^{-1}(1 - z)^{1/4} \tag{II.9}$$

$n = n_0 + 3/4$:

$$f_n(z) = \frac{4}{3} + \frac{4}{7} + \frac{4}{11} + \cdots + \frac{4}{4n_0 + 3} - 3\log 2 + \frac{\pi}{2}$$

$$- 4(1 - z)^{3/4} \left[\frac{1}{3} + \frac{1 - z}{7} + \frac{(1 - z)^2}{11} + \cdots + \frac{(1 - z)^{n_0}}{4n_0 + 3} \right]$$

$$+ \log[1 + (1 - z)^{1/2}] + 2\log[1 + (1 - z)^{1/4}]$$

$$- 2\tan^{-1}(1 - z)^{1/4} \tag{II.10}$$

Calculation methods

One of the aims of the work was to calculate $x = V_R/V_m$ when $y = K_{d_r}V_b/V_m$ was obtained experimentally. But Eq. (II.4) is implicit according to x. Therefore y must be calculated for a given x and the relation between y and x examined graphically. Before these calculations are made the order of the variables must be established in Eq. (II.4). The values of n are in many cases probably lower than 6. Thus n was considered to lie in the range of 0·25–6·00, at intervals of 0·25. The values of c were chosen to be between 0·5 and 1·0 at intervals of 0·1. If $c < 0.5$, we obtain gradients with too small an elution efficiency. The values of x were taken in the range of 0–5·00 at intervals of 0·1. For $x > 5.00$ the concentration of the gradient is too small.

Ohashi and Kogushi, using an ALGOL-H computer, have computed 7200 values of y for the combinations of the above-mentioned variables.

In this way, 2400 values of the function $f_n(z)$ were computed for different values of z (Table II.1). The values of z were chosen between 0 and 1·00, at intervals of 0·01 as the limits of z are between 0 and 1.

The authors suggest two practical and easy methods for the calculation of the approximate value of x. The two methods depend on whether y is bigger or smaller than 0·7.

Method 1, $y > 0.7$.

When y is big enough, $f_n(ce^{-x})$ is very small compared to $x - f_n(c)$ in Eq. (II.4). Consequently, a first approximate value of x is obtained from Eq. (II.4) in which the term $f_n(ce^{-x})$ given by Eq. (II.11) is neglected.

$$x_1 = y + f_n(c) \tag{II.11}$$

By means of Newton's method, the second approximate value is given by the following relationship:

$$x_2 = x_1 - \frac{G(x_1) - y}{G'(x_1)} = x_1 - \frac{f_n(ce^{-x_1})}{(1 - ce^{-x_1})^n} \tag{II.12}$$

By using the data for the functions $f_n(z)$ given in Table II.1 we can calculate the second approximate value of x_2. For the cases in which n is between two of the values given in Table II.1, the value of $f_n(z)$ is obtained by interpolation.

As seen in Table II.2 the errors for the values of x_2 obtained by this method are always less than 1 per cent when y is bigger than 0·7. When $y < 0.7$ this method gives large errors for x_2, especially when n is large, and the authors have created another method for these cases.

Method 2, $y < 0.7$.

When $y < 0.7$, the relationship between x and y is graphically illustrated in Figs. II.1–II.6 for different values of c. From these figures a first approximate value of x_1 can be obtained for an experimental value of y.

The second approximate value of x_2 is obtained by means of Newton's method, using the equation

$$x_2 = x_1 - \frac{x_1 - f_n(c) + f_n(ce^{-x_1}) - y}{(1 - ce^{-x_1})^n} \tag{II.13}$$

The first approximate value of x_1 may be read on the graph to ± 0.02, and the errors of x_2 are under 3 per cent for the cases shown in Table II.3.

TABLE II.1. Values of $f_n(z)$

z	n 0.25	0.50	0.75	1.00	1.25	1.50	1.75	2.00	2.25	2.50	2.75	3.00
0.00	0.0000	0.0000	0.0000	0.0000	0.0000	0.0000	0.0000	0.0000	0.0000	0.0000	0.0000	0.0000
0.01	0.0025	0.0050	0.0075	0.0100	0.0125	0.0150	0.0175	0.0200	0.0224	0.0249	0.0274	0.0299
0.02	0.0050	0.0100	0.0150	0.0200	0.0250	0.0299	0.0349	0.0398	0.0447	0.0496	0.0545	0.0594
0.03	0.0075	0.0151	0.0225	0.0300	0.0374	0.0448	0.0522	0.0596	0.0669	0.0742	0.0814	0.0887
0.04	0.0101	0.0201	0.0301	0.0400	0.0499	0.0597	0.0695	0.0792	0.0889	0.0985	0.1081	0.1176
0.05	0.0126	0.0252	0.0376	0.0500	0.0623	0.0745	0.0867	0.0988	0.1107	0.1227	0.1345	0.1463
0.06	0.0152	0.0302	0.0452	0.0600	0.0747	0.0893	0.1038	0.1182	0.1325	0.1466	0.1607	0.1747
0.07	0.0177	0.0353	0.0527	0.0700	0.0871	0.1041	0.1209	0.1376	0.1541	0.1704	0.1867	0.2028
0.08	0.0203	0.0404	0.0603	0.0800	0.0995	0.1188	0.1379	0.1568	0.1755	0.1941	0.2124	0.2306
0.09	0.0229	0.0455	0.0679	0.0900	0.1119	0.1335	0.1548	0.1759	0.1968	0.2175	0.2379	0.2581
0.10	0.0255	0.0506	0.0755	0.1000	0.1242	0.1481	0.1717	0.1950	0.2180	0.2407	0.2632	0.2853
0.11	0.0281	0.0558	0.0831	0.1100	0.1365	0.1627	0.1885	0.2139	0.2390	0.2638	0.2882	0.3123
0.12	0.0307	0.0609	0.0907	0.1200	0.1489	0.1773	0.2052	0.2328	0.2599	0.2867	0.3130	0.3390
0.13	0.0333	0.0661	0.0983	0.1300	0.1611	0.1918	0.2219	0.2515	0.2870	0.3094	0.3376	0.3654
0.14	0.0360	0.0713	0.1060	0.1400	0.1734	0.2063	0.2385	0.2702	0.3013	0.3319	0.3620	0.3915
0.15	0.0386	0.0765	0.1136	0.1500	0.1857	0.2207	0.2551	0.2887	0.3218	0.3543	0.3861	0.4174
0.16	0.0413	0.0817	0.1213	0.1600	0.1979	0.2351	0.2715	0.3072	0.3422	0.3764	0.4100	0.4430
0.17	0.0440	0.0869	0.1289	0.1700	0.2102	0.2495	0.2879	0.3255	0.3624	0.3984	0.4337	0.4683
0.18	0.0466	0.0922	0.1366	0.1800	0.2224	0.2638	0.3043	0.3438	0.3825	0.4202	0.4572	0.4933
0.19	0.0493	0.0974	0.1443	0.1900	0.2346	0.2781	0.3205	0.3619	0.4024	0.4419	0.4805	0.5181
0.20	0.0520	0.1027	0.1520	0.2000	0.2468	0.2923	0.3367	0.3800	0.4222	0.4633	0.5035	0.5427
0.21	0.0548	0.1080	0.1597	0.2100	0.2598	0.3065	0.3529	0.3980	0.4419	0.4846	0.5263	0.5669
0.22	0.0575	0.1133	0.1674	0.2200	0.2711	0.3207	0.3689	0.4158	0.4614	0.5058	0.5489	0.5909
0.23	0.0602	0.1186	0.1752	0.2300	0.2832	0.3348	0.3849	0.4336	0.4808	0.5267	0.5713	0.6147

TABLE II.1 (*continued*)

z	n 0.25	0.50	0.75	1.00	1.25	1.50	1.75	2.00	2.25	2.50	2.75	3.00
0.24	0.0630	0.1239	0.1829	0.2400	0.2953	0.3489	0.4008	0.4512	0.5001	0.5475	0.5935	0.6382
0.25	0.0658	0.1293	0.1907	0.2500	0.3074	0.3629	0.4167	0.4688	0.5192	0.5681	0.6155	0.6615
0.26	0.0685	0.1346	0.1984	0.2600	0.3195	0.3769	0.4325	0.4862	0.5382	0.5885	0.6372	0.6845
0.27	0.0713	0.1400	0.2062	0.2700	0.3315	0.3909	0.4482	0.5035	0.5570	0.6088	0.6588	0.7072
0.28	0.0741	0.1454	0.2140	0.2800	0.3436	0.4048	0.4638	0.5208	0.5758	0.6289	0.6801	0.7297
0.29	0.0770	0.1508	0.2218	0.2900	0.3556	0.4187	0.4794	0.5379	0.5944	0.6488	0.7013	0.7520
0.30	0.0798	0.1563	0.2296	0.3000	0.3676	0.4325	0.4949	0.5550	0.6128	0.6685	0.7222	0.7740
0.31	0.0827	0.1617	0.2375	0.3100	0.3796	0.4463	0.5104	0.5719	0.6311	0.6881	0.7429	0.7958
0.32	0.0855	0.1672	0.2453	0.3200	0.3915	0.4600	0.5258	0.5888	0.6493	0.7075	0.7635	0.8173
0.33	0.0884	0.1727	0.2532	0.3300	0.4035	0.4738	0.5411	0.6055	0.6674	0.7268	0.7838	0.8386
0.34	0.0913	0.1782	0.2610	0.3400	0.4154	0.4874	0.5563	0.6222	0.6853	0.7459	0.8039	0.8597
0.35	0.0942	0.1837	0.2689	0.3500	0.4273	0.5010	0.5714	0.6387	0.7031	0.7648	0.8239	0.8805
0.36	0.0971	0.1893	0.2768	0.3600	0.4392	0.5146	0.5865	0.6552	0.7208	0.7835	0.8436	0.9012
0.37	0.1001	0.1948	0.2847	0.3700	0.4510	0.5281	0.6016	0.6715	0.7383	0.8021	0.8631	0.9215
0.38	0.1030	0.2004	0.2926	0.3800	0.4629	0.5416	0.6165	0.6878	0.7557	0.8206	0.8825	0.9417
0.39	0.1060	0.2060	0.3006	0.3900	0.4747	0.5551	0.6314	0.7039	0.7730	0.8388	0.9016	0.9616
0.40	0.1050	0.2117	0.3085	0.4000	0.4865	0.5685	0.6462	0.7200	0.7902	0.8569	0.9206	0.9813
0.41	0.1120	0.2173	0.3165	0.4100	0.4983	0.5818	0.6609	0.7359	0.8072	0.8749	0.9394	1.0008
0.42	0.1150	0.2230	0.3245	0.4200	0.5101	0.5952	0.6756	0.7518	0.8241	0.8927	0.9579	1.0201
0.43	0.1181	0.2287	0.3324	0.4300	0.5218	0.6084	0.6902	0.7675	0.8408	0.9103	0.9763	1.0392
0.44	0.1211	0.2344	0.3405	0.4400	0.5336	0.6217	0.7047	0.7832	0.8574	0.9278	0.9945	1.0580
0.45	0.1242	0.2401	0.3485	0.4500	0.5453	0.6348	0.7192	0.7987	0.8739	0.9451	1.0126	1.0766
0.46	0.1273	0.2459	0.3565	0.4600	0.5570	0.6480	0.7336	0.8142	0.8903	0.9623	1.0304	1.0950
0.47	0.1304	0.2516	0.3646	0.4700	0.5686	0.6611	0.7479	0.8295	0.9065	0.9793	1.0481	1.1133
0.48	0.1335	0.2574	0.3726	0.4800	0.5803	0.6741	0.7621	0.8448	0.9227	0.9961	1.0655	1.1313
0.49	0.1367	0.2632	0.3807	0.4900	0.5919	0.6871	0.7763	0.8599	0.9386	1.0128	1.0828	1.1491

TABLE II.1 (*continued*)

z	n 0.25	0.50	0.75	1.00	1.25	1.50	1.75	2.00	2.25	2.50	2.75	3.00
0.50	0.1399	0.2691	0.3888	0.5000	0.6035	0.7001	0.7904	0.8750	0.9545	1.0293	1.0999	1.1667
0.51	0.1430	0.2750	0.3969	0.5100	0.6151	0.7130	0.8044	0.8899	0.9702	1.0475	1.1169	1.1841
0.52	0.1463	0.2809	0.4051	0.5200	0.6266	0.7258	0.8183	0.9048	0.9858	1.0620	1.1336	1.2013
0.53	0.1495	0.2868	0.4132	0.5300	0.6382	0.7386	0.8322	0.9195	1.0013	1.0781	1.1502	1.2183
0.54	0.1527	0.2927	0.4214	0.5400	0.6497	0.7514	0.8460	0.9342	1.0167	1.0940	1.1666	1.2351
0.55	0.1560	0.2987	0.4296	0.5500	0.6612	0.7641	0.8597	0.9487	1.0319	1.1098	1.1829	1.2517
0.56	0.1593	0.3047	0.4378	0.5600	0.6726	0.7768	0.8734	0.9632	1.0470	1.1254	1.1990	1.2681
0.57	0.1627	0.3107	0.4460	0.5700	0.6841	0.7894	0.8869	0.9775	1.0620	1.1409	1.2149	1.2844
0.58	0.1660	0.3168	0.4542	0.5800	0.6955	0.8020	0.9004	0.9918	1.0768	1.1563	1.2306	1.3004
0.59	0.1694	0.3229	0.4625	0.5900	0.7069	0.8145	0.9139	1.0059	1.0916	1.1714	1.2462	1.3163
0.60	0.1728	0.3290	0.4708	0.6000	0.7183	0.8270	0.9272	1.0200	1.1062	1.1865	1.2616	1.3320
0.61	0.1762	0.3351	0.4790	0.6100	0.7296	0.8394	0.9405	1.0340	1.1207	1.2014	1.2768	1.3475
0.62	0.1787	0.3413	0.4874	0.6200	0.7410	0.8518	0.9537	1.0478	1.1350	1.2162	1.2919	1.3628
0.63	0.1831	0.3475	0.4957	0.6300	0.7523	0.8641	0.9668	1.0615	1.1493	1.2308	1.3068	1.3780
0.64	0.1866	0.3537	0.5040	0.6400	0.7636	0.8764	0.9799	1.0752	1.1634	1.2453	1.3216	1.3930
0.65	0.1902	0.3600	0.5124	0.6500	0.7748	0.8886	0.9928	1.0887	1.1774	1.2596	1.3362	1.4078
0.66	0.1937	0.3663	0.5208	0.6600	0.7860	0.9008	1.0057	1.1022	1.1913	1.2738	1.3507	1.4224
0.67	0.1973	0.3726	0.5292	0.6700	0.7972	0.9129	1.0186	1.1155	1.2050	1.2879	1.3640	1.4369
0.68	0.2010	0.3790	0.5377	0.6800	0.8084	0.9250	1.0313	1.1238	1.2186	1.3018	1.3791	1.4512
0.69	0.2046	0.3854	0.5461	0.6900	0.8196	0.9370	1.0440	1.1419	1.2322	1.3156	1.3931	1.4654
0.70	0.2083	0.3918	0.5546	0.7000	0.8307	0.9490	1.0565	1.1550	1.2455	1.3292	1.4069	1.4793
0.71	0.2121	0.3983	0.5631	0.7100	0.8418	0.9609	1.0691	1.1679	1.2588	1.3427	1.4206	1.4932
0.72	0.2158	0.4048	0.5716	0.7200	0.8529	0.9727	1.0815	1.1808	1.2720	1.3561	1.4342	1.5068
0.73	0.2196	0.4114	0.5802	0.7300	0.8639	0.9845	1.0938	1.1935	1.2850	1.3694	1.4475	1.5203
0.74	0.2235	0.4180	0.5888	0.7400	0.8749	0.9963	1.1061	1.2062	1.2979	1.3825	1.4608	1.5337

TABLE II.1 (continued)

z	n 0.25	0.50	0.75	1.00	1.25	1.50	1.75	2.00	2.25	2.50	2.75	3.00
0.75	0.2274	0.4246	0.5974	0.7500	0.8859	1.0080	1.1183	1.2188	1.3107	1.3955	1.4739	1.5469
0.76	0.2313	0.4313	0.6060	0.7600	0.8969	1.0196	1.1304	1.2312	1.3234	1.4083	1.4869	1.5599
0.77	0.2353	0.4381	0.6147	0.7700	0.9078	1.0312	1.1425	1.2435	1.3360	1.4210	1.4997	1.5728
0.78	0.2393	0.4448	0.6234	0.7800	0.9187	1.0427	1.1544	1.2558	1.3484	1.4336	1.5124	1.5856
0.79	0.2433	0.4517	0.6321	0.7900	0.9296	1.0542	1.1663	1.2679	1.3608	1.4461	1.5250	1.5982
0.80	0.2475	0.4586	0.6408	0.8000	0.9405	1.0656	1.1781	1.2800	1.3730	1.4584	1.5374	1.6107
0.81	0.2516	0.4655	0.6496	0.8100	0.9513	1.0769	1.1898	1.2919	1.3851	1.4707	1.5497	1.6230
0.82	0.2558	0.4725	0.6584	0.8200	0.9620	1.0882	1.2014	1.3038	1.3971	1.4827	1.5618	1.6352
0.83	0.2601	0.4795	0.6673	0.8300	0.9728	1.0995	1.2130	1.3155	1.4090	1.4947	1.5738	1.6472
0.84	0.2645	0.4867	0.6761	0.8400	0.9835	1.1107	1.2244	1.3272	1.4208	1.5066	1.5857	1.6592
0.85	0.2689	0.4938	0.6850	0.8500	0.9942	1.1218	1.2358	1.3388	1.4324	1.5183	1.5975	1.6710
0.86	0.2734	0.5011	0.6940	0.8600	1.0048	1.1328	1.2471	1.3502	1.4440	1.5299	1.6091	1.6826
0.87	0.2779	0.5084	0.7030	0.8700	1.0155	1.1438	1.2583	1.3615	1.4554	1.5414	1.6206	1.6942
0.88	0.2825	0.5158	0.7120	0.8800	1.0260	1.1547	1.2695	1.3728	1.4667	1.5527	1.6320	1.7056
0.89	0.2873	0.5232	0.7211	0.8900	1.0366	1.1656	1.2805	1.3839	1.4779	1.5640	1.6433	1.7168
0.90	0.2921	0.5308	0.7302	0.9000	1.0471	1.1764	1.2915	1.3950	1.4890	1.5751	1.6544	1.7280
0.91	0.2970	0.5384	0.7393	0.9100	1.0576	1.1871	1.3023	1.4059	1.5000	1.5861	1.6655	1.7390
0.92	0.3020	0.5462	0.7486	0.9200	1.0680	1.1978	1.3131	1.4168	1.5109	1.5970	1.6764	1.7500
0.93	0.3072	0.5540	0.7578	0.9300	1.0784	1.2083	1.3238	1.4275	1.5217	1.6078	1.6872	1.7608
0.94	0.3125	0.5620	0.7671	0.9400	1.0887	1.2189	1.3344	1.4382	1.5324	1.6185	1.6979	1.7715
0.95	0.3179	0.5701	0.7765	0.9500	1.0990	1.2293	1.3449	1.4487	1.5429	1.6291	1.7085	1.7820
0.96	0.3236	0.5783	0.7860	0.9600	1.1093	1.2397	1.3553	1.4592	1.5534	1.6396	1.7189	1.7925
0.97	0.3295	0.5868	0.7955	0.9700	1.1195	1.2500	1.3657	1.4695	1.5638	1.6499	1.7293	1.8029
0.98	0.3356	0.5954	0.8051	0.9800	1.1296	1.2602	1.3759	1.4798	1.5740	1.6602	1.7396	1.8131
0.99	0.3423	0.6043	0.8148	0.9900	1.1397	1.2703	1.3861	1.4899	1.5842	1.6703	1.7497	1.8233
1.00	0.3498	0.6137	0.8247	1.0000	1.1498	1.2804	1.3961	1.5000	1.5942	1.6804	1.7598	1.8333

TABLE II.1 (*continued*)

z	n 3·25	3·50	3·75	4·00	4·25	4·50	4·75	5·00	5·25	5·50	5·75	6·00
0·00	0·0000	0·0000	0·0000	0·0000	0·0000	0·0000	0·0000	0·0000	0·0000	0·0000	0·0000	0·0000
0·01	0·0323	0·0348	0·0372	0·0397	0·0422	0·0446	0·0471	0·0495	0·0519	0·0544	0·0568	0·0593
0·02	0·0643	0·0651	0·0740	0·0788	0·0836	0·0884	0·0932	0·0980	0·1028	0·1076	0·1123	0·1171
0·03	0·0959	0·1031	0·1102	0·1173	0·1244	0·1315	0·1386	0·1456	0·1526	0·1596	0·1665	0·1734
0·04	0·1271	0·1365	0·1459	0·1553	0·1646	0·1738	0·1830	0·1922	0·2013	0·2104	0·2194	0·2284
0·05	0·1580	0·1696	0·1812	0·1927	0·2041	0·2154	0·2267	0·2379	0·2490	0·2601	0·2711	0·2821
0·06	0·1885	0·2023	0·2159	0·2295	0·2429	0·2563	0·2695	0·2827	0·2958	0·3087	0·3216	0·3344
0·07	0·2187	0·2345	0·2502	0·2658	0·2812	0·2964	0·3116	0·3266	0·3415	0·3563	0·3709	0·3854
0·08	0·2486	0·2664	0·2840	0·3015	0·3188	0·3359	0·3529	0·3697	0·3863	0·4028	0·4191	0·4353
0·09	0·2781	0·2978	0·3173	0·3367	0·3558	0·3747	0·3934	0·4118	0·4301	0·4482	0·4662	0·4839
0·10	0·3072	0·3288	0·3502	0·3713	0·3922	0·4128	0·4331	0·4532	0·4731	0·4927	0·5121	0·5313
0·11	0·3361	0·3595	0·3826	0·4054	0·4280	0·4502	0·4721	0·4938	0·5151	0·5362	0·5570	0·5776
0·12	0·3645	0·3897	0·4146	0·4391	0·4632	0·4870	0·5104	0·5335	0·5563	0·5788	0·6009	0·6228
0·13	0·3927	0·4196	0·4461	0·4722	0·4978	0·5231	0·5480	0·5725	0·5966	0·6204	0·6438	0·6669
0·14	0·4206	0·4491	0·4772	0·5048	0·5319	0·5586	0·5848	0·6107	0·6361	0·6611	0·6857	0·7099
0·15	0·4481	0·4782	0·5078	0·5369	0·5654	0·5935	0·6210	0·6481	0·6748	0·7009	0·7266	0·7519
0·16	0·4753	0·5069	0·5380	0·5685	0·5984	0·6278	0·6566	0·6849	0·7126	0·7399	0·7667	0·7930
0·17	0·5021	0·5353	0·5678	0·5996	0·6309	0·6614	0·6914	0·7209	0·7497	0·7780	0·8058	0·8330
0·18	0·5287	0·5633	0·5972	0·6303	0·6628	0·6945	0·7257	0·7562	0·7860	0·8153	0·8440	0·8722
0·19	0·5550	0·5909	0·6261	0·6605	0·6942	0·7271	0·7593	0·7908	0·8216	0·8518	0·8814	0·9104
0·20	0·5809	0·6182	0·6547	0·6903	0·7250	0·7590	0·7923	0·8247	0·8565	0·8876	0·9180	0·9477
0·21	0·6065	0·6451	0·6828	0·7196	0·7554	0·7904	0·8246	0·8580	0·8906	0·9225	0·9537	0·9842
0·22	0·6319	0·6717	0·7106	0·7484	0·7853	0·8213	0·8564	0·8907	0·9241	0·9568	0·9886	1·0198
0·23	0·6569	0·6980	0·7379	0·7768	0·8147	0·8516	0·8876	0·9227	0·9569	0·9903	1·0228	1·0546
0·24	0·5816	0·7238	0·7649	0·8048	0·8436	0·8814	0·9182	0·9541	0·9890	1·0231	1·0563	1·0886

TABLE II.1 (*continued*)

z	n 3·25	3·50	3·75	4·00	4·25	4·50	4·75	5·00	5·25	5·50	5·75	6·00
0·25	0·7061	0·7494	0·7915	0·8324	0·8721	0·9107	0·9483	0·9849	1·0205	1·0552	1·0890	1·1219
0·26	0·7302	0·7746	0·8177	0·8595	0·9001	0·9395	0·9778	1·0151	1·0514	1·0866	1·1210	1·1544
0·27	0·7541	0·7995	0·8435	0·8862	0·9276	0·9678	1·0068	1·0448	1·0816	1·1174	1·1523	1·1862
0·28	0·7777	0·8241	0·8690	0·9125	0·9547	0·9956	1·0353	1·0738	1·1112	1·1476	1·1829	1·2173
0·29	0·8010	0·8483	0·8941	0·9385	0·9814	1·0230	1·0633	1·1024	1·1403	1·1771	1·2129	1·2477
0·30	0·8240	0·8722	0·9189	0·9640	1·0076	1·0498	1·0907	1·1304	1·1688	1·2061	1·2423	1·2774
0·31	0·8467	0·8959	0·9433	0·9891	1·0334	1·0762	1·1177	1·1578	1·1967	1·2344	1·2710	1·3065
0·32	0·8692	0·9192	0·9674	1·0139	1·0588	1·1022	1·1442	1·1848	1·2241	1·2622	1·2992	1·3350
0·33	0·8914	0·9422	0·9911	1·0383	1·0838	1·1277	1·1702	1·2112	1·2510	1·2894	1·3267	1·3628
0·34	0·9133	0·9648	1·0145	1·0623	1·1083	1·1528	1·1957	1·2372	1·2773	1·3161	1·3537	1·3901
0·35	0·9350	0·9872	1·0375	1·0859	1·1325	1·1775	1·2208	1·2627	1·3032	1·3423	1·3801	1·4168
0·36	0·9563	1·0093	1·0602	1·1092	1·1563	1·2017	1·2455	1·2877	1·3285	1·3679	1·4061	1·4429
0·37	0·9775	1·0311	1·0827	1·1322	1·1798	1·2256	1·2697	1·3123	1·3534	1·3931	1·4314	1·4685
0·38	0·9984	1·0527	1·1047	1·1547	1·2028	1·2490	1·2935	1·3364	1·3778	1·4177	1·4563	1·4936
0·39	1·0190	1·0739	1·1265	1·1770	1·2255	1·2721	1·3169	1·3601	1·4017	1·4419	1·4807	1·5182
0·40	1·0394	1·0949	1·1480	1·1989	1·2478	1·2948	1·3399	1·3834	1·4253	1·4656	1·5046	1·5423
0·41	1·0595	1·1155	1·1692	1·2205	1·2698	1·3171	1·3625	1·4062	1·4483	1·4889	1·5281	1·5659
0·42	1·0794	1·1359	1·1900	1·2418	1·2914	1·3390	1·3847	1·4287	1·4710	1·5117	1·5511	1·5890
0·43	1·0990	1·1561	1·2106	1·2628	1·3127	1·3606	1·4066	1·4507	1·4932	1·5342	1·5736	1·6117
0·44	1·1184	1·1760	1·2309	1·2834	1·3337	1·3818	1·4280	1·4724	1·5151	1·5561	1·5957	1·6339
0·45	1·1375	1·1956	1·2509	1·3037	1·3543	1·4027	1·4491	1·4937	1·5365	1·5777	1·6174	1·6557
0·46	1·1565	1·2149	1·2706	1·3238	1·3746	1·4232	1·4699	1·5146	1·5576	1·5989	1·6388	1·6771
0·47	1·1752	1·2340	1·2901	1·3435	1·3946	1·4435	1·4903	1·5352	1·5783	1·6197	1·6597	1·6981
0·48	1·1936	1·2529	1·3092	1·3630	1·4143	1·4634	1·5103	1·5554	1·5986	1·6402	1·6802	1·7188
0·49	1·2118	1·2715	1·3281	1·3822	1·4337	1·4829	1·5301	1·5753	1·6186	1·6603	1·7004	1·7390

APPENDIX II

TABLE II.1 (*continued*)

z	n 3·25	3·50	3·75	4·00	4·25	4·50	4·75	5·00	5·25	5·50	5·75	6·00
0·50	1·2299	1·2898	1·3468	1·4010	1·4528	1·5022	1·5495	1·5948	1·6383	1·6800	1·7202	1·7589
0·51	1·2476	1·3079	1·3652	1·4197	1·4716	1·5212	1·5686	1·6140	1·6576	1·6994	1·7396	1·7784
0·52	1·2652	1·3258	1·3833	1·4380	1·4901	1·5393	1·5874	1·6329	1·6765	1·7185	1·7587	1·7975
0·53	1·2826	1·3434	1·4012	1·4561	1·5084	1·5582	1·6059	1·6515	1·6952	1·7372	1·7775	1·8164
0·54	1·2997	1·3608	1·4188	1·4739	1·5263	1·5763	1·6241	1·6698	1·7136	1·7556	1·7960	1·3349
0·55	1·3166	1·3780	1·4362	1·4915	1·5440	1·5941	1·6420	1·6878	1·7316	1·7737	1·8141	1·8530
0·56	1·3334	1·3950	1·4534	1·5088	1·5615	1·6117	1·6596	1·7055	1·7494	1·7915	1·8320	1·8709
0·57	1·3499	1·4117	1·4703	1·5258	1·5787	1·6290	1·6770	1·7229	1·7669	1·8090	1·8495	1·8885
0·58	1·3662	1·4282	1·4870	1·5427	1·5956	1·6460	1·6941	1·7400	1·7841	1·8263	1·8668	1·9058
0·59	1·3823	1·4446	1·5034	1·5592	1·6123	1·6628	1·7109	1·7569	1·8010	1·8432	1·8838	1·9228
0·60	1·3982	1·4606	1·5197	1·5756	1·6287	1·6793	1·7275	1·7736	1·8176	1·8599	1·9005	1·9395
0·61	1·4139	1·4765	1·5357	1·5917	1·6449	1·6955	1·7438	1·7899	1·8340	1·8763	1·9170	1·9560
0·62	1·4295	1·4922	1·5515	1·6076	1·6609	1·7116	1·7599	1·8060	1·8502	1·8925	1·9331	1·9722
0·63	1·4448	1·5077	1·5671	1·6233	1·6767	1·7274	1·7757	1·8219	1·8661	1·9084	1·9491	1·9882
0·64	1·4600	1·5230	1·5825	1·6388	1·6922	1·7430	1·7914	1·8376	1·8818	1·9241	1·9648	2·0039
0·65	1·4749	1·5381	1·5977	1·6540	1·7075	1·7583	1·8068	1·8530	1·8972	1·9396	1·9803	2·0193
0·66	1·4897	1·5530	1·6127	1·6691	1·7226	1·7735	1·8219	1·8682	1·9124	1·9548	1·9955	2·0346
0·67	1·5043	1·5677	1·6274	1·6839	1·7375	1·7884	1·8369	1·8832	1·9274	1·9698	2·0105	2·0496
0·68	1·5188	1·5822	1·6420	1·6986	1·7522	1·8031	1·8516	1·8979	1·9422	1·9846	2·0253	2·0644
0·69	1·5330	1·5966	1·6564	1·7130	1·7667	1·8176	1·8662	1·9125	1·9568	1·9992	2·0399	2·0790
0·70	1·5471	1·6107	1·6707	1·7273	1·7810	1·8320	1·8805	1·9268	1·9711	2·0135	2·0542	2·0934
0·71	1·5610	1·6247	1·6847	1·7414	1·7951	1·8461	1·8946	1·9410	1·9853	2·0277	2·0684	2·1075
0·72	1·5748	1·6385	1·6986	1·7553	1·8090	1·8600	1·9086	1·9549	1·9992	2·0417	2·0824	2·1215
0·73	1·5883	1·6522	1·7122	1·7690	1·8227	1·8738	1·9224	1·9687	2·0130	2·0555	2·0962	2·1353

TABLE II.1 (continued)

z	n 3.25	3.50	3.75	4.00	4.25	4.50	4.75	5.00	5.25	5.50	5.75	6.00
0.74	1.6018	1.6656	1.7258	1.7825	1.8363	1.8873	1.9359	1.9823	2.0266	2.0691	2.1098	2.1489
0.75	1.6150	1.6790	1.7391	1.7959	1.8497	1.9007	1.9493	1.9957	2.0400	2.0825	2.1232	2.1623
0.76	1.6281	1.6921	1.7523	1.8091	1.8629	1.9140	1.9626	2.0089	2.0533	2.0957	2.1364	2.1756
0.77	1.6411	1.7051	1.7653	1.8221	1.8759	1.9270	1.9756	2.0220	2.0663	2.1088	2.1495	2.1886
0.78	1.6539	1.7179	1.7782	1.8350	1.8888	1.9399	1.9885	2.0349	2.0792	2.1217	2.1624	2.2015
0.79	1.6666	1.7306	1.7909	1.8477	1.9015	1.9526	2.0013	2.0476	2.0920	2.1344	2.1751	2.2143
0.80	1.6791	1.7431	1.8034	1.8603	1.9141	1.9652	2.0138	2.0602	2.1045	2.1470	2.1877	2.2269
0.81	1.6914	1.7555	1.8158	1.8727	1.9265	1.9776	2.0262	2.0726	2.1170	2.1594	2.2001	2.2393
0.82	1.7036	1.7678	1.8280	1.8849	1.9388	1.9899	2.0385	2.0849	2.1292	2.1717	2.2124	2.2516
0.83	1.7157	1.7798	1.8401	1.8970	1.9509	2.0020	2.0506	2.0970	2.1413	2.1838	2.2245	2.2637
0.84	1.7277	1.7918	1.8521	1.9090	1.9629	2.0140	2.0626	2.1090	2.1533	2.1958	2.2365	2.2756
0.85	1.7395	1.8036	1.8639	1.9208	1.9747	2.0258	2.0744	2.1208	2.1652	2.2076	2.2483	2.2875
0.86	1.7511	1.8153	1.8756	1.9325	1.9864	2.0375	2.0861	2.1325	2.1768	2.2193	2.2600	2.2992
0.87	1.7627	1.8269	1.8872	1.9441	1.9979	2.0490	2.0977	2.1441	2.1884	2.2309	2.2716	2.3107
0.88	1.7741	1.8383	1.8986	1.9555	2.0094	2.0605	2.1091	2.1555	2.1998	2.2423	2.2830	2.3222
0.89	1.7854	1.8496	1.9099	1.9668	2.0207	2.0718	2.1204	2.1668	2.2111	2.2536	2.2943	2.3335
0.90	1.7966	1.8607	1.9211	1.9780	2.0218	2.0829	2.1316	2.1780	2.2223	2.2648	2.3055	2.3446
0.91	1.8076	1.8718	1.9321	1.9890	2.0429	2.0940	2.1426	2.1890	2.2334	2.2758	2.3165	2.3557
0.92	1.8185	1.8827	1.9430	2.0000	2.0538	2.1049	2.1536	2.2000	2.2443	2.2867	2.3275	2.3666
0.93	1.8293	1.8935	1.9538	2.0108	2.0646	2.1157	2.1644	2.2108	2.2551	2.2976	2.3383	2.3774
0.94	1.8400	1.9042	1.9645	2.0215	2.0753	2.1264	2.1751	2.2215	2.2658	2.3083	2.3490	2.3881
0.95	1.8506	1.9148	1.9751	2.0320	2.0859	2.1370	2.1857	2.2320	2.2764	2.3188	2.3596	2.3987
0.96	1.8611	1.9253	1.9856	2.0425	2.0964	2.1475	2.1961	2.2425	2.2868	2.3293	2.3700	2.4092
0.97	1.8714	1.9356	1.9960	2.0529	2.1067	2.1578	2.2065	2.2529	2.2972	2.3397	2.3804	2.4195
0.98	1.8817	1.9459	2.0062	2.0631	2.1170	2.1681	2.2167	2.2631	2.3075	2.3499	2.3907	2.4298
0.99	1.8918	1.9560	2.0164	2.0733	2.1271	2.1783	2.2269	2.2733	2.3176	2.3601	2.4008	2.4399
1.00	1.9019	1.9661	2.0264	2.0833	2.1372	2.1883	2.2369	2.2833	2.3277	2.3701	2.4109	2.4500

Appendix II

Table II.2. Errors of x_2 calculated by Method 1

n	c	x	y	x_1	x_2	$100(x_2 - x)/x$
0·25	0·5	0·700	0·6254	0·7653	0·700	0·0
0·25	0·5	1·000	0·9078	1·0487	1·000	0·0
0·25	1·0	0·900	0·6612	1·0110	0·901	0·1
0·25	1·0	1·000	0·7497	1·0995	1·001	0·1
6·00	0·5	1·000	0·1285	1·8874	1·221	22·1
6·00	0·5	2·000	0·6148	2·3737	2·023	1·2
6·00	0·5	2·100	0·6819	2·4408	2·119	0·9
6·00	0·5	3·000	1·3860	3·1449	3·002	0·1
6·00	1·0	2·000	0·2400	2·6900	2·127	6·4
6·00	1·0	2·600	0·5569	3·0069	2·629	1·1
6·00	1·0	2·900	0·6213	3·0713	2·722	0·8
6·00	1·0	3·000	0·8309	3·2809	3·010	0·3

Table II.3. Errors of x_2 calculated by Method 2

n	c	x	y	x_1	x_2	$100(x_2 - x)/x$
1·00	1·0	0·10	0·0048	0·08	0·102	2·0
1·00	1·0	0·10	0·0048	0·12	0·101	1·0
2·00	1·0	0·20	0·0023	0·18	0·206	3·0
2·00	1·0	0·20	0·0023	0·22	0·202	1·0
3·00	1·0	0·30	0·0014	0·28	0·301	0·3
3·00	1·0	0·30	0·0014	0·32	0·305	1·7
4·00	1·0	0·40	0·0011	0·38	0·400	0·0
4·00	1·0	0·40	0·0011	0·42	0·406	1·5
5·00	1·0	0·50	0·0009	0·48	0·505	1·0
5·00	1·0	0·50	0·0009	0·52	0·502	0·4
6·00	1·0	0·60	0·0009	0·58	0·608	1·3
6·00	1·0	0·60	0·0009	0·62	0·610	1·7

Appendix II

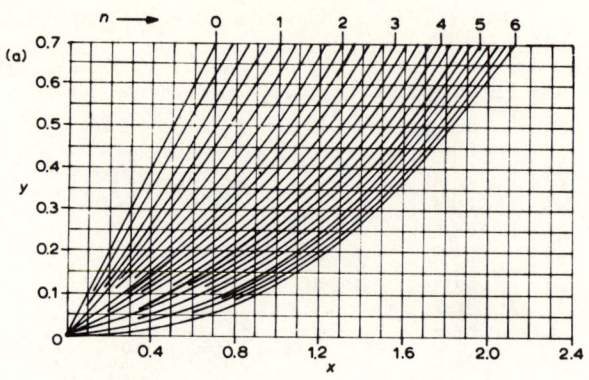
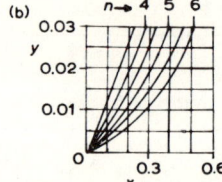

Fig. II.1

Method 2: $c = 0.5$; n at intervals of (a) 0.25, (b) 0.50

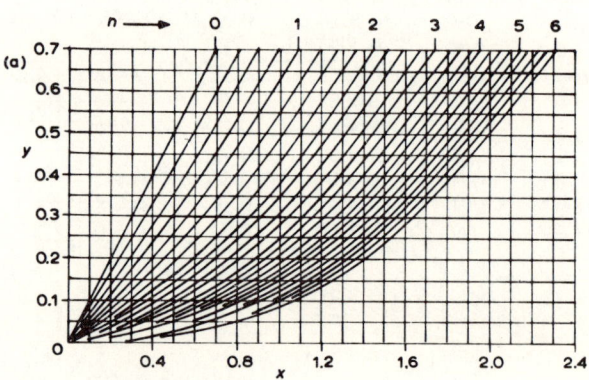
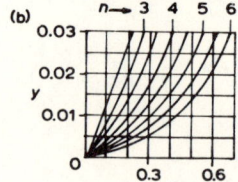

Fig. II.2

Method 2: $c = 0.6$; n at intervals of (a) 0.25, (b) 0.50

APPENDIX II   331

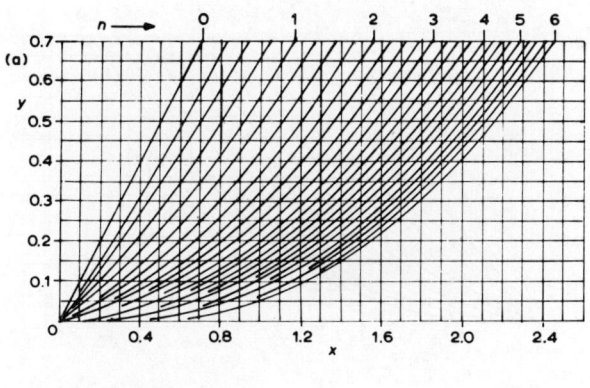

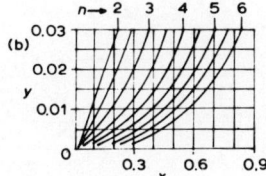

Fig. II.3

Method 2: $c = 0{\cdot}7$; n at intervals of (a) 0·25, (b) 0·50

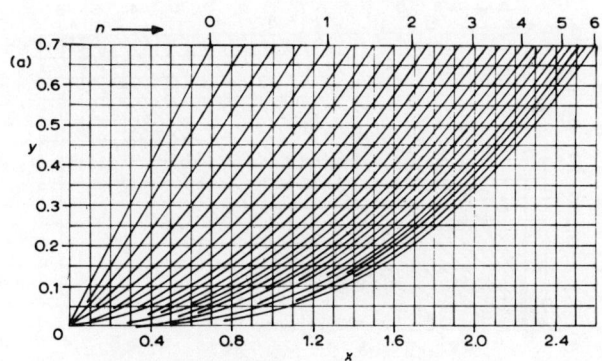

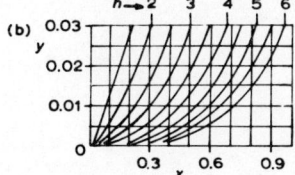

Fig. II.4

Method 2: $c = 0{\cdot}8$; n at intervals of (a) 0·25, (b) 0·50

332   APPENDIX II

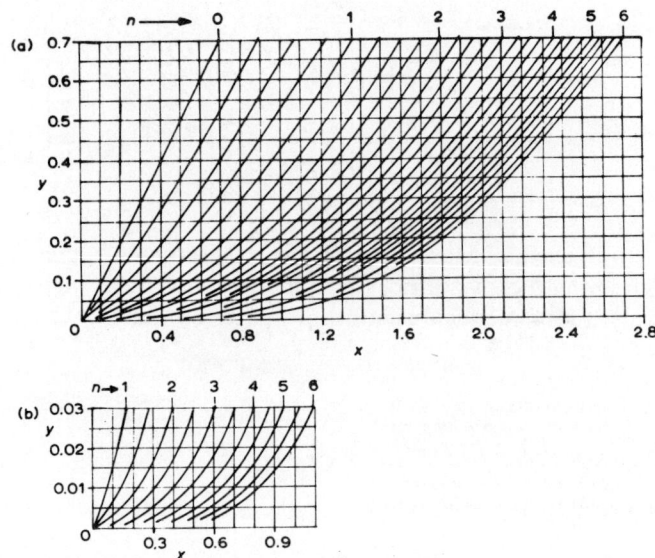

Fig. II.5

Method 2: $c = 0.9$; n at intervals of (a) 0·25, (b) 0·50

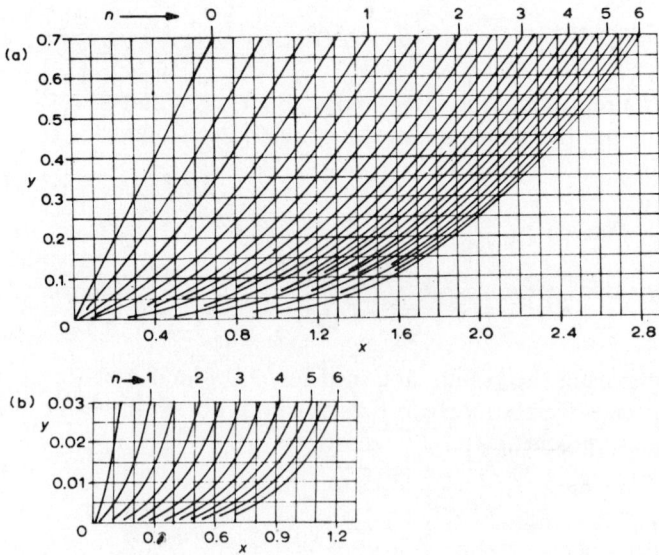

Fig. II.6

Method 2: $c = 1.0$; n at intervals of (a) 0·25, (b) 0·50

APPENDIX III

The method used by Massart and Bossaert ([166] in Chapter 7) consists in bracketing the root K'_d of Eq. (7.185) between two near limits z_L and z_U obtained with four identical significant figures so that:

$$\varphi(z_L) < 0, \qquad \varphi(z_U) > 0$$

Because of the monotonic character of φ, we may conclude that:

$$z_L < K'_d < z_U$$

and this procedure therefore implies the numerical integration of the Eq. (7.186).

Therefore the integration interval $(0, V_R)$ is divided into n equal intervals:

$$\int_0^V f(V)\,dV = \int_0^k f(V)\,dV + \int_k^{2k} f(V)\,dV + \cdots \int_{jk}^{(j+1)k} f(V)\,dV$$

$$+ \cdots \int_{(n-1)k}^{nk} f(V)\,dV$$

where $f(V)$ represents the integrand in Eq. (7.186) and $k = V_R/n$.

By applying the Newton-Cotes formula, for instance, for six intervals, we obtain:

$$\int_{jk}^{(j+1)k} f(V)\,dV \sim \frac{k}{288}[19f(jk) + 75f(jk + \delta) + 50f(jk + 2\delta)$$

$$+ 50f(jk + 3\delta) + 75f(jk + 4\delta) + 19f(j+1)k]$$

where $\delta = k/5$.

By comparing the results obtained for $n = 8$ and $n = 16$, we see that the division into eight intervals is sufficient to obtain the value of K'_d with a four-figure precision.

INDEX

Acids, 265
   amino, 43, 81, 84, 105, 107, 109, 110.
      126, 240, 263, 284, 294, 295, 312
   fatty, 107, 187
   organic, 107, 125, 204, 206, 234
Activity gradients, 126, 129, 262
Adsorbents, activity, 93
   porous-coated, 96
Alanine, 81, 103, 109
Alkali metals, 43, 107, 108, 209
Alkaline earth metals, 43
Alkaloids, 265, 267
   steroid, 265
Alkyl oleates, 47
Amidone oligosaccharides, 43
Amino-acids, 43, 81, 84, 105, 107, 109,
      110, 126, 240, 263, 284, 294, 295,
      312
p-Aminoazobenzene, 98, 274
Ampholytes, 265
Anions, separation, 126, 211
Antibiotics, 250
Antidiagonal gradients, 130, 265, 268
Antiparallel gradients, 130, 258, 263, 268,
      271, 272, 276, 277, 280, 289, 292,
      301, 302
Aspartic acid, 109, 283
Azobenzene, 98

Bases, 265
Bismuth, 284
Broadening of zones, 3, 57, 58, 59, 68, 70
Butter Yellow, 263, 264, 267, 300

Caffeine, 41
Carbohydrates, 110, 126, 326
Cations, separation, 40, 42, 44, 125, 185
Cephalosporins, 41
Chlorinated hydrocarbon residues, 48
Chlorophenols, 238

Cholesterol, 265
   esters, 47, 48
Cholesteryl stearate, 265
Classification of gradients, 126, 217
Cobalt, 43, 107, 116, 262, 276, 277
Column dimensions, 91, 105
   packing, 95
   structure of bed, 95
Combined gradients, 129, 306
Complexing agents in ion-exchange, 111
Compositional gradients, 126, 129, 258,
      262
Compression factor, 223, 232
Compound gradients, 128, 151, 169, 185
Concave gradients, 128, 137, 158, 185
Concentration gradients, 20, 125, 137
Continuous gradients, 128, 157, 258
Convection, 57
Convex gradients, 128, 137, 185
Copper, 43
Cross-sectional gradients, 129

Darcy's law, 10, 20, 22, 23, 25, 94
Dehomogenisation of eluent, 19, 46, 229,
      291
Diagonal gradients, 130
Diaminobutyric acid, 84
Diffusion, coefficients, 15
   eddy, 57, 67
      coupled, 70
   longitudinal, 57, 64, 67
   molecular, 57, 64, 67
p-Dimethylazobenzene, 48
Dimethyl phthalate, 99
Dionyl phthalate, 99, 275
Dinoseb, 302
Discontinuous gradients, 133, 185, 258
Distribution coefficient, 63
Dyes, 289
   as reference compounds, 39

Eddy diffusion, 57, 67
  coupled, 70
Efficiency, 55, 67, 85, 97, 99, 206, 232
Eluent, critical parameters, 41, 60, 94, 95, 105, 110
  migration, and non-uniform concentration, 13, 19
  and porosity, 22
  and surface tension, 6
  and viscosity, 6
  ascendant, 4
  capillary models, 14, 20
  closed column, 20
  descendant, 7
  diffusion models, 5
  horizontal, 9
  open column, 29
  statistical models, 21
  temperature effect, 5
Eluotropic series, 111
Elution, calculated curve, 61, 198
  gradient, 125, 126, 133, 185
    stepwise, 128
  pressure, 100
  temperature, 99
  volume, in ion-exchange, 211
Environmental gradients, 129, 270
Enzymes, 126
Eosin, 268
Essential oils, 303
Ethyl Orange, 100
Evaporation gradients, 126
Exponential gradients, 137, 158, 185

Fatty acids, 107, 187
Fick's law, 15
Flow, factors controlling, 22
  gradients, 126, 129
  laminar, 24, 25
  mechanisms, 14
  turbulent, 24

Gaussian distribution, 56
Glucosamine, 109
Glucosides, 289
Glutamic acid, 84, 109
Glycine, 81, 103, 109
Gradients, activity, 126, 129, 258, 265
  antidiagonal, 130, 265, 268

antiparallel, 130, 258, 263, 268, 271, 272, 276, 277, 280, 289, 292, 301, 302
classification, 126, 129
closed column, 133, 198, 270
combined, 129, 306
compositional, 126, 129, 258, 262
compound, 128, 151, 169, 185
concave, 128, 137, 185
concentration, 20, 125, 137
continuous, 128, 157, 258
convex, 128, 137, 185
cross-sectional, 129
diagonal, 130
discontinuous, 133, 185, 258
elution, 125, 133, 185
environmental, 129, 370
evaporation, 126
exponential, 127, 158, 185
flow, 126, 129
grain size, 129
impregnation, 126, 129, 258, 264
ion-exchange column, 208
ionic strength, 129, 137, 265, 312, 313
layer thickness, 129, 301
linear, 128, 156, 185
mobile phase, 133, 185, 195, 306, 313
nomenclature, 129
open column, 185, 228, 278
orthogonal, 130, 258, 263, 264, 265, 266, 267, 272, 277, 278, 282, 289, 300
paper chromatography, 185, 228
parabolic, 159
parallel, 130, 258, 262, 263, 267, 268, 272, 278, 283, 284, 292, 294, 295, 296, 297, 298, 299
pH, 125, 129, 151, 179, 210, 258, 265, 267, 312, 313
polarity, 129, 171, 275, 295, 299
programmed, 179
salt, 179
selection, 234
stationary phase, 40, 126, 129, 258
stepwise, 128, 157
temperature, 7, 20, 32, 126, 129, 270, 306, 312
thin-layer chromatography, 189, 228
vapour phase, 126, 129, 297, 298, 299

vapour pressure, 129
velocity, 302
Grain size gradients, 129

Hafnium, 284
HETP, 55
  and adsorbent, 94
  and column dimensions, 92
  and flow-rate, 94
  and packing, 95
  and particle size, 93
  and resolution, 78, 83
  and R_f value, 39
  in ion-exchange, 63, 102
    controlling factors, 64
  temperature effect, 107, 290
Histidine, 264, 283
Humidity, effect on R_f, 45, 115
Hypnotics, 295

Impregnation gradients, 126, 129, 258, 264
Indoles, 267
Indophenol, 263, 264, 267, 300
  blue, 48
Iodide, 262
Ion-exchange, column dimensions, 105
  complexing agents, 111
  elution parameters, 105
  elution volume, 211
  gradients, 208
  HETP, 63, 102
  on chromatographic paper, 40
  resins, cross-linking, 103
    particle size, 64, 102
    quantity, 103
  resolution, 102
  temperature, 107
Ionic strength gradients, 129, 137, 265, 312, 313
  effect on resolution, 110
Iron, 262

Kinetics, ascending chromatography, 4
  descending chromatography, 7
  eluent migration, 4, 18, 20
  horizontal chromatography, 9
  isothermal, 5, 29
  surface tension, 6, 7

temperature gradient, 7
viscosity, 6, 7
zone migration, 29
Kozeny–Carman equation, 24
'KS-Vario' chambers, 45, 298

Layer thickness gradients, 129, 301
Lead, 43, 116, 284
Linear gradients, 128, 156, 185
Lipids, 136, 244

Manganese, 283, 284
Mass transfer effects, 26, 57, 105
p-Methoxyazobenzene, 274
Methyl oleate, 274
Methyl Orange, 100
Methyl palmitate, 99, 275
Mobile phase, choice of, 111
  effect on R_f, 41
  gradients, 133, 185, 195, 306, 313
  optimum velocity, 60

Navier–Stokes equations, 22
N-chambers, 294
Nickel, 43, 262, 283, 284, 286
p-Nitroaniline, 300
Nomenclature of gradients, 129
Nucleic acids, 126, 247
Nucleotide bases, 43
Nucleotides, 136

Oleates, alkyl, 47
Oligosaccharides, 43, 126
Open column gradients, 185, 228, 278
Optimisation of chromatography, 91
Orthogonal gradients, 130, 258, 263, 264, 265, 266, 267, 272, 277, 278, 282, 289, 300

Packing of column, 95
Paper chromatography, gradients, 185, 228
Parabolic gradients, 159
Parallel gradients, 130, 258, 262, 263, 267, 268, 272, 278, 283, 284, 292, 294, 295, 296, 297, 298, 299
Particle shape, 60
Particle size, 22, 24, 60, 64, 93, 102
Partition coefficient, 70
  effect on R_f, 35

Peak capacity, 206
Peak position, 198, 209, 211, 212, 214
Peclet number, 105
Penicillin, 41
Peptides, 126
pH, effect on resolution, 110
　gradients, 125, 129, 151, 179, 210, 258, 265, 267, 312, 313
　indicators, 265
　optimum, 113
Phosphorus oxyacids, 214, 216, 252
Plate theory, 56
Poisson distribution, 56
Polarity gradients, 129, 171, 275, 295, 299
Polyethylene glycol, 239
Polymers, 126, 307, 308, 312
Polypropylene glycol, 239
Polystyrenes, 277
Porosity of particles, 24
Porous media, models, 14, 20
　statistical, 21
Pressure, efficiency, 100
　ion-exchange, 105
　specific retention volume, 101
Programmed gradients, 179
Proteins, 126, 136, 187, 240

Rare earths, 43, 103, 104, 107, 108, 209, 213, 218, 251, 252, 284
Resolution, and analysis time, 101
　closed column, 102
　efficiency, 85
　gradients, 221, 230, 260
　index, 86
　ion-exchange, 102
　open column, 83
　power, 82
　R_f values, 72, 83
　temperature, 99, 290
Retention time, 55
　relative, 74
Retention volume, 66, 74, 99, 101, 200
Reynolds numbers, 25, 105
R_f values, and adsorbent, 40, 94
　adsorbent structure, 41, 95
　ambient gas phase, 44
　column parameters, 92
　composition of mobile phase, 41
　composition of stationary phase, 40
　eluent migration, 32

　esterification, 32
　factors affecting, 39
　HETP, 39
　humidity, 45, 115
　ion-exchange, 40
　medium, 42
　mobile phase parameters, 41, 94, 95
　particle size, 46, 93
　partition coefficient, 35
　pH, 113
　pressure, 100
　reproducibility, 39, 46, 47
　role of paper, 40
　sample size, 93
　selectivity, 72
　temperature, 42, 44, 97, 116, 283
　time, 101
　vapour phase, 44

Saturated chambers, 44, 300
Salt gradients, 179
Sample size, 93
Sandwich chambers, 44, 283
Schmidt number, 105
Selection of gradients, 234
Selectivity, 70, 72, 74, 108
　coefficient, 72
Separation, 65
　efficiency, 55, 67, 98, 100
　factor, 72
　function, 82, 86
　relative retention, 74
Serine, 109
SK chambers, 46
Specific retention volume, 101
Squalane, 99, 274, 275
Squalene, 265
Stationary phase, choice of, 111
　effect of structure, 68
　on R_f, 40
　gradients, 40, 126, 129, 258
Statistical moments, and HETP, 58
　and resolution, 79
Stepwise gradients, 128, 257
Steroid alkaloids, 265
Steroids, 126, 248
Streptothricins, 250
Sudan II, 98
Sudan III, 98, 274
Sudan Red, 48, 263, 264, 267, 300

Sugars, 110, 126, 236
Sulphonamides, 300
Sulphoxides, 115
Supercritical-fluid chromatography, 99
Surface porosity, controlled, 97

Temperature, and ion-exchange, 107
  and partition coefficient, 31
  and plate height, 107
  and resolution, 107, 108
  and R_f values, 42, 97, 116, 283
  and selectivity, 108
  and separation efficiency, 98
  gradients, 7, 20, 32, 126, 129, 270, 306, 312
  programmes, 270
Terpene alcohols, 264
Theobromine, 41
Theophylline, 41
Theoretical plates, 55
  in ion-exchange, 63, 102
  number of, 56, 63, 67, 81
Thin-layer chromatography gradients, 189, 228

Threonine, 84
Time of analysis, 101
Tortuosity factor, 57, 110
Tristearine, 275

Unsaturated chambers, 44, 294

Valine, 109
Van Deemter equation, 58, 67, 94
Vapour phase gradients, 126, 129, 297, 298, 299
Vapour pressure gradients, 129
'Varigrad', 162, 171, 175
Velocity gradients, 302
Velocity of mobile phase, optimum, 61
Viscosity of eluent, 95

Zinc, 117, 276
Zirconium, 284
Zone broadening, 3, 57, 58, 59, 68, 70
  migration, 31, 33, 34